Passive Acoustic Monitoring of Cetaceans

Passive acoustic monitoring (PAM) is used increasingly by the scientific community to study, survey and census marine mammals, especially cetaceans, many of which are easier to hear than to see. PAM is also used to support efforts to mitigate potential negative effects of human activities such as ship traffic, military and civilian sonar and offshore exploration.

Walter Zimmer provides an integrated approach to PAM, combining physical principles, discussion of technical tools and application-oriented concepts of operations. In addition, relevant information and tools necessary to assess existing and future PAM systems are presented, with MATLAB® code used to generate figures and results so that readers can reproduce data and modify code to analyse the impact of changes. This allows the principles to be studied while discovering potential difficulties and side effects. Aimed at graduate students and researchers, the book provides all information and tools necessary to gain a comprehensive understanding of this interdisciplinary subject.

Walter M. X. Zimmer holds a Ph.D. in theoretical physics from the University of Regensburg, Germany. He is currently a Scientist in the Applied Research Department of the NATO Undersea Research Centre (NURC) in La Spezia, Italy, and a Guest Investigator at the Biology Department of the Woods Hole Oceanographic Institution (WHOI), Woods Hole, MA, USA.

Passive Acoustic Monitoring of Cetaceans

WALTER M. X. ZIMMER

NATO Undersea Research Centre

CAMBRIDGE
UNIVERSITY PRESS

CAMBRIDGE
UNIVERSITY PRESS

University Printing House, Cambridge CB2 8BS, United Kingdom

Cambridge University Press is part of the University of Cambridge.

It furthers the University's mission by disseminating knowledge in the pursuit of
education, learning and research at the highest international levels of excellence.

www.cambridge.org
Information on this title: www.cambridge.org/9781107428386

First published 2011
First paperback edition 2014

A catalogue record for this publication is available from the British Library

Library of Congress Cataloguing in Publication data
Zimmer, Walter M. X., 1949–
Passive Acoustic Monitoring of Cetaceans / Walter MX Zimmer.
   p.   cm
Includes bibliographical references and index.
ISBN 978-0-521-19342-9
1. Cetacea – Monitoring.   2. Cetacea – Effect of noise on.   3. Dolphin sounds.
4. Whale sounds.   I. Title.
QL737.C4Z53   2011
599.5072'3–dc22

                                                                        2010052636

ISBN 978-0-521-19342-9 Hardback
ISBN 978-1-107-42838-6 Paperback

Additional resources for this publication at www.cambridge.org/9780521193429

Für Giannina

Contents

Acknowledgements

The data used within this book were in part made available by Gianni Pavan (CIBRA, Università degli Studi di Pavia, Italy, http://www.unipv.it/cibra), Carmen Bazua-Duran (Universidad Nacional Autonoma de Mexico, Mexico), Denise Risch (Northeast Fisheries Science Center, NOAA, USA), Robert Dziak, Sharon Nieukirk and Dave Mellinger (Oregon State University and NOAA, USA), Douglas Gillespie (University of St Andrews, UK), Chris Clark (Cornell University, USA), Anna Moscop (International Fund for Animal Welfare, www.ifaw.org/sotw) and Peter L. Tyack (Woods Hole Oceanographic Institution, WHOI, USA).

In 1998, when I became interested in cetacean research, I had worked for some time in underwater research, but knew little about whales and dolphins. Without Gianni Pavan (CIBRA, Università degli Studi di Pavia, IT) and the whole CIBRA team, Peter L. Tyack and Mark Johnson, both of Woods Hole Oceanographic Institution (WHOI, USA), and Peter T. Madsen (University of Aarhus, DK) who shared with me their knowledge about whales and dolphins and who engaged in a continuing and extremely fruitful collaboration, I would have never learned so much about these fantastic and extraordinary creatures.

I would also like to thank Arnold B-Nagy (NURC), John Harwood, Len Tomas and Tiago Marques (CREEM, UK) for discussions on population estimation, general ecological modelling and distance sampling. Piero Guerrini, Vittorio Grandi and Luigi Troiano (NURC) were always available for discussions about hydrophones, electronics and system implementation.

Special thanks to David Hughes (NURC) and Peter Tyack who undertook the huge task of reading and commenting on the whole manuscript. Tiago Marques and Gianni Pavan provided useful input on parts of the manuscript.

Finally, I would like to express my gratitude to copy-editor Lynn Davy and to Martin Griffiths, Lynette Talbot and Abigail Jones from the Cambridge University Press editorial and production team for helping me to implement an idea that was simmering for quite some time.

Introduction

Since antiquity humans have been fascinated by whales and dolphins. They were attracted by the social behaviour and apparent curiosity of dolphins, which resulted in frequent interaction with humans. Myths were generated, especially around large whales, which behaved differently from other sea fauna. Even if early research on whales may have been driven by the economic interests of the whaling industry, the widespread recognition that whales, dolphins and porpoises are marine mammals significantly changed the scientific impetus. In fact, cetaceans (whales, dolphins and porpoises), together with the sirenians (manatees and dugongs), are the only mammals that live all their life in the sea. For an air-breather, living in the water is a continuous challenge and as such marine mammals deserve our respect and our protection. Consequently, in addition to a pure interest in knowledge, scientific research increasingly studies marine mammals to support their conservation and protection.

Studying the life of marine mammals is a challenge for scientists, who may prefer a laboratory environment to the sometimes hostile sea, and most research in the past was limited to observations of surface behaviour. Only a few scientists were properly equipped to study the underwater behaviour of marine mammals. Studying pinnipeds (seals, sea lions and walruses) and sirenians was somewhat easier as these animals are sometimes accessible near shore: pinnipeds spend some time on land and sirenians live in very shallow water or rivers. Most cetacean species, however, occupy the vast areas of the oceans of the world. Living all their life in water, cetaceans have not only adapted their lifestyle, but have also modified the way in which they interact with each other and with the environment. In particular, cetaceans' sound generation and auditory systems evolved and adapted to their new environment, basing their daily life mainly on acoustics. Apart from being essential for cetaceans, this use of sound allows human researchers to eavesdrop and to study cetacean behaviour from a distance.

Although cetaceans are found in all oceans, they are highly mobile and most species that are of special concern to environmental management or are of public concern could be considered as rare or elusive. Classic cetacean surveys have used visual (sighting) methods to detect the animals, but there is growing recognition that many species of interest are easier to hear than to see. Owing to technological progress, there is now increasing awareness in the biology community about the usefulness of passive acoustic monitoring (PAM) for the study of cetaceans in their natural environment. PAM is a good technique for surveying and studying cetaceans, not only because they frequently use sound for their day-to-day activities, but also because acoustics is so far the only tool that allows the study of submerged animals that are not visible to human observers and does not interfere with the animals' behaviour if properly implemented. PAM is expected to improve researchers' overall capability to monitor the temporal and spatial behaviour of cetaceans and therefore their habitat usage.

Monitoring cetaceans acoustically requires that the animals be detected acoustically. Cetaceans can be detected acoustically, not only by listening for sounds emitted by the animals (passive acoustic detection) but also by using whale-finding sonar to listen for echoes reflecting from the animal (active acoustic detection). Whale-finding sonar systems do not require the whale to make a sound, but to be effective they must produce substantial sound energy to obtain detectable echoes. This is because the sound has to travel twice the distance from the sonar to the whale and the whale reflects only a part of the sound energy back to the active sonar. While some success in detecting baleen whales with active sonar has been reported (e.g. Lucifredi and Stein, 2007), the feasibility of active acoustic detection has not yet been demonstrated for deep-diving whales and the increased sound energy required to detect these species may generate additional risks to the well-being of the animals. Passive acoustic monitoring (PAM), on the other hand, is based on listening to the acoustic output from whales without interfering with the animals' behaviour.

PAM is not only important to survey and census of marine mammals, but also an essential ingredient in efforts to mitigate potential negative effects of human activities (ship traffic, offshore exploration, military and civilian sonar, etc.) on marine mammals. One of the expectations for successful marine mammal risk mitigation is that PAM becomes an affordable technology, allowing a more or less continuous survey of acoustically active cetaceans, especially in remote or hostile marine environments.

The successful application of PAM requires both appropriate technology and operational concepts. Although the impact of technology or system parameters (e.g. self-noise, array gain, processing bandwidth and gain) is easy to assess, the impact of operational concepts is more difficult to quantify, because it depends on the behaviour of whales, dolphins and porpoises as well as environmental characteristics. Other operational constraints include the mobility of the PAM system (whether it should be moored, allowed to float, or be towed from a ship). A range of biological and oceanographic parameters will influence the number of sensors necessary to obtain the required success rate and confidence level. The design of the sonar systems will also depend on the objective of PAM: a system designed for abundance estimation may have very different requirements from one that is tuned for risk mitigation, in which failure to detect a whale that is present could constitute a failure of the PAM system.

The initial stimulus for compiling this book resulted from my involvement in passive acoustic detection/monitoring of deep-diving toothed whales, especially beaked whales, to support the risk mitigation of anthropogenic activities. Although PAM as application is rather new in cetacean research, the use of passive acoustics is well established in ocean engineering. As an interdisciplinary subject, successful PAM combines physics, technology and biology and requires an integrated presentation.

The object of this book is therefore to provide an integrated approach to PAM, combining the required physical principles with discussions of the available technological tools and an analysis of application-oriented concepts of operations. In particular, the reader of the book should be enabled to understand the physical basics behind PAM, its technological implementation and its operational use. The book is aimed at students, researchers and professionals who are interested in cetaceans and want to understand the concepts behind PAM, or who may need to implement PAM. As such, the book provides all relevant information and tools necessary to assess existing and future PAM systems. By addressing most aspects of PAM systems, the book may also function as a framework for alternative approaches.

The book is divided into ten chapters organized into three parts that correspond roughly to three major academic disciplines:

- Underwater acoustics
- Signal processing
- Ecology

Underwater acoustics is a well-documented subject (e.g. Urick, 1983; Medwin and Clay, 1998; Lurton, 2002; Medwin *et al.*, 2005) and this book tries to synthesize the subject matter by relating the presentation to PAM.

Signal processing is also the subject of a vast list of books and publications; the recent book by Au and Hastings (2008) on bioacoustics addresses some key methods. Here, I try to synthesize the signal processing techniques that are relevant to PAM, with special attention to techniques that have recently been proposed for the detection and classification of marine mammals. The text is based on some examples of sound recordings, and develops different signal processing algorithms. The reader is, however, invited to consult the literature to learn about alternative techniques or to use the data for his or her own signal processing schemes.

The applications of PAM to detect cetaceans are part of ecology and the description of PAM operation follows closely what is published in this field. Of particular help were the textbooks by Southwood and Henderson (2006) on ecological methods, by Thompson (2004) on sampling rare or elusive species, and the standard texts on distance sampling by Buckland *et al.* (2001, 2004). Here I try to present some of the methods that are, or may become, relevant to PAM applications.

A common feature of all chapters in this book is the explicit use of Matlab code (Matlab® version 7.5.0) to generate the various figures and results. Although this is not a book about programming Matlab, I tried, nevertheless, to include Matlab code fragments throughout the book with the purpose of complementing the information presented (plain text, equations, figures) with realistic and practical examples. In fact, the reader should not only be put in the position of being able to reproduce the presented results, but also be enabled to modify the code and to analyse the impact of his or her changes. This type of 'learning by doing' will allow the reader to study the principles and to discover potential difficulties and side effects of the method presented. To facilitate this hands-on approach, both data and Matlab code that are reproduced in the book, together with some additional scripts, are available for download from the book's website: www.cambridge.org/9780521193429. When reading a particular chapter of the book, the reader is therefore invited to locate the related MATLAB® script and to try to understand the different MATLAB® instructions and to compare them with the mathematical formulation; it is always advisable to utilize the built-in MATLAB® help system when confronted with new matlab instructions. The reader may further check www.wmxz.eu for additional information, new datasets, updated MATLAB® scripts, discussions, and future PAM developments.

The data used in this book were provided by different marine mammal researchers for the scope of this book. Any further use of the data requires the consent of the scientist who originated the data.

Part I

Underwater acoustics (the basics)

Part I should provide the basic knowledge of underwater acoustics that is needed for the remaining part of the book. Although the word *acoustics* was originally associated with the sound properties of rooms, its usage is here broadened to include wave phenomena in media other than air, and frequencies other than those audible to the human ear.

The first chapter summarizes the notations and basic concepts of underwater sound with a bias towards the needs of the remaining chapters. The purpose of the second chapter is twofold, namely to introduce the sounds of interest, i.e. the sounds made by different cetaceans, and to develop techniques that are suited to describing the different sound categories. The third chapter presents and discusses all the components of the sonar equation, which is the workhorse of sonar design and performance analysis; I focus on the passive sonar equation, which is the version that is relevant for PAM.

1 Principles of underwater sound

The objective of this chapter is to provide the notation and basic concepts of underwater sound that may be found useful for understanding both the description of cetacean sound and the use of the sonar equation. The approach should be complete enough without going into too much detail, but should provide the basis, in terms of concept and notation, to support the discussion of the remainder of the book. As this chapter can be considered as a general introduction to underwater sound, it synthesizes various textbooks and presents the information in the context of PAM, covering the following subjects:

Sound as a pressure wave and the wave equation
Measuring underwater sound (the decibel scale)
Sound velocity models and profiles
Sound propagation
Sound as an information carrier, disturbance or noise

1.1 Sound as a pressure wave

Historically, the term *sound* was used to describe pressure waves in air and that are audible by humans (Randall, 1951), but in this book, I will follow the recent custom in underwater acoustics and use the word sound to describe all pressure waves that are generated by an initial pressure fluctuation irrespective of frequency and media in which the sound waves propagate.

1.1.1 Wave equation

The wave equation is one of the most important equations in physics. It is of such importance that it merits a detailed derivation. In fact, all textbooks in physics and most books in oceanography go into lengthy derivations of the wave equation (e.g. Randall, 1951; Kinsler *et al.*, 2000; Medwin and Clay, 1998). The derivation of the wave equation from more basic physical principles depends mainly on the media in which the waves propagate and may be straightforward or somewhat complicated. However, it is the beauty of the wave equation that in the end its notation is independent of the physical phenomenon (acoustic wave, sea surface waves, electromagnetic waves, etc.) and the media in which the wave is propagating (gas, liquid, solid, etc.).

The propagation of waves is related to one of the fundamental principles in physics, Newton's second law of motion, which states that 'a change in motion is proportional to the motive force impressed and takes place along the straight line in which the force is impressed' (Newton, quoted after Crease, 2008).

In mathematical terms, a modern form of this equation of motion of an object is given by

$$\frac{d}{dt}(mu) = F \tag{1.1}$$

where m is the mass (measured in kg) of the object, which in principle could vary with time (e.g. rockets), u is the speed of the object (measured in m/s), and F is the total force acting on the object (measured in kg m/s^2). The term $\frac{d}{dt}(mu)$ denotes the temporal change of the product of mass m and speed u; the product mu is also known as the momentum of the object. For constant m, Newton's second law is better known in the form $ma = F$, where $a = \frac{d}{dt}u$ represents the derivative of speed with respect to time, i.e. the acceleration of the object.

Equation 1.1 is written in a way that indicates that both u and F are simple numbers, or scalars. This is appropriate if the force F may be described by a single number. Typically, such a scalar notation is used when the force is acting along a single dimension of the real world, say only vertically, or in a single horizontal direction. In the case of an arbitrary three-dimensional description of the force, that is, where we have the force described by components in x, y, z directions (for a Cartesian co-ordinate system), we should have a set of three different equations. However, in such cases it is common to adopt a vector notation combining all directional equations so that assuming constant mass m Equation 1.1 may be written as

$$\frac{d}{dt}\mathbf{u} = \frac{1}{m}\mathbf{F} \tag{1.2}$$

where $\mathbf{u} = (u_x, u_y, u_z)$ and $\mathbf{F} = (F_x, F_y, F_z)$ denoting the x, y, z components of speed vector $\mathbf{u}$ and force vector $\mathbf{F}$. We use bold characters to describe vectors to distinguish from scalars.

In other words, Equation 1.2 is nothing more than the compact notation of the three equations of motion that are needed to describe the response of an object to a force that is given in Cartesian co-ordinates by

$$\frac{d}{dt}u_x = \frac{1}{m}F_x$$
$$\frac{d}{dt}u_y = \frac{1}{m}F_y \tag{1.3}$$
$$\frac{d}{dt}u_z = \frac{1}{m}F_z$$

If we were to describe the force in spherical co-ordinates, that is, we measure the force in the radial direction R, along azimuth and elevation angles (θ, φ):

$\mathbf{F} = (F_R, F_\theta, F_\varphi)$, then we obtain

$$\frac{\mathrm{d}}{\mathrm{d}t} u_R = \frac{1}{m} F_R$$
$$\frac{\mathrm{d}}{\mathrm{d}t} u_\theta = \frac{1}{m} F_\theta \qquad (1.4)$$
$$\frac{\mathrm{d}}{\mathrm{d}t} u_\varphi = \frac{1}{m} F_\varphi$$

By adapting a vector notation in Equation 1.2, we obtain not only a more compact formula but also a notation that does not depend on the underlying implementation, that is, the notation does not depend on the way in which we measure the forces and speeds in reality.

Newton's second law is valid for all moving objects. Therefore, it also applies to the motion of small particles that are displaced by some forces in a given medium, keeping the surroundings constant. In order to displace particles in this way the medium must be compressible. This is the case for gas, but also for liquids and even solids, although, of course, gases are more easily compressed than liquids or solids.

Without any forces applied to particles, none of the particles within a gas should change its speed, according to Newton's second law. We empirically know that actual gases cannot exist without any forces, as the gas molecules in a given volume are continuously changing their direction due to collisions and that there are gravitational forces acting not only on the gas, but also between the gas molecules, holding the gas together. The collisions between the gas molecules are typically described by the pressure of the gas, where high pressure indicates high collision rates. Real gases are therefore characterized by quantities such as pressure, volume and temperature, which build the basis of thermodynamics, a very successful physical discipline.

To generate a sound wave we have to disturb the pressure equilibrium by exerting an additional force P, measured in N/m^2 (newtons per square metre). By forcefully displacing particles in a medium, say a gas, we create in general a situation where the pressure of the gas has been locally changed. That is, we create a pressure gradient within the medium. If the displacement force is removed, then we expect the gas particles to return to their (dynamic) equilibrium.

This restoring force of a pressure gradient is given by

$$\mathbf{F} = -V \nabla P \qquad (1.5)$$

where $V = m\rho$ is the volume of a small parcel of air with density ρ on which the pressure gradient ∇P acted. The operator ∇ is called the Nabla operator and describes the spatial gradient, which in our case states how the pressure P varies in x, y, z directions, that is $\nabla P = \left(\frac{\mathrm{d}P}{\mathrm{d}x}, \frac{\mathrm{d}P}{\mathrm{d}y}, \frac{\mathrm{d}P}{\mathrm{d}z} \right)$.

With Equation 1.5, Newton's second law becomes, in terms of the pressure gradient,

$$\frac{\mathrm{d}}{\mathrm{d}t} \mathbf{u} = -\frac{1}{\rho} \nabla P \qquad (1.6)$$

that is, the particle velocity $\mathbf{u}$ is in opposite direction to the pressure gradient ∇P.

Combining the equation of motion (1.6) with the equation of continuity, which is given in Cartesian co-ordinates by

$$\frac{1}{\rho}\frac{d}{dt}\rho + \frac{d}{dx}u_x + \frac{d}{dy}u_y + \frac{d}{dz}u_z = 0 \qquad (1.7)$$

we obtain an equation that relates the changing gas density to the changing pressure:

$$\frac{d^2}{dt^2}\rho = \frac{d^2}{dx^2}P + \frac{d^2}{dy^2}P + \frac{d^2}{dz^2}P \qquad (1.8)$$

As $\frac{d}{dt}\rho$ is the rate at which the density is changing due to external forces and $\frac{d}{dx}P$ is the pressure gradient along the x-axis, Equation 1.8 says simply that the temporal change in the density rate is given by the spatial variation of the pressure gradient.

Equation 1.8 is presented in Cartesian co-ordinates and, similar to Equation 1.5, we introduce a notation free of the co-ordinate system by denoting

$$\nabla^2 P = \frac{d^2}{dx^2}P + \frac{d^2}{dy^2}P + \frac{d^2}{dz^2}P \qquad (1.9)$$

where ∇^2 is also called the Laplacian operator, so that

$$\frac{d^2}{dt^2}\rho = \nabla^2 P \qquad (1.10)$$

In order to complete the derivation of the wave equation we need a relation between pressure and density. Without going into the specifics of the different media (gas, liquid), we express the pressure in the medium as a function f of the density ρ

$$P = f(\rho) \qquad (1.11)$$

and assume that variations (denoted by the symbol δ) in pressure are linearly proportional to variations in density

$$\delta P = c^2(\delta\rho) \qquad (1.12)$$

We use c^2 to indicate that the proportionality constant is positive and that the pressure always increases with increasing density. Consequently, we obtain

$$\frac{d^2}{dt^2}P = c^2\frac{d^2}{dt^2}\rho \qquad (1.13)$$

and after inserting Equation 1.13 in Equation 1.10 we obtain the wave equation, which in terms of pressure reads

$$\frac{d^2}{dt^2}P = c^2\nabla^2 P \qquad (1.14)$$

Equation 1.14 is the general wave equation, which relates the temporal variation of the local pressure to the spatial differences in the surrounding pressure field. The spatial differences are described by the Laplacian operator ∇^2, the form of which depends on the co-ordinate system chosen for the application.

In cases where there is complete spherical symmetry around the location of interest we obtain the wave equation by using the Laplacian operator in spherical coordinates, which, maintaining only derivatives with respect to radius vector r, becomes

$$\nabla^2 = \left(\frac{\partial^2}{\partial r^2} + \frac{2}{r} \frac{\partial}{\partial r} \right) \tag{1.15}$$

yielding, after some manipulations, the spherical wave equation

$$\frac{d^2(rP)}{dt^2} = c^2 \frac{\partial^2(rP)}{\partial r^2} \tag{1.16}$$

This spherical wave equation plays an important role in underwater acoustics as the ocean is in general very large compared with the sound source and over reasonable distances complete spherical symmetry applies. In addition, the spherical wave equation is a one-dimensional wave equation (depending only on range r), simplifying the analysis significantly.

1.1.2 The solution of the wave equation

To solve the wave equation we consider the spherical wave equation

$$\frac{d^2(rP)}{dt^2} = c^2 \frac{\partial^2(rP)}{\partial r^2} \tag{1.17}$$

and note that its general solution is given by the relation

$$rP = f(ct \pm r) \tag{1.18}$$

for any function f that depends purely on the argument $(ct \pm r)$.

That Equation 1.18 is the solution of Equation 1.17 can be easily verified by differentiating the function f twice with respect to time t and radius r:

$$\frac{d^2}{dt^2} f(ct \pm r) = c^2 f(ct \pm r)$$
$$\frac{\partial^2}{\partial r^2} f(ct \pm r) = f(ct \pm r) \tag{1.19}$$

which, after insertion into Equation 1.17, obviously solves this equation.

This general solution (Equation 1.18) includes both outgoing (minus sign) and incoming waves (plus sign). Considering only outgoing or diverging waves we write, for the pressure,

$$P = \frac{1}{r} f(ct - r) \tag{1.20}$$

and note that, independent of function f, the decrease of pressure P is inversely proportional to the range r.

The solution of the wave equation relates distance r to time t via the constant c, indicating that this constant c describes the speed at which the disturbance travels in the acoustic medium. For this reason, it is appropriate to call c the wave velocity or sound speed; Equation 1.12, which introduces c, may therefore be considered as the defining equation of the sound speed.

Caveat

The wave equation, as given in Equation 1.14, describes the propagation of a disturbance within an acoustic medium; it does not describe the generation of sound and the coupling of the sound generator to the acoustic medium. For a mathematically correct generic description, the wave equation should be augmented by an additional term describing an external force that disturbs the acoustic medium. In other words, the wave equation 1.14 describes the physics of the pressure wave only after external forces have been switched off.

Example In the following example, we demonstrate how an arbitrary disturbance propagates through an acoustic medium. For this, we assume that the pressure disturbance f may be approximated by a Gaussian function

$$f(r - ct) = P_0 \exp\left\{ -\frac{1}{2}\left(\frac{ct - r}{\sigma}\right)^2 \right\}$$ (1.21)

where P_0 is the maximum amplitude and σ is a measure of the width of the Gaussian disturbance. To 'see' how the disturbance propagates through an acoustic medium we simulate the whole process on the computer. For this simulation, we use a short Matlab script to model the dynamics of Equations 1.20 and 1.21.

Matlab script

```
%Scr1_1
r=1:0.1:100; % range in km
t=10:10:50; % time steps in seconds
c=1.5; %sound speed in units of km/s
%
P0=1;
s=2;
%
P=zeros(length(r),length(t)); %pre-allocate pressure vector
%
for ii=1:length(t)
    P(:,ii)=P0*exp(-0.5*((c*t(ii)-r)/s).^2)./r; %Eq 1.21
end
%
figure(1)
plot(r,P,'k')
```

```
xlabel('Distance r [km]')
ylabel('Pressure P [rel]')
text(20,0.065,'t=10 s')
text(30,0.036,'t=20 s')
text(45,0.025,'t=30 s')
text(60,0.020,'t=40 s')
text(75,0.016,'t=50 s')
```

This script defines first a range vector r that varies in steps of 0.1 km from 1 to 100 km. It defines further a time vector t for which we wanted to plot the pressure of the outgoing disturbance. As we measure the range in km and the time in seconds, we define next the wave velocity in km/s. Here we use a value of c = 1.5 km/s, which is a typical value for sound speed in seawater.

After defining the peak value at unit range (1 km) and setting the sigma of the Gaussian function, called s in the script, we allocate the memory for the different pressure vectors. Looping next through all time steps we estimate the pressure for all ranges. The script ends with plotting the results and labelling the axes and curves.

Executing the script Scr1_1 results in Figure 1.1.

From Figure 1.1 we note that the different Gaussian curves peak at regular spatial intervals of 15 km, which is to be expected for a sound speed of 1.5 km/s and a temporal separation of 10 s between consecutive curves. We note further that the peak amplitude of the curves decreases as time goes on, or the further the curve is from the origin. While, by

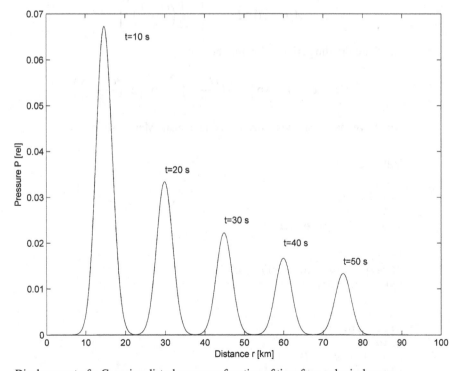

Figure 1.1   Displacement of a Gaussian disturbance as a function of time for a spherical wave.

definition, the peak amplitude at a range of 1 km is at 1.0 (and too large to be plotted in Figure 1.1), it reduced to about 0.033 after 20 s, where it will be at a range of 30 km and will be further reduced to 0.01 after 66 s at a range of 100 km. This reduction in pressure is obviously due to the $1/r$ term in Equation 1.20 for spherical waves.

1.1.3   Periodic solution to the wave equation

In the previous section, we have seen that the wave equation describes the motion of a single disturbance of the ambient pressure as a function of time. Next, we introduce the periodic solution to the wave equation. This solution is appropriate to describe waves, for example sound waves in air and in water. A standard way to describe a periodic solution is to introduce a periodic function, for example a cosine function as shown in the next equation

$$f(r - ct) = A(ct - r)\cos\left\{2\pi\frac{ct - r}{\lambda}\right\} \tag{1.22}$$

where λ describes the periodicity, that is, it measures the distance for which the function $f(ct - r)$ repeats itself and is therefore called the wavelength of the periodic function. The factor $A(ct - r)$ is the amplitude of the pressure wave and must also be a function of $(ct - r)$.

For example: a Gabor pulse is characterized by a Gaussian amplitude function

$$A(r - ct) \equiv P_0 \exp\left\{-\frac{1}{2}\left(\frac{ct - r}{\sigma}\right)^2\right\} \tag{1.23}$$

and therefore the periodic function becomes

$$f(r - ct) \equiv P_0 \exp\left\{-\frac{1}{2}\left(\frac{ct - r}{\sigma}\right)^2\right\}\cos\left\{2\pi\left(\frac{ct - r}{\lambda}\right)\right\} \tag{1.24}$$

which we may easily visualize by using a small Matlab script

Matlab script
```
% Scr1_2
r=0:0.1:35;
c=1.5;
lam=10; %wavelength [km]
sig=4;
t=10;
P0=exp(-1/2*((c*t-r)/sig).^2);
p=P0.*cos(2*pi*(c*t-r)/lam);

figure(1)
plot(r,p,'k')
grid on
xlabel('Range [km]')
ylabel('Pressure [rel]')
```

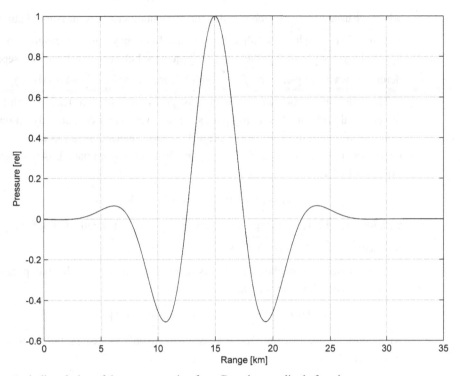

Figure 1.2   Periodic solution of the wave equation for a Gaussian amplitude function

The script differs from the previous script not only in having a different expression for the pressure function but also in that it estimates the sound pressure only for a single time $t = 10$ s but at different ranges.

Executing the script, we obtain Figure 1.2.

Figure 1.2 shows a cosine wave that is modulated with a Gaussian (bell-shape) function. The pressure is oscillating and is reaching its maximum value at 15 km. The distance between the two negative peaks is nearly 9 km, somewhat short of the nominal wavelength, and this is due to the Gaussian amplitude function that modifies the cosine waveform. As all range units are in km, the modelled wavelength λ is 10 km.

The Gabor pulse plays an important role in underwater bioacoustics as bottlenose dolphins (*Tursiops truncatus*) emit echolocation clicks that present a similar shape and for which Gabor pulses seem to be reasonable models (Au, 1993). Only the wavelength is different.

1.1.4   Wavelength frequency relation

When we inspect Equation 1.22, we note that the argument of the cosine term is dimensionless, as it should be, but it is written using the wavelength as divisor. An alternative notation replaces the division by a multiplication

$$f(r - ct) = A(ct - r)\cos(\omega t - kr) \tag{1.25}$$

where we used $k = \frac{2\pi}{\lambda}$, which is called the wave number and is measured in units of m^{-1}, and $\omega = 2\pi\left(\frac{c}{\lambda}\right)$, which is called the angular frequency and is measured in radians/s. Typically, we will not use the angular frequency but keep the factor 2π separate and define the wave frequency by $f = \frac{c}{\lambda}$, which is measured in s^{-1} or Hz (Hertz).

The solution of the wave equation as given in Equation 1.25 is valid for one-dimensional problems (including pure spherical waves that depend only on one dimension, the distance from the wave origin). A more general solution allows k and r to be vectors (e.g. k_x, k_y, k_z and r_x, r_y, r_z in Cartesian co-ordinates), that is, we replace Equation 1.25 by

$$f(ct - r) = A(ct - r)\cos(\omega t - \mathbf{kr}) \tag{1.26}$$

with $\mathbf{kr} = k_x r_x + k_y r_y + k_z r_z$ (again in Cartesian co-ordinates) and $\mathbf{r}$ being the range vector with length $r = |\mathbf{r}| = \sqrt{r_x^2 + r_y^2 + r_z^2}$

Very often, it is mathematically more convenient to describe the periodic wave solution in complex vector notation

$$f(ct - r) = A(ct - r)\exp\{i(\omega t - \mathbf{kr})\} \tag{1.27}$$

where the acoustic pressure P is obtained by using only the real part of the wave solution $f(ct - r)$.

1.1.5   Acoustic impedance

Let us consider the general solution of an outgoing planar wave propagating in the x direction, that is, the y and z components of the wave equation are assumed to be zero. Let us further assume that the local particle speed u satisfies the wave equation, that is: $u = u(ct - x)$, then by differentiating the solution of the wave equation with respect to time and range and after combining the two terms we obtain

$$\frac{\mathrm{d}}{\mathrm{d}t}u(ct - x) = -c\frac{\mathrm{d}}{\mathrm{d}x}u(ct - x) \tag{1.28}$$

As the one-dimensional version of Equation 1.6 may be written as

$$\rho\frac{\mathrm{d}}{\mathrm{d}t}u = -\frac{\partial}{\partial x}P \tag{1.29}$$

we obtain

$$\frac{\partial}{\partial x}P = \rho c\frac{\mathrm{d}}{\mathrm{d}x}u \tag{1.30}$$

and after integration

$$P = Zu \tag{1.31}$$

where we define $Z = \rho c$.

In analogy to Ohm's law, with the particle velocity replacing the electric current and the pressure replacing the electric voltage, the quantity Z may be called the acoustic impedance and is measured in $N\,s/m^3$ (Newton seconds per cubic metre).

The acoustic impedance may in general be a complex number and is only a real number if pressure and particle velocity are in phase, that is, if a change in pressure results immediately in a change in particle velocity. This is typically the case for plane waves where we have $P = (\rho_0 c)u$ and therefore

$$Z_0 = \rho_0 c \qquad (1.32)$$

Z_0 is called the characteristic acoustic impedance and describes the characteristics of the propagation medium for plane waves. ρ_0 is the mean density of the propagation medium.

Example Let us consider two media where sound propagation is of general importance, air and seawater. In air at 20 °C we find the characteristic acoustic impedance to be $Z_{0\_air} = 415\,N\,s/m^3$ and in seawater at 13 °C the value is $Z_{0\_water} = 1.540 \times 10^6\,N\,s/m^3$, that is, the characteristic acoustic impedance in seawater is about 3710 times the value in air.

In terms of Equation 1.31, we can also say that for a given particle velocity the acoustic pressure in seawater will be about 3710 times higher than in air, or equivalently, to reach the same acoustic pressure the particle velocity in seawater can be 3710 times smaller than in air. Air is a soft and water is a harder medium and, using the electric analogy again, water is a better conductor of sound waves than air is.

1.1.6 Plane wave approximation to spherical waves

Recalling Equation 1.6 and applying it to spherical waves represented by $P = \dfrac{A}{r}\exp\{i(\omega t - kr)\}$, we obtain by integration for the particle velocity

$$u = \frac{1}{\rho_0}\left(\frac{1}{r} + ik\right)\frac{P}{i\omega} = \frac{1}{\rho_0 c}\left(1 - \frac{i}{kr}\right)P \qquad (1.33)$$

We note that the acoustic impedance is now a complex number, but that the imaginary part vanishes for $kr \gg 1$, indicating that for values of r that are significantly larger than the wavelength λ, spherical waves also assume the characteristics of a plane wave, as the particle velocity becomes directly proportional to the sound pressure.

1.1.7 Acoustic intensity

The acoustic intensity I, measured in W/m^2, is the acoustic power passing perpendicular through a unit area, and is the product of acoustic pressure and particle velocity. For a one-dimensional wave equation this becomes

$$I(x, t) = P(x, t)u(x, t) \qquad (1.34)$$

As sound pressure and particle velocity vary as a function of position x and time t, so does the acoustic intensity.

If we consider only plane waves then we may use Equation 1.31 to replace the particle velocity and obtain

$$I(x, t) = \frac{P(x, t)^2}{\rho_0 c} \qquad (1.35)$$

As the acoustic impedance in seawater is about 3710 times the acoustic impedance in air, we conclude from Equation 1.35 that for the same pressure the resulting acoustic intensity of a plane wave in water is 3710 times lower than the acoustic intensity in air.

1.1.8   Acoustic energy flux

The acoustic energy flux is the amount of acoustic energy flowing perpendicularly through a unit area. It is measured in terms of W s/m^2 (Watt seconds per square metre) and is obtained by integrating the acoustic intensity $I(x,t)$ over the duration of the sound wave. Using Equation 1.35 for the acoustic intensity, we obtain for the acoustic energy flux $F(x)$

$$F(x) = \frac{1}{\rho_0 c} \int_{-\infty}^{\infty} P^2(x, t) \mathrm{d}t \qquad (1.36)$$

The energy flux is a function of the distance x from the sound source and the integral covers in general all times from minus to plus infinity. This may or may not be a problem. If the signal of interest is limited in time, say $t_1 < t < t_2$, then it makes sense to limit the integration to these times as the signal outside these time windows becomes zero by definition. In this case, the integral is finite and the acoustic energy flux is well defined. In cases, however, where the signal of interest is continuous the integral will in general become infinite, that is, the acoustic energy as estimated by Equation 1.36 becomes useless.

Example In cases of short transients of acoustic waves, for example Gaussian pulses (Equation 1.24), the energy flux is estimated by integrating over all times (minus infinity to plus infinity)

$$F(x) = \frac{P_0^2}{\rho_0 c} \int_{-\infty}^{\infty} \exp\left\{ -\left(\frac{ct - x}{\sigma}\right)^2 \right\} \cos^2\left\{ 2\pi\left(\frac{ct - x}{\lambda}\right) \right\} \mathrm{d}t \qquad (1.37)$$

The integral in Equation (1.37) may be estimated in closed form by noting that $\cos^2 x = \frac{1}{2}(1 + \cos(2x))$ and with the following integral

$$\int_{-\infty}^{\infty} \exp\left\{-\left(\frac{ct-x}{\sigma}\right)^2\right\} \cos\left\{4\pi\left(\frac{ct-x}{\lambda}\right)\right\} dt$$

$$= \frac{\sqrt{\pi}\sigma}{c} \exp\left\{-\left(\frac{\lambda}{8\pi\sigma}\right)^2\right\}$$

$$(1.38)$$

and using

$$S_0 = \int_{-\infty}^{\infty} \exp\left\{-\left(\frac{ct-x}{\sigma}\right)^2\right\} dt = \frac{\sqrt{\pi}\sigma}{c} \qquad (1.39)$$

we obtain for the energy flux

$$F(x) = F_0 = \frac{P_0^2}{2\rho_0 c} S_0 \left[1 + \exp\left\{-\left(\frac{\lambda}{8\pi\sigma}\right)^2\right\}\right] \qquad (1.40)$$

S_0 is measured in seconds and is the integral over the shape factor (amplitude squared) of the waveform, that is, it is proportional to the pulse duration.

1.1.9 Mean acoustic intensity

As said before, the acoustic energy flux is not well defined for cases where the sound is continuous, as then the time integral (Equation 1.36) becomes infinite. In such situations, it is more convenient to characterize the sound not by an energy value but by a time-averaged, mean intensity value that, using an appropriate time window of length T, is estimated as

$$\bar{I}(x) = \frac{F(x)}{T} = \frac{1}{\rho_0 c} \frac{1}{T} \int_0^T A^2(ct-x) \cos^2(\omega t - kx) dt \qquad (1.41)$$

where the pressure squared is integrated over the time of interest T. The mean intensity $\bar{I}(x)$ will in general be a function of the location x.

Example Let us now assume that the acoustic wave is purely sinusoidal, that is, the amplitude function $A(ct-x) = P_0$ is assumed to be constant

$$P(x,t) = P_0 \cos(\omega t - kx) \qquad (1.42)$$

As the acoustic wave is a purely periodic function, it is sufficient to integrate over one cycle of the waveform $t = [0 \cdots 2\pi]$ and divide by the length of the period $T = 2\pi$.

$$\bar{I}(x) = \bar{I}_0 = \frac{P_0^2}{\rho_0 c} \frac{1}{2\pi} \int_0^{2\pi} \cos^2(\omega t - kx) dt \qquad (1.43)$$

Carrying out the integration, we obtain

$$\bar{I}_0 = \frac{1}{2}\frac{P_0^2}{\rho_0 c} = \frac{P_{\mathrm{RMS}}{}^2}{\rho_0 c} \tag{1.44}$$

The mean acoustic intensity of a sinusoidal wave is independent of the position x and equal to one-half of the peak intensity. P_{RMS} is then the classical root mean squared (RMS) pressure value.

We note that the energy flux of a Gaussian wave pulse (Equation 1.40) is related to the mean intensity of a pure sinusoidal wave (Equation 1.44) by

$$F_0 = \bar{I}_0 S_0 \left[1 + \exp\left\{ -\left(\frac{\lambda}{8\pi\sigma} \right)^2 \right\} \right] \tag{1.45}$$

That is, the energy flux of a Gaussian pulse is numerically only a little larger than the mean acoustic density of a purely sinusoidal wave multiplied by the effective duration S_0 of the Gaussian pulse. This is especially true for cases where σ is of the same order or larger than the wavelength λ.

Mathematical note

The following integrals were used in the previous examples:

$$\int\limits_{-\infty}^{\infty} \exp(-a^2 x)\mathrm{d}x = \frac{\sqrt{\pi}}{a} \tag{1.46}$$

$$\int\limits_{-\infty}^{\infty} \exp(-a^2 x)\cos(bx)\mathrm{d}x = \frac{\sqrt{\pi}}{a}\exp\left\{ -\frac{a^2}{4b^2} \right\} \tag{1.47}$$

1.1.10   Total radiated acoustic energy

Emitting sound requires energy. A good indication of the energetic effort for sound generation is the total radiated acoustic energy, which is obtained from the energy flux by integrating over the complete sphere surrounding the sound source.

$$E_A = \int\limits_{0}^{2\pi} \mathrm{d}\varphi \int\limits_{0}^{\pi} F(\varphi, \vartheta, r) \sin \vartheta \mathrm{d}\vartheta \tag{1.48}$$

where $F(\varphi, \vartheta, r)$ is the acoustic energy flux measured now in terms of spherical coordinates φ, ϑ, r (azimuth, elevation, and range). For a constant energy flux $F(\varphi, \vartheta, r) = F_0$ we obtain

$$E_A = 4\pi F_0 \tag{1.49}$$

1.2   Measuring underwater sound (the decibel scale)

The decibel scale is an important tool in acoustics and has been introduced to describe naturally occurring sound intensities in a convenient way. In fact, sound intensities that are audible to humans range from a just audible 10^{-12} W/m^2 to a painful 10 W/m^2. In addition, if we take, for example, the solution to the spherical wave equation we note that the sound intensity should decrease with range squared, that is

$$I(r, t) = \frac{I(1, t)}{r^2} \qquad (1.50)$$

or

$$\frac{I(r, t)}{I(1, t)} = r^{-2} \qquad (1.51)$$

When increasing the range from 1 m to 1 km (10^3 m), the intensity of a spherical spreading sound wave should therefore decrease by a factor of 1 million or to 10^{-6} of the initial value. It was therefore intuitive to introduce a logarithmic scale as a more convenient way of describing sound intensity as such a scale avoids the use of exponents that are common in scientific notations.

1.2.1   Formal definition of the decibel scale

One decibel (dB) is one tenth of a Bel, which itself is the base 10 logarithm of a sound intensity ratio.

$$[\mathrm{dB}] = 10 \log\left(\frac{I}{I_{\mathrm{ref}}}\right) \qquad (1.52)$$

Likewise, the inversion is given by

$$I = I_{\mathrm{ref}} 10^{([\mathrm{dB}]/10)} \qquad (1.53)$$

From the formal definition, we note that the decibel scale is only properly defined if the reference intensity is also given. Unfortunately, there is no unique convention for the selection of the reference intensity. In addition, while the use of intensity as reference would correspond to the formal definition of the decibel scale ('deci' stands for 'a tenth of') it is custom to refer the decibel scale to a sound pressure as reference. While one could see an inconsistency in this convention, there should in practice be no problem as the conversion from intensity ratio to pressure ratio is trivial and shown in Equation 1.54:

$$10 \log\left(\frac{I}{I_{\mathrm{ref}}}\right) = 10 \log\left(\frac{P^2}{P_{\mathrm{ref}}^2}\right) = 20 \log\left(\frac{P}{P_{\mathrm{ref}}}\right) \qquad (1.54)$$

The standard reference for airborne sound is 10^{-12} W/m^2, which is roughly the intensity of a barely audible 1000 Hz pure tone and may easily be expressed in equivalent (RMS) reference pressure of about 2×10^{-5} N/m^2 or 20 μPa.

While a reference sound pressure of 20 μPa has some practical reasons for human air acoustics, it has not been adapted for underwater acoustics, where instead a reference sound pressure of 1 μPa has been selected.

As there are different reference pressures for airborne and waterborne sound, special attention is needed not to confuse airborne and waterborne decibel values. In addition, when referring the dB scale to pressure, the same numerical values result in different sound intensities due to the dependency of the acoustic impedance on density and sound speed. This may be of importance where sound intensities are the relevant sound characteristics.

1.2.2   Sound pressure level

Using the decibel scale, we now can express the sound pressure more conveniently as sound pressure level *SPL*, which is the sound pressure on the decibel scale

$$SPL = 20 \log\left(\frac{P}{P_{\text{ref}}}\right) \tag{1.55}$$

For underwater sound we express the sound pressure level in [dB re 1 μPa], that is, in dB relative to an RMS pressure of 1 μPa, which in water is equivalent to a sound intensity of 6.5×10^{-19} W/m².

How important the use of the reference pressure and the specification of the propagation media are is shown by the following example: a sound pressure of 120 dB re 1 μPa is in water equivalent to 6.5×10^{-7} W/m², but a sound pressure of 120 dB re 20 μPa is in air equivalent to 1 W/m². The same dB value (120 dB) results from a sound intensity that in water is 6.5×10^{-7} times the sound intensity in air.

Conversely a sound intensity of 1 W/m², which in air is equivalent to 120 dB re 20 μPa, corresponds in water to a sound pressure of 182 dB re 1 μPa.

This example is of importance when comparing or discussing sound effects in water and in air, as it says that more energy is required to produce the same sound pressure in water than in air, or in other words a pressure wave in water carries more energy than the same pressure wave in air. The difference is mainly due to the different impedance, or resistance to particle motion.

1.2.3   dB arithmetic made simple

Arithmetic in dB follows two rules:

- addition in dB corresponds to a multiplication of ratios
- subtraction in dB corresponds to a division of ratios

For most practical cases, we only have to remember a few numbers:

0 dB is equivalent to an intensity ratio of 1
10 dB is equivalent to an intensity ratio of 10
3 dB is equivalent to an intensity ratio of 2
5 dB is equivalent to an intensity ratio of about 3

Table 1.1 Conversion from dB to intensity ratio

dB	0	1	2	3	4	5	6	7	8	9	10
Ratio	1	1.25	1.6	2	2.5	3.2	4	5	6.4	8	10

Table 1.2 Conversion from intensity ratio to dB

Ratio	1	2	3	4	5	6	7	8	9	10
dB	0	3	4.8	6	7	7.8	8.5	9	9.5	10

All other values may easily be constructed by the basic rules of addition/multiplication or by straight interpolation.

Some examples of converting intensity ratios to/from dB values

a ratio of 16 is expressed as $16 = 2^4$, equivalent to 4×3 dB $= 12$ dB;

a ratio of 2.5 is expressed as $2.5 = 10/4$, equivalent to 10 dB $- (2 \times 3)$ dB $= 4$ dB

20 dB is equivalent to a ratio of 10^2

33 dB is equivalent to a ratio of 2×10^3 (33 dB $= (3 + 30)$ dB or 2×10^3)

37 dB is equivalent to a ratio of 5×10^3 (37 dB $= (-3 + 40)$ dB or $\frac{1}{2} \times 10^4$)

1.3   Sound velocity

The sound velocity or sound speed is formally defined as a constant relating pressure changes to density changes in the acoustic media (Equation 1.12)

$$c^2 \equiv \frac{\partial P}{\partial \rho} \tag{1.56}$$

If large pressure variations are needed to obtain small density changes, as in liquids or solids, then the sound speed is high; if on the other hand one needs only small pressure variation to change the density, as in gases, the sound speed is low.

1.3.1   Sound velocity in air

As terrestrial acoustics is mainly interested in sound propagation in air and as the sound velocity in air is a classical example in physics, it may be interesting to follow the derivation of a sound speed formula.

The density of a gas depends on its pressure and temperature

$$\rho = \frac{P}{RT} \tag{1.57}$$

where R is the gas constant.

In the case of constant temperature $T =$ const, the Boyle–Mariotte law holds

$$PV = \text{const} \tag{1.58}$$

Sound propagation, however, does not keep the temperature constant as the pressure variation is too fast to exchange heat with the surrounding gas, that is, the process is adiabatic instead of isothermic and the following Poisson equation holds

$$PV^\gamma = \text{const} \tag{1.59}$$

Using this Poisson equation, we obtain the sound speed by noting that

$$\frac{\mathrm{d}}{\mathrm{d}\rho}(PV^\gamma) = 0 \tag{1.60}$$

that is

$$\frac{\partial P}{\partial \rho} + P\gamma \frac{1}{V}\frac{\partial V}{\partial \rho} = 0 \tag{1.61}$$

and

$$\frac{\partial V}{\partial \rho} = \frac{\partial}{\partial \rho}\left(\frac{m}{\rho}\right) = -\frac{1}{\rho}\left(\frac{m}{\rho}\right) = -\frac{V}{\rho} \tag{1.62}$$

so that

$$c^2 \equiv \frac{\partial P}{\partial \rho} = \frac{\gamma P}{\rho} = \gamma RT \tag{1.63}$$

Inserting the appropriate values for dry air at 0 °C ($\gamma = 1.4$, $R = 287$ J/kg/K, $T = 273$ K) we obtain $c = 331.2$ m/s, which is very close to the empirical measurements of the speed of sound in air.

1.3.2 Sound velocity in oceans

To obtain the speed of sound in water one needs again a relation between pressure and density. In general we can say that for all compressible materials, that is also for liquids, the volume V reacts proportionally to small pressure variations

$$\mathrm{d}V = -\kappa V \mathrm{d}P \tag{1.64}$$

with a finite compressibility κ.

Differentiating with respect to density and using Equation 1.62 we get

$$\frac{\partial V}{\partial \rho} = -\frac{V}{\rho} = -\kappa V \frac{\partial P}{\partial \rho} \tag{1.65}$$

or

$$c^2 \equiv \frac{\partial P}{\partial \rho} = \frac{1}{\kappa \rho} \tag{1.66}$$

For pure water at 20 °C we have $\kappa = 0.46\ 10^{-9}$ Pa^{-1} and $\rho = 1000$ kg/m^3 and obtain $c = 1474$ m/s, similar to the measured speed of sound in water.

As no closed theory exists to describe the speed of sound in real seawater, empirical methods are needed to obtain realistic values for the sound speed. Sound speed may be measured directly with reusable or expendable sound-velocity meters, or by using empirical expressions that easily fit measured physical quantities (temperature, salinity, depth) to measured sound speeds. These formulas are all approximations to measurements and a variety of such formulas may be found in oceanographic textbooks.

Leroy *et al.* (2008, 2009) gives a new formula that can be used in all oceans, which includes in addition to temperature, salinity and depth, also the geographic latitude, and which may conveniently be expressed as

$$
\begin{aligned}
c = {} & 1402.5 + 5T - \left(\frac{T}{4.288}\right)^2 + \left(\frac{T}{16.8}\right)^3 \\
& + \left[1.33 - \left(\frac{T}{81.3}\right) + \left(\frac{T}{107.2}\right)^2\right] S \\
& + \left(\frac{Z}{64.1}\right) + \left(\frac{Z}{1980.3}\right)^2 + \left(\frac{Z}{5155}\right)^3 \\
& + \left(\frac{Z}{18519}\right)\left(\frac{\Phi}{45} - 1\right) - \left(\frac{Z}{10172}\right)^3 T \\
& + \left[\left(\frac{T}{57.74}\right)^2 + \left(\frac{S}{69.93}\right)\right]\left(\frac{Z}{1000}\right)
\end{aligned}
\tag{1.67}
$$

where T is the temperature in °C, S is salinity in PSU (practical salinity units), Φ is the geographic latitude in degrees, and Z is depth in metres.

To obtain the sound speed one has therefore to measure both temperature and salinity as functions of depth. This is typically done with an oceanographic instrument called a CTD (conductivity–temperature–depth) profiler. Very often, however, only the temperature is measured, with a device called an XBT (expandable bathythermograph) and the salinity is either taken from databases or assumed to be constant. Using temperature or salinity data from databases is only justified if the required accuracy is limited or one is only interested in typical oceanographic scenarios.

Example For this example, we use the temperature and salinity obtained from the National Oceanographic Data Center (NODC) '*World Ocean Atlas 2009*' climatology datasets, which can be downloaded from http://www.nodc.noaa.gov. If we choose August as month and take as position 41°N and 5°E, which is somewhere west of Corsica in the western Mediterranean Sea, we obtain Figure 1.3.

We note that the temperature is decreasing continuously from the surface to the selected maximum depth of 1500 m. The salinity is approximately 38 PSU and relatively constant. The resulting sound speed reaches nearly 1535 m/s at the surface, shows a well-defined minimum at a depth of 100 m and increases again with depth to nearly 1530 m/s at 1500 m.

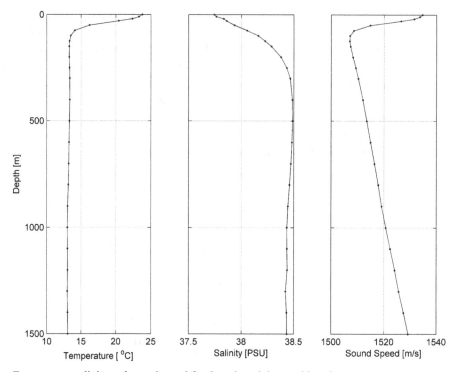

Figure 1.3   Temperature, salinity and sound speed for the selected time and location.

This decrease in temperature close to the surface is also called the thermocline; the lower end of the thermocline typically correlates with the minimum sound speed, the depth of which is also called the depth of the sound channel.

From this example, we may conclude that the sound speed is, in general, not constant in the world's oceans. From Leroy's sound speed equation, we see that even for constant temperature and salinity the sound speed increases with depth by about 1 m/s every 64 m. When assuming non-constant sound speed it is convenient to realize that in general, ocean water is horizontally stratified and, with some exceptions, it is safe to assume that the sound speed has a predominant vertical gradient. Close to the surface, sound speed is dominated by temperature variations; temperature itself tends to be horizontally stratified, with warmer water being found closer to the surface. The temperature of the surface water is a direct consequence of the solar energy falling on it, and will therefore show a strong diurnal variation. At depth, where seawater temperature tends to be less variable, the sound speed is more a function of water pressure. Despite the dominance of vertical variations in sound speed, some minor horizontal sound speed gradients may exist, for example due to varying sea surface temperature. In the following, we assume for simplicity that the sound speed has only a depth-dependent component, that is, $c = c(z)$.

Table 1.3 Typical salinities in world oceans and seas

Ocean basin	Salinity [PSU]
North Sea	35.4
Baltic Sea	13
Mediterranean Sea	38
Black Sea	22
Atlantic	35
Pacific	34.5
Arctic	33

Source: Urban (2002).

Matlab script

The following function assumes that the NODC datasets are already loaded into a local directory '../climatology'. If not already done, the function extracts temperature and salinity from the files containing climatological data and stores them in a temporary Matlab data file. It uses then Leroy's formula to estimate the sound speed at a selected location.

```
function [sv,T,S]=getSoundSpeed(month,lat0,lng0)
%month=8;
%lat0=41;
%lng0=5;
%
root0='../climatology/';
%
Leroy = @(T,S,Z,L) ...
        1402.5 + 5*T - (T/4.288).^2 + (T/16.8).^3 ...
        + (1.33 - (T/81.3) + (T/107.2).^2).*S ...
        + (Z/64.1) + (Z/1980.3).^2 - (Z/5155).^3 ...
        + (Z/18519).*(L/45-1) -T.*(Z/10172).^3 ...
        + ((T/57.74).^2 + (S/69.93)).*(Z/1000);
%
lfn=sprintf('ts_%02d_1d',month);
if exist([lfn '.mat'],'file')~=2
    % extract data from climatology
    %
    temp=[];
    root1=sprintf('t_%02d_1d/',month);
    dirs=dir([root0 root1 '*.csv']);
    %
    for ii=1:length(dirs)
        fname=[root0 root1 dirs(ii).name];
        fprintf('loading %s\n',fname)
        temp(ii).dat=xlsread(fname,'A:D');
    end

    sal=[];
    root1=sprintf('s_%02d_1d/',month);
    dirs=dir([root0 root1 '*.csv']);
```

```
%
for ii=1:length(dirs)
    fname=[root0 root1 dirs(ii).name];
    fprintf('loading %s\n',fname)
    sal(ii).dat=xlsread(fname,'A:D');
end
% save data to local mat file
save(lfn,'temp','sal')
end
%
%load local mat file to access data
load(lfn)
%
%extract temperature and salinity vector
T=zeros(length(temp),1);
D=0*T;
S=0*T;

for ii=1:length(temp)
    data=temp(ii).dat;
    gd=find(abs(data(:,1)-lat0)<=1 & abs(data(:,2)-lng0)<=1);
    %
    % temperature
    T(ii)=mean(data(gd,4));
    % depth
    D(ii)=mean(data(gd,3));
    % salinity
    S(ii)=mean(sal(ii).dat(gd,4));
end

%estimate sound speed
sv=[D,Leroy(T,S,D,lat0)];
```

The Leroy formula of the sound speed is here given as what is called an anonymous function. If the function is used in different functions or scripts then it becomes convenient to program the functionality as a regular Matlab function. The script tries to use climatological data stored in a Matlab data file; if this file does not exist, the script will generate the file.

The following script loads the sound speed profile and generates the graphics shown in Figure 1.3.

```
%Scr1_3
month=8;
lat0=41;
lng0=5;
%
[sv,T,S]=getSoundSpeed(month,lat0,lng0);
%
figure(1)
subplot(131)
plot(T,sv(:,1),'k.-')
axis ij
```

```
grid on
ylabel ('Depth [m] ')
xlabel ('Temperature [ ^oC] ')

subplot (132)
plot (S, sv (:,1) , 'k.-')
set (gca, 'yticklabel', [])
axis ij
grid on
xlabel ('Salinity [PSU] ')

subplot (133)
plot (sv (:,2) , sv (:,1) , 'k.-')
set (gca, 'yticklabel', [])
axis ij
grid on
xlabel ('Sound Speed [m/s] ')
```

1.4   Sound propagation

The solution of the wave equation (e.g. Equation 1.20 for a spherical wave) describes the propagation of sound through an acoustic medium, such as water. It is intuitive to realize that, in an ideal scenario where the sound speed is constant and where there are no boundaries, a spherical sound wave will continue to expand as a sphere of increasing radius. In the following section we try to understand how sound will propagate in a real environment where the sound speed is not constant and where boundaries are present.

1.4.1   Snell's law

Here we consider a particular solution of the sonar equation $f(ct - r)$ where $r = ct$. We may interpret this equation as describing the location of the centre of the propagating wave. This equation tells us that the range rate of the propagating wave is given by the now depth-dependent sound speed:

$$\frac{dr}{dt} = c(z) \tag{1.68}$$

As, by assumption, the sound speed varies only in the vertical direction and not in the horizontal plane, so the range rate changes only in a vertical direction and not within the horizontal layers. In Figure 1.4 we have plotted $dr = cdt$ so that the direction of the propagating sound wave is oriented downwards and its propagating direction forms an angle ϑ with the (vertical) sound speed gradient.

We observe that in a horizontally layered medium the horizontal increment of a sound wave must be constant

$$dx = \frac{dr}{\sin \vartheta} = \text{const} \tag{1.69}$$

With $dr = c(z)dt$ and $\vartheta = \vartheta(z)$ we obtain from this requirement the general Snell's law

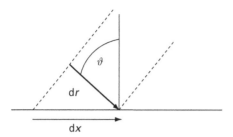

Figure 1.4   Geometry of wave propagation in a horizontal sound speed layer.

$$\frac{c(z)}{\sin \vartheta(z)} = \text{const} \tag{1.70}$$

The angle ϑ, which describes the direction in which the sound wave is propagating, therefore varies as a function of the sound speed. In particular

$$\sin \vartheta(z) = \frac{c(z)}{c(z_0)} \sin \vartheta(z_0) \tag{1.71}$$

where z_0 is the depth at which the sound wave was generated and $\vartheta(z_0)$ is the initial angle for which the wave propagation is analysed.

1.4.2   Reflection at a plane interface

Figure 1.4 describes a special case of the more general situation where the sound wave arrives at a discontinuity in the sound speed. Such a discontinuity may be the sea surface or the bottom of the ocean. We assume, as shown in Figure 1.5, that the sound wave is propagating in medium 1 with the incident angle ϑ_1, that at the interface a part of this sound wave is reflected back into medium 1 with angle ϑ_1 and that a third part of the sound wave is penetrating the interface and continues to propagate in medium 2 with angle ϑ_2.

In general, one can state that the total pressure of the different waves at the interface in both media must be the same

$$P_i + P_r = P_t \tag{1.72}$$

where P_i is the pressure of the incident wave, P_r is the pressure of the wave reflected from the interface, and P_t is the pressure of the wave transmitted through the interface.

Furthermore, the sum of the particle velocity normal to the interface must be the same on both sides of the interface.

$$u_{zi} - u_{zr} = u_{zt} \tag{1.73}$$

where u_{zi}, u_{zr} and u_{zt} are the vertical particle velocity (i.e. normal to the interface) of the incident, reflected and transmitted wave, respectively; we use the fact that the vertical particle velocity component for the reflected wave is inverted relative to particle velocity of the incident wave.

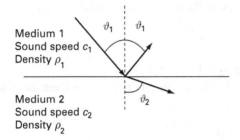

Figure 1.5   Reflection/refraction of a sound wave on a plane interface.

From Figure 1.4 we note that $u_z = u \cos \vartheta$ and from Equation 1.31 we have $u = \dfrac{P}{\rho c}$ and therefore obtain

$$\frac{P_i}{\rho_1 c_1} \cos \vartheta_1 - \frac{P_r}{\rho_1 c_1} \cos \vartheta_1 = \frac{P_t}{\rho_2 c_2} \cos \vartheta_2$$

$$\frac{P_i}{\rho_1 c_1} \cos \vartheta_1 - \frac{P_r}{\rho_1 c_1} \cos \vartheta_1 = \frac{P_i + P_r}{\rho_2 c_2} \cos \vartheta_2$$

and finally

$$P_i \left(\frac{\cos \vartheta_1}{\rho_1 c_1} - \frac{\cos \vartheta_2}{\rho_2 c_2} \right) = P_r \left(\frac{\cos \vartheta_1}{\rho_1 c_1} + \frac{\cos \vartheta_2}{\rho_2 c_2} \right)$$

The reflection coefficient R is the quotient of the pressure of the reflected wave to the pressure of the incident wave

$$R = \frac{P_r}{P_i} = \frac{\rho_2 c_2 \cos \vartheta_1 - \rho_1 c_1 \cos \vartheta_2}{\rho_2 c_2 \cos \vartheta_1 + \rho_1 c_1 \cos \vartheta_2} \tag{1.74}$$

The transmission coefficient T is the relative pressure of the wave transmitted through the interface and is obtained from Equation 1.72 as $T = 1 + R$ or with Equation 1.74

$$T = \frac{2 \rho_2 c_2 \cos \vartheta_1}{\rho_2 c_2 \cos \vartheta_1 + \rho_1 c_1 \cos \vartheta_2} \tag{1.75}$$

The relation between ϑ_2 and ϑ_1 is given by Snell's law:

$$\sin \vartheta_2 = \frac{c_2}{c_1} \sin \vartheta_1 \tag{1.76}$$

Total reflection may occur if the sound speed in the second medium exceeds the following relation

$$c_2 > \frac{c_1}{\sin \vartheta_1} \tag{1.77}$$

or equivalently, if the incident angle of the incident wave satisfies

$$\sin \vartheta_1 > \frac{c_1}{c_2} \tag{1.78}$$

as then no real solution exists for ϑ_2 in Equation 1.76.

The critical angle at which total reflection occurs is given by the ratio of the sound speeds

$$\sin \vartheta_C = \frac{c_1}{c_2} \tag{1.79}$$

Total reflection results in a phase shift of the reflected sound wave according to the relation

$$\tan \varphi = \frac{\rho_1}{\rho_2} \frac{\sqrt{\sin^2 \vartheta_1 - \sin^2 \vartheta_C}}{\cos \vartheta_1} \tag{1.80}$$

which varies from zero at the critical angle, where total reflection starts to occur, to a phase angle of 90° at an incident angle of 90°. We will come back in the next section to the phase shift of bottom-reflected sound after we learn more about typical bottom characteristics.

Example: vertical incidence Vertical incidences are a special case of Equation 1.75, yielding very instructive results when applied to air–water interfaces. Vertical incidences (Figure 1.4) occur when $\cos \vartheta_1 = 1$; and under the assumption of a plain interface we obtain from Equation 1.75

$$T = \frac{2\rho_2 c_2}{\rho_2 c_2 + \rho_1 c_1}$$

Sound transmission air to water

$$T_{\mathrm{A2W}} = 2 \frac{1.54 \times 10^6}{(1.54 \times 10^6 + 415)} \approx 2$$

The sound pressure in water is, for vertical incidence, about twice the sound pressure in air.

Sound transmission water to air

$$T_{\mathrm{W2A}} = 2 \frac{415}{(1.54 \times 10^6 + 415)} \approx \frac{1}{2000}$$

The sound pressure in air decreases, for vertical incidence, by about 2000 times.

It is therefore easier for marine life to hear airborne sound than it is for humans to hear, say, whales and dolphins, in the air. Because sound barely transmits from water into air, it is therefore fair to say that, in air, the sound pressure of waterborne sound is nearly zero. The

Table 1.4 Typical bottom characteristics

Type	ρ [kg/m^3]	c_p [m/s]	a_p [dB/λ_p]	c_s [m/s]	a_s [dB/λ_p]
Silt	1700	1575	1	$80\,z^{0.3}$	1.5
Gravel	2000	1800	0.6	$180\,z^{0.3}$	1.5
Limestone	2400	3000	0.1	1500	0.2
Basalt	2700	5250	0.1	2500	0.2

Notation: ρ is the density; c_p and c_s are the compressional and shear speeds respectively; a_p and a_s are the attenuation for compressional and shear waves, respectively; z is the thickness of the silt and gravel layer. The attenuation coefficient is given in relation to the wavelength to reflect the frequency-dependence of the attenuation.
Source: Jensen *et al.* (2000).

water–air interface is therefore also called a pressure release or soft interface. Assuming now $P_t = 0$, Equation 1.72 indicates $P_r = -P_i$, that is the pressure function of the surface-reflected wave is inverted relative to the incoming signal (180° out of phase).

1.4.3   Bottom interface

Sound propagation in seawater is typically bounded not only by an air interface but also by a bottom interface. Whereas the air interface is, in most cases, treated as a 'vacuum', or completely reflecting, the bottom interface must be considered more specifically.

In general, the bottom characteristics are described by bottom density, sound speed and attenuation of pressure and shear waves (Table 1.4). Shear waves are typical of solid materials, where sound propagation may also be due to oscillations that are transverse to the direction of propagation. Shear waves do not exist in liquids and gases.

It should be noted that the values given in Table 1.4 are merely indicative. Especially for long ranges, where sound propagation interacts increasingly with the bottom, a site-specific geo-acoustic model should be established combined with sufficient ground truth information.

Example Equation 1.80 indicates that under certain circumstances interfaces may result in total reflection and associated phase shift. Using the bottom characteristics of limestone, it is instructive to determine the onset of total reflection and this phase shift.

Matlab code

```
%Scr1_4
th1=0:90;
%water
rho1=1000;
c1=1500;
% Limestone
```

```
rho2=2400;
c2=3000;
%
arg1=real(rho1*sqrt((c2*sin(th1*pi/180)).^2-c1.^2));
arg2=rho2*c2*cos(th1*pi/180);
phi=atan2(arg1,arg2)*180/pi;

figure(1)
plot(th1,phi,'k')
xlabel('Incident angle [^o]')
ylabel('Phase shift [^o]')
title('Limestone')
```

From Figure 1.6 we note that total reflection occurs for incidence angles (ϑ in Figure 1.4) greater than 30° and that a phase shift of 45° occurs at an incidence angle of 70°. For incidence angles less than 30°, sound is partially reflected and the remaining sound energy enters the second layer as a transmitted wave. In this regime, the reflected wave does not suffer any phase shift, but with the onset of total reflection (incident angle > 30°), a phase shift in the reflected wave is observed. As the critical angle is related to the critical range, one consequence is that, at distances greater than this critical range, reflected waves experience a variable degree of phase shift, resulting in more and more randomization of any initially deterministic signal.

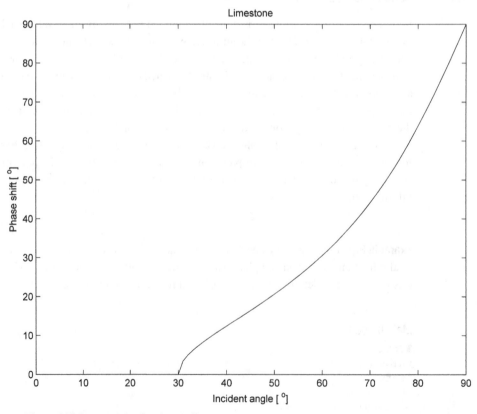

Figure 1.6   Phase shift due to total reflection on limestone.

1.4.4   Sound absorption

Sound waves that travel through the ocean will always suffer an absorption loss, that is, sound energy is transformed from mechanical into other forms of energy. Absorption loss in seawater is mainly due to the finite amount of time that real fluids require for responding to pressure waves. This process is called relaxation and may be due to both the shear viscosity of water and the ionic dissociation of chemical molecules (magnesium sulphate and boric acid) in salt water.

Extensive research has been put into empirical formulas to describe seawater sound absorption as a function of frequency, salinity, acidity and water depth. The most complete such formula is given by Francois and Garrison (1982a, b), which presents the attenuation of sound in seawater in dB/km according to

$$\alpha_f = A_1 P_1 \frac{f_1 f^2}{f_1^2 + f^2} + A_2 P_2 \frac{f_2 f^2}{f_2^2 + f^2} + A_3 P_3 f^2 \tag{1.81}$$

where the frequency f is given in kHz.
The contribution of boric acid $B(OH)_3$ is given by

$$A_1 = \frac{8.86}{c} 10^{(0.78pH - 5)}$$

$$P_1 = 1$$

$$f_1 = 2.8 \sqrt{\frac{S}{35}} 10^{\left(4 - \frac{1245}{273+T}\right)}$$

The contribution of magnesium sulphate $Mg(SO)_4$ is given by

$$A_2 = 21.44 \frac{S}{c}(1 + 0.025T)$$

$$P_2 = 1 - 1.37 \times 10^{-4}z + 6.2 \times 10^{-9}z^2$$

$$f_2 = \frac{8.17 \times 10^{\left(8 - \frac{1990}{273+T}\right)}}{1 + 0.0018(S - 35)}$$

The contribution of pure water viscosity is given by

$$A_3 = 4.937 \times 10^{-4} - 2.59 \times 10^{-5}T + 9.11 \times 10^{-7}T^2$$
$$- 1.50 \times 10^{-8}T^3 \qquad (T \le 20\,^{\circ}C)$$

$$A_3 = 3.964 \times 10^{-4} - 1.146 \times 10^{-5}T + 1.45 \times 10^{-7}T^2$$
$$- 6.6 \times 10^{-10}T^3 \qquad (T > 20\,^{\circ}C)$$

$$P_3 = 1 - 3.83 \times 10^{-5}z + 4.9 \times 10^{-10}z^2$$

Some frequencies of interest in PAM are 100 Hz, 1 kHz, 15 kHz, 40 kHz and 100 kHz, where the attenuation coefficients become 0.0007, 0.05, 1.6, 9.5 and 41 dB/km, respectively (Matlab script Sct1_5).

Matlab code to estimate absorption coefficients

```
%Scr1_5
f=[0.1 1 15 40 100]; %frequency in kHz
%
T=20; % Temperature in deg C
S=38; % Salinity in parts per thousand
Z=0; % Depth in m
L=42; % Latitude in deg
pH=8; % pH value of sea water
%
% sound speed in m/s
c=Leroy(T,S,Z,L);
%
% attenuation in dB/km
FrancoisGarrison(f,T,S,Z,c,pH)
```

Constraining salinity, temperature and depth, one may obtain a simpler formula for the frequency dependence of the sound attenuation. As an example, we obtain, for salinity of 38 PSU, depth of zero, and 20 °C

$$\alpha_f = \frac{0.27f^2}{2.7 + f^2} + \frac{106f^2}{17400 + f^2} + 2.2 \times 10^{-4}f^2 \tag{1.82}$$

where the frequency f is given in kHz.

1.5   Signals, noise and interference

So far, we have treated sound as something neutral, independent of any application. However, when considering passive acoustic monitoring (PAM) one should realize that sound may be composed of signals, noise and interference.

Signals are sound messages that we want to receive with our PAM system. Noise and interference are sounds that we do not want to receive, and which in general will degrade our capability to recognize our signals of interest; interferences are signal-like sounds but of no interest to the PAM application. The discrimination between signals and interference is highly application-dependent. In the case of PAM of cetaceans, we can consider as signals all cetacean vocalizations (e.g. clicks, whistles, calls) and as interference all non-cetacean sounds, be they biological (e.g. snapping shrimps) or anthropogenic (e.g. echo-sounders).

The term *noise* is in general not very well defined in acoustics and sometimes includes all unwanted sounds. Here, the term *noise* is used to describe only randomly varying sound and is therefore in contrast to interferences. This differentiation is justified, as different signal processing techniques are used to detect signals against noise, and to classify signals in the presence of interferences. Random noise is always there, but interference may or may not be present. Interferences can become a major problem, and depending on the PAM objectives it may be possible that even cetacean sounds are considered as interferences. For example, the classification performance of a beaked whale detector may degrade in the presence of dolphin clicks. Some PAM implementation may even consider sound from conspecifics as interferences reducing system performance. This intrinsic ambiguity requires a careful differentiation between signal and interference.

1.5.1   Signals

Suppressing the range-dependent component of the wave function, we may write the sound pressure of a plane wave as

$$P(t) = A(t) \cos\{2\pi f(t)\, t + \varphi(t)\} \tag{1.83}$$

Signals are sound waves that carry information, which may be coded in the amplitude $A(t)$, in the frequency $f(t)$, or in the phase function $\varphi(t)$.

Typically $A(t)$ will be limited in time, allowing us to assume, without loss of generality, that

$$A(t) = 0 \text{ for } t < 0 \text{ and } t > T \tag{1.84}$$

The signal frequency $f(t)$ may also carry information and is assumed to be positive and limited

$$f(t) > 0 \text{ and } f(t) < F_{\max} \tag{1.85}$$

where $F_{\max}$ is the maximum frequency of the signal.

The signal phase $\varphi(t)$ is limited between 0 and 2π, and may be used for information-dependent phase modulation.

Signals in which the information is encoded in the amplitude $A(t)$ are called amplitude-modulated or AM signals; signals in which the information is encoded in the signal frequency $f(t)$ are called frequency-modulated or FM signals. Although pure implementations of AM and FM signals are known from artificial sonar systems, biological sounds are rarely so easy classified; in nearly all cases, both amplitude and frequency modulations are present. Cetacean sounds cover a large variety of AM and FM signals and will be described in more detail in Chapter 2. A further scheme to encode information is the phase modulation, which, however, may be expressed in terms of frequency modulation.

1.5.2   Noise

Noise is considered to be randomly varying sound against which signals must be detected. Noise can be created by randomly varying amplitude $A(t)$, frequency $f(t)$, or phase $\varphi(t)$ fluctuations. One important feature of naturally occurring noise is that, in general, its origin cannot be attributed to a single sound source; hence noise may be considered as a very large sum of individual sound waves with varying amplitudes, frequencies and phases. Such noise fluctuations are mostly due to the vast number of different noise sources, repeated interaction of the propagating sound waves with ocean boundaries and spatial variations of the sound speed profiles.

As noise is best described as a random process, it is intuitive to describe it by basic statistical distributions. Considering the importance of the Gauss distribution

$$f(a) = \frac{1}{\sqrt{2\pi}\sigma}\exp\left\{-\frac{1}{2}\left(\frac{a - a_0}{\sigma}\right)^2\right\} \tag{1.86}$$

in general statistics, it is no surprise that this distribution is also of interest for the description of noise phenomena, even if its merits have more to do with its easy use than with its accurate description of noise processes. Empirical noise is never really perfectly Gaussian.

With respect to narrowband noise, a convenient statistical distribution of noise amplitude is given by explicitly considering noise as a large sum of sinusoidal waves modulated by random phase

$$P(t) = A_0 \exp\{i(\omega t + \varphi_0)\} = \sum_{n=1}^{N} A_n \exp\{i(\omega t + \varphi_n)\} \tag{1.87}$$

where we use the complex (exponential) form of the sound wave and consider φ_n as a random phase equally distributed between 0 and 2π.

Suppressing the temporal term $\exp\{i\omega t\}$, e.g. by setting $t = 0$, we obtain

$$P(0) = P_1 + iP_2 \tag{1.88}$$

with

$$P_1 = \sum_{n=1}^{N} A_n \cos\varphi_n$$
$$P_2 = \sum_{n=1}^{N} A_n \sin\varphi_n \tag{1.89}$$

As the number of sources N becomes very large then both quantities P_1 and P_2 tend toward zero-centred Gaussian variables with variance $\sigma^2 = \sum_{n=1}^{N} A_n^2$.

The variable $P^2 = |P(0)|^2 = P_1^2 + P_2^2$ follows a chi-squared law with two degrees of freedom and consequently the distribution of the amplitude $P = \sqrt{P_1^2 + P_2^2}$ reads

$$f(P) = \frac{P}{\sigma^2}\exp\left\{-\frac{1}{2}\left(\frac{P}{\sigma}\right)^2\right\} \tag{1.90}$$

which is also known as the Rayleigh distribution.

The most probable amplitude value (the peak of the Rayleigh distribution) is $P_{\text{peak}} = \sigma$; the mean amplitude value is given by $<P> = \sigma\sqrt{\frac{\pi}{2}}$ and the variance of the amplitude is $\text{var}(P) = \sigma^2\left(2 - \frac{\pi}{2}\right)$, resulting in a coefficient of variation CV

$$CV = \frac{\text{var}(P)}{<P>^2} = \frac{4 - \pi}{\pi}$$

The Rayleigh distribution is widely used in underwater acoustics to describe amplitude fluctuations of noise and will play an important role in tuning the detection performance of a PAM system.

2 Cetacean sounds

The objective of this chapter is twofold: to introduce the sounds of interest, i.e. the sounds made by different cetaceans, and to develop techniques suited to describe the different sound categories. I interlace sound descriptions and processing techniques to allow the reader to experiment with the Matlab scripts, maybe to change some parameters, or to visualize the results differently, etc.

I initially classify the sounds made by cetaceans into two categories, according to their observed functionality. For the purposes of PAM, cetacean sounds are conveniently divided into 'echolocation clicks' and 'communication signals'.

Typical echolocation clicks are made by toothed whales during foraging. They are short pulses of significant intensity, highly directional and used for locating food.

Typical communication signals are whistles or complex call sequences emitted by dolphins and baleen whales. They are relatively tonal or pulsed signals with a varying degree of spectral variability and with potentially low directionality. This differentiation does not suggest that echolocation signals are not useful for communication, but it emphasizes a difference in the primary function of the sound.

Cetacean clicks are next described both in time and in the frequency domain; a time–frequency description is developed for tonal and pulsed cetacean vocalizations. This chapter also discusses the directivity of echolocation signals and the expected source levels of cetacean sounds.

2.1 Classification of cetaceans

All whales, dolphins and porpoises form the taxonomic order Cetacea, which is divided into two suborders, Odontoceti (toothed whales) and Mysticeti (baleen whales). These are further divided into families, e.g. Delphinidae (ocean dolphins) or Ziphiidae (beaked whales). Within families, cetaceans are classified into genera and species. For example, common bottlenose dolphins carry the scientific name *Tursiops* (genus) *truncatus* (species) and belong to the family Delphinidae, suborder Odontoceti. Although the taxonomy tries to group and link together related species, ongoing research, e.g. in genetics, continuously challenges the relation between the species. For examples of work on genetic classification, see Henshaw *et al.* (1997), Dalebout *et al.* (2001), Arnason *et al.* (2004) or Morisaka and Connor (2007).

The visual observation of wild cetaceans is in most cases detailed enough to allow classification to the species level. In some cases, it may further be possible to identify the individuals, i.e. recognizing the specific individual as already detected previously. Such identification requires features or marks that are unique to an individual, allowing later re-identification, and is done mostly by photographic documentation (photo-ID). Sometimes, however, the classification to species level is difficult and it is then common practice to classify to genus or family level only (e.g. unidentified beaked whale).

2.2   Classification of cetacean sounds

Cetaceans generate a variety of sounds for a variety of purposes. Even if the taxonomy of cetaceans is continuously challenged by new scientific discoveries (e.g. those provided by genetic analysis), the overall approach tends to produce unique classifications. The same cannot be said for the classification of cetacean sounds. There are moans, grunts, shrieks, knocks, thumps, shot-guns, creaks, buzzes, clicks, pulses, up or down calls, ratchets, trumpets, etc. All these terms seem more to reflect the imagination of the listener than to be a useful definition of cetacean sound. Of course, one can say that sounds made by cetaceans are most likely species-specific, so in the end it should not matter what name the sound has, as for classification purposes every species should be treated separately.

If we set the descriptive approach aside, we may classify the sound either by its function or by its acoustic appearance. The most intuitive use of sound is for communication within the same species or between different species; consequently, communication sounds are a major functional class. Cetaceans are also known to use sound for echolocation, that is, to detect, classify or inspect objects, be it prey or other submerged items in their vicinity, indicating that echolocation signals are a second major functional class. Approaching the sound classification from an acoustics standpoint, sound can be considered either periodic, e.g. tonal, or aperiodic and therefore pulse-like. Interestingly, there is no one-to-one relation between function and acoustic properties; there are pulses used for communication and short tonal sounds used for echolocation.

2.2.1   General characteristics of echolocation signals

Echolocation is the use of sound, or better, the use of echoes from an emitted animal sound to estimate the location, range and direction of an object. The term was coined in the 1950s by Griffin (1958) to describe the use of ultrasonic clicks by bats and is now also adapted for cetaceans. So far, only toothed whales have a generally accepted capacity for echolocation.

In order to be useful for localization, echolocation sounds should satisfy some basic requirements; in particular the sound must support the estimation of range to the object but also the estimation of the angle at which the objects can be found (both in azimuth and in elevation).

Successful echolocation requires the object to reflect incoming sound back to the sound emitter. How much sound is backscattered depends mainly on the size and acoustic properties of the object and the frequency used. Typically, only a fraction of incoming sound energy is reflected back from the object to the echolocating animal. The ratio of reflected to incoming sound intensity is then called the target strength.

Estimation of range is typically done by measuring the time a sound pulse needs to travel from the animal to the object and back. Obviously, the animal must be able to receive the returning sound and it is fair to assume that the animal is silent again before the echo returns from the object. That this is not strictly necessary is demonstrated by horseshoe bats, which emit constant frequency (CF) echolocation signals that are so long as to suggest the onset of echo processing while the animal is still emitting the CF signal. This 'listening while calling' capability, however, has not been demonstrated in cetaceans. Neither has it been demonstrated that cetaceans are able to estimate the range of distant objects beyond the range that is related to the interval between two consecutive clicks.

Angular estimation may be done in three ways by: using an omni-directional transmitter and a directional receiver, using a directional transmitter and an omni-directional receiver, or using both a directional transmitter and a directional receiver. Like all mammals, cetaceans are binaural and therefore possess a directional receiver, the angular resolution of which is limited by the distance between the ears. All echolocating cetaceans therefore improve their overall angular resolution by emitting a narrow sound beam, the directionality of which is again limited by the size of the cetacean head.

2.2.2   General characteristics of communication signals

Communication is intuitively the most obvious use of sound, but care must be taken to avoid anthropomorphic projections. Communication is generally understood as the transfer of information from one being to another for the advantage of the sender, the receiver, or both. Although in the context of acoustic communication it seems obvious that the information is encoded in sound signals, the amount of information transferred by sound is for cetaceans completely unknown and subject to scientific debate. The information content will vary with the contexts of the signal exchange and the behavioural states of the animals; the expectations, or intentions, of the sender or receiver of signals may further determine the information content.

Communication in a noisy or complex environment may require the animal to increase the complexity of the transmitted signals to obtain sufficient redundancy to convey the information to the receiver. As this approach may fail for very long-range communication, an alternative strategy is then to slow down the information rate and to transmit very simple signals over long periods of time to reduce the required bandwidth of the communication channel.

An important consequence for PAM systems is that the type and occurrence of communication signals is unpredictable; they vary from short pulses, suited for determining presence or absence only, to stereotyped whistles emitted for social cohesion, to multi-themed songs whose functional purposes are still not fully understood. PAM systems that are based on communication signals for detecting and monitoring cetaceans must therefore be flexible and robust enough to recognize all types of cetacean communication sound.

From the acoustics point of view, one would expect communication signals to be less directional than echolocation signals, because highly directional signals allow mainly private peer-to-peer communication, where sender and receiver face each other. In a normal three-dimensional ocean scenario, sender and receiver are more or less arbitrarily oriented; this suggests that, for communication signals to be successful, they should be emitted with little or no directional preference. A weak forward–backward directionality of, say, less than 6 dB (intensity ratio of 4) is, however, a plausible observation, because sound that is generated at the head of the animal may experience some attenuation when propagating backwards through the body of the animal. A second necessity for sound to be useful as a communication signal is that it allows the encoding of information, as discussed in Section 1.5.1. For example, a continuous tone with constant frequency and amplitude does not convey information other than its presence.

2.2.3   General presentation of cetacean sounds

The variety of terms used to characterize cetacean sounds suggests that there will not be a single way to describe and visualize cetacean sounds. Of course, cetacean sounds may be considered to carry information and therefore expressed in terms of Equation. 1.83, that is:

$$P(t) = A(t)\cos\{2\pi f(t)\,t\} \tag{2.1}$$

Here the explicit phase term $\varphi(t)$ has been dropped, as it may be easily expressed in terms of frequency $f(t)$.

In Chapter 1 we constructed the sound pressure $P(t)$ by choosing specific functions of $A(t)$ and $f(t)$, that is, we synthesized the sound (e.g. Gabor pulse, Equation 1.24). To describe cetacean sounds, we are confronted with the opposite problem: we have sound recordings, that is, we have $P(t)$, and want to analyse these sounds in terms of $A(t)$ and $f(t)$.

2.3   Cetacean sound presentation in the time domain

The easiest way to describe cetacean sound is to plot directly the sound pressure measurements, that is, we plot $P(t)$ as a time series, as done in Figure 1.2 for a Gaussian pulse. This type of presentation is the most common, and is appropriate for signals that exhibit large amplitude variations. All cetacean clicks fall into this category.

2.3.1   Beaked whale clicks

Figure 2.1 shows a sequence of cetacean clicks in the time domain. It presents the received sound pressure (vertical or amplitude axis, measured in arbitrary units) as

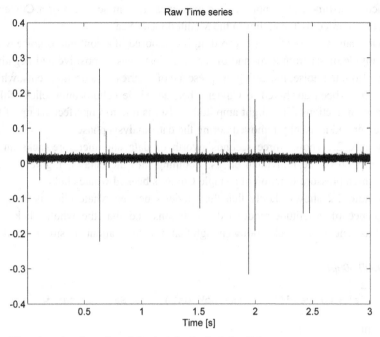

Figure 2.1   Time domain of a series of Cuvier's beaked whale clicks.

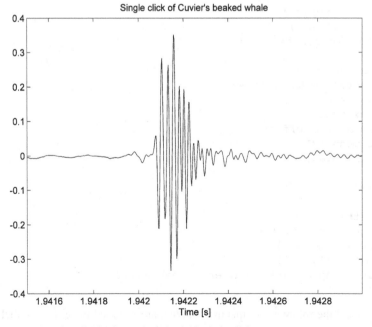

Figure 2.2   Time domain of a single Cuvier's beaked whale click.

function of time (horizontal or time axis, measured in seconds) of a Cuvier's beaked whale recorded by the author in the Mediterranean Sea.

One can easily see that the recording is composed of a continuous noise with a slight amplitude offset from zero, and of recurring instances of positive and negative deflections from the noise, indicating the presence of a series of short transients, which in this case have been attributed to Cuvier's beaked whale echolocation clicks followed by surface reflections. The slight amplitude offset is due to imperfect set-up of the sound receiver and is usually removed during the data analysis phase.

Figure 2.1 shows three seconds' worth of data and therefore lacks any detailed description of the individual clicks. Zooming into the dataset, Figure 2.2 shows the measured pressure function of a single Cuvier's beaked whale click.

Figure 2.2 shows clearly that the Cuvier's beaked whale click is composed of a sequence of amplitude-modulated oscillations and that the whole click lasts about 200 μs. The whole click is long enough that it features about six strong oscillations.

Matlab script

```
%Scr2_1
[xx,fs] = wavread('../Pam_book_data/zifio Select Scan.wav');
%
% construct time vector
tt = (1:length(xx))/fs;

figure(1)
clf,
plot(tt,xx,'k');
set(gca,'fontsize',12)
xlabel('Time [s]')
xlim(tt([1 end]))
title('Raw Time series')

% zoom into data
tsel=tt>1.9415 & tt<1.9430;
tt1=tt(tsel);
xx1=xx(tsel);
% remove bias (offset)
bias=mean(xx1);
xx1=xx1-bias;

figure(2)
clf
plot(tt1,xx1,'k')
set(gca,'fontsize',12)
xlabel('Time [s]')
xlim(tt1([1 end]))
title('Single click of Cuvier''s beaked whale')
```

This and the following scripts use data that are assumed to be in a parallel directory called Pam_book_data , that is, the following Matlab instruction dir('../

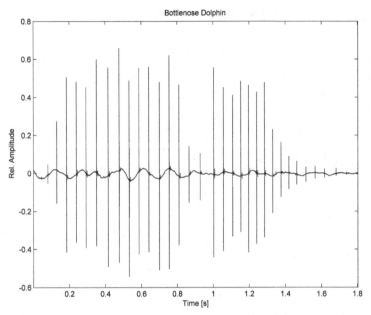

Figure 2.3   Click train of a bottlenose dolphin (courtesy C. Bazua Duran).

`pam_book_data'`) results in a list of files. If necessary, this part of the script should be adapted to reflect the location of the data files.

2.3.2   Dolphin clicks

Cuvier's beaked whale clicks are a relatively new observation; in fact, they were only recently discovered by means of acoustic archival tags (DTAG; see Johnson and Tyack, 2003) that were placed on the back of Cuvier's beaked whales (Zimmer *et al.*, 2005a). Dolphin echolocation clicks have been studied extensively in the past; mostly because dolphins have long been kept in captivity and trained to perform acoustic tests in a laboratory environment (see, e.g. Au, 1993). Figure 2.3 shows a click train of a free-ranging bottlenose dolphin. The Matlab script `Scr2_2` is not reproduced in the text as it is redundant to `Scr2_1`. The time sequence in Figure 2.3 shows about 16 clicks per second with varying amplitude, where amplitude variations of up to a factor of 10 may be observed. Recalling that the amplitude is proportional to the received sound pressure, the observed amplitude variation corresponds to a received level variation of 20 dB. The reason for this variation cannot be determined without additional information about the behaviour of the dolphin. The variation may be due to changes in source level or due to changing geometry, resulting in differences in the received sound level. Figure 2.3 shows further a slowly oscillating background noise with about 8–9 cycles per second, which is mostly due to very low-frequency vibration present in the experimental setup (e.g. boat engine). Such unwanted low-frequency disturbances are very frequent in field experiments and at a later stage we will implement filters to remove them.

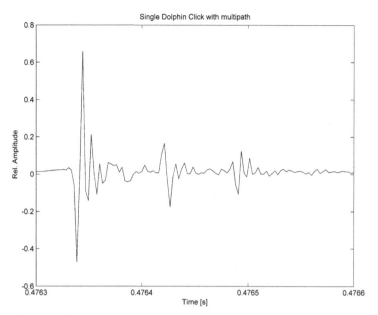

Figure 2.4   Single dolphin click/pulse with multi-path.

Figure 2.4 is a zoom into Figure 2.3 showing a very short snippet of 300 μs around what in Figure 2.3 appears to be a single click. In reality there is a strong click followed by a sequence of weaker clicks. All clicks are very short, with only 2–3 oscillations lasting about 30 μs. The bottlenose dolphin clicks are much shorter than the clicks presented in Figure 2.2 for Cuvier's beaked whales and are therefore best described as short pulses, similar to the Gabor pulse (Figure 1.2). The second pulse in Figure 2.4 is not only smaller in amplitude but also phase-inverted: that is, whereas for the first (main) pulse the deflections are first negative and then positive, for the second pulse the deflections are first positive and then negative. This phase reversal is typical of a surface reflection as mentioned in Section 1.4.2. The shape of the third pulse is again in phase with the main pulse, indicating a bottom reflection. The short interval between direct arrival (first pulse), surface and bottom reflections indicates finally that the bottlenose dolphin data were recorded in shallow waters, where the boundary reflections follow the original clicks at short intervals.

Descriptions of dolphin clicks constitute a significant amount of underwater bioacoustic literature, which is of no surprise when considering that there is often easy access to these animals (see, e.g. Au, 1993).

2.3.3   Sperm whale echolocation clicks

Sperm whales are, after dolphins, a favourite subject for passive acoustics as they emit long sequences of audible echolocation or sonar clicks. Figure 2.5 shows such a click sequence, which is sometimes also referred to as a click train. Again, the script Scr2_3

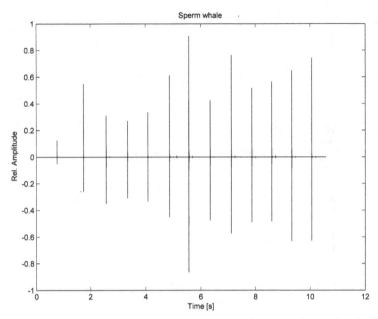

Figure 2.5   Sperm whale click train.

is omitted from the text as it is similar to `Scr2_1`. The data were recorded with a digital archiving tag (DTAG; Johnson and Tyack, 2003) attached on the back of a sperm whale in the Mediterranean Sea.

Figure 2.6 is a zoom into Figure 2.5 and shows that each sperm whale click is composed of a sequence of short pulses separated by 3.7 ms. This multi-pulse structure is typical of sperm whale clicks and is a consequence of the particular sound production of sperm whales, also called bent-horn sound generation (Norris and Harvey, 1972; Møhl, 2001; Zimmer *et al.*, 2005b, c), which we will discuss in more detail in the next section. For the time being we note that the distance between consecutive pulses within a single sperm whale click is called the inter-pulse interval (IPI) and is determined by the time the sound needs to traverse twice the spermaceti organ of the sperm whale. The IPI of sperm whales is therefore related to the length of the animal, for which two formulas may be found in the literature:

- Formula of Gordon (1987) for smaller animals (IPI < 5 ms)

$$L = 4.833 + 1.453(\text{IPI}) - 0.009(\text{IPI})^2$$

- Formula of Rhinelander and Dawson (2004) for larger animals (IPI > 5 ms)

$$L = 17.12 - 2.189(\text{IPI}) + 0.251(\text{IPI})^2$$

where L and IPI are given in m and ms, respectively.

Here we apply the formula of Gordon and find the overall length of the sperm whale to be 10.2 m, which could be considered as a juvenile.

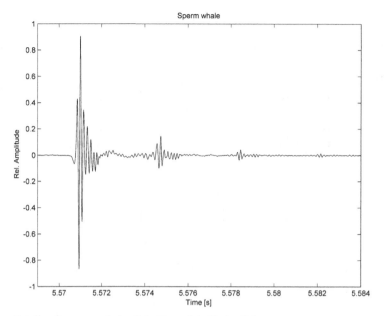

Figure 2.6   Details of a sperm whale click. Zoom into Figure 2.5.

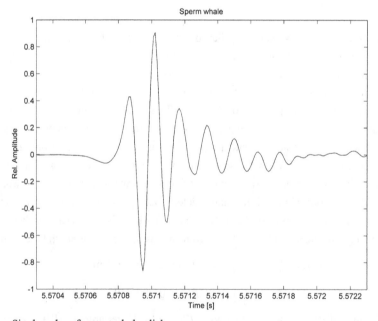

Figure 2.7   Single pulse of sperm whale click.

If we zoom in to the first pulse of Figure 2.6, we obtain what is shown in Figure 2.7. We note that the single pulse consists of some four oscillations lasting over a period of 1 ms, with elevated amplitudes for the first two oscillations, similar to the dolphin click (Figure 2.4).

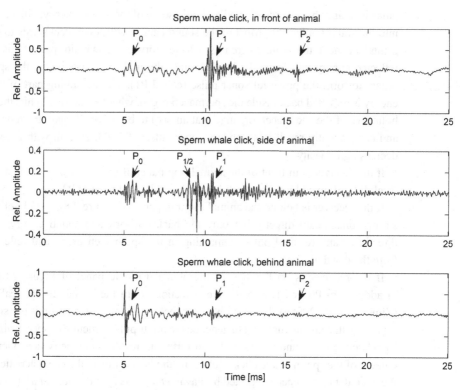

Figure 2.8   Sperm whale clicks recorded at distance under different aspects.

In contrast to Figure 2.4 where a single dolphin pulse is followed by multiple boundary reflections, the multi-pulse structure of a sperm whale click (Figure 2.6) is emitted by the whale itself.

2.3.4   Remotely received sperm whale clicks

Figures 2.6 and 2.7 show a single sperm whale click and pulse, but these data are recorded by using the DTAG attached to the animal dorsally behind the head and the data are not necessarily characteristic of the sound one may record with a distant hydrophone.

Figure 2.8 shows three sperm whale clicks recorded at different times from a single animal at distances that exceed 1 km. The top click was recorded in front of the animal, the middle click was recorded from the side of the animal, and the bottom clids was recorded from behind the animal. Although there is no doubt that this sperm whale also emits its click with a multi-pulse structure, some explanation is required of the details of Figure 2.8.

Without at this point going into details of sperm whale sound production, Figures 2.6 and 2.8 confirm that the multi-pulse click generation is complicated but not mysterious. According to the actual accepted bent-horn theory (Norris and Harvey, 1972; Møhl, 2001), a single pulse is generated below the blowhole at the tip of the nose. This original pulse is mainly backward-oriented and only some of the sound energy will leak directly into the water (called P0). Most of the sound energy will travel backwards towards the frontal air sac at the skull, where it is reflected forward. The major part of this forward-reflected energy will leave the head through the junk, which is below the spermaceti organ, to form the powerful sonar pulse (called P1). The remaining part of the sound energy is reflected back inside the spermaceti organ, where it once again hits the forward boundary of the spermaceti organ (distal air sac) to be reflected again towards the skull and consequently result in secondary sonar pulses, P2, P3, etc., all with progressively decreasing intensity.

If the receiver is in front of the animal (top panel of Figure 2.8), it will first sense the weak P0, followed by the powerful sonar beam P1, the weaker P2, etc.

If the receiver is behind the animal (bottom panel of Figure 2.8), it will not sense the strong sonar beam directly but only the backward-oriented sound energy, either P0 directly or any residual pulses remaining in the spermaceti organ and reflected back from the distal air sac.

If the receiver is on the side of the animal (middle panel of Figure 2.8), it will, in addition to P0 or P1, also receive reflections from the frontal air sac (called P1/2) (Zimmer et al., 2005b). It should be clear that the boundaries that limit the spermaceti organ and that are instrumental to generate the multi-pulse structure of sperm whale clicks by reflecting the sound waves back and forth will result in geometry-dependent energy content of the sperm whale click. Robust methods to estimate the characteristic IPI from the click data have been suggested by Pavan et al. (1997) and Teloni et al. (2007).

2.3.5   Sperm whale coda

Sperm whales not only emit echolocation clicks but are also known to produce stereo-typed click sequences, called coda, which are short click sequences and which are very often repeated, as shown in Figure 2.9.

Sperm whale coda are thought to serve as communication between sperm whales, as so far no coda have been registered for solitary whales. Figure 2.9 shows what is called a 3+1 coda from the Mediterranean Sea (Pavan et al. 2000). As is expected for communication signals, the actual pattern may change for different geographic regions (Watkins and Schevill, 1977; Weilgart and Whitehead, 1997).

Figure 2.10 confirms that coda clicks also show the typical multi-pulse structure seen in the regular clicks in Figure 2.6. Although the inter-pulse interval (IPI) is again 3.7 ms there are two differences that are worth noting: first, the amplitude difference between consecutive pulses is less for coda clicks; and second, between the clear visible pulses we find additional acoustic energy. Both effects are due to the increased reverberation within the spermaceti organ of sperm whales, similar to the internal reflection described for Figure 2.8. This increase in reverberation, which also gives sperm whale coda clicks a metallic quality, may also be due to intentional changes in the geometry of the spermaceti organ to improve the communication functionality of sperm whale coda, i.e. to reduce the directionality of the sound emission.

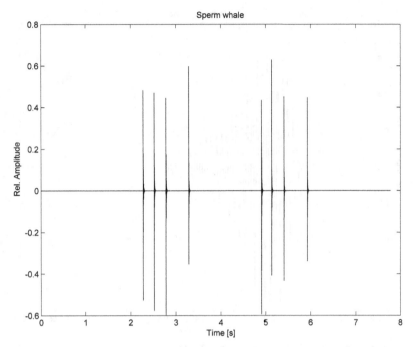

Figure 2.9 Sperm whale coda.

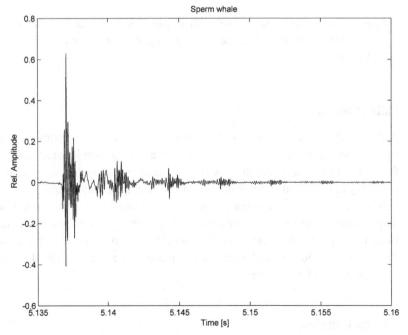

Figure 2.10 Pulse structure of single sperm whale coda click.

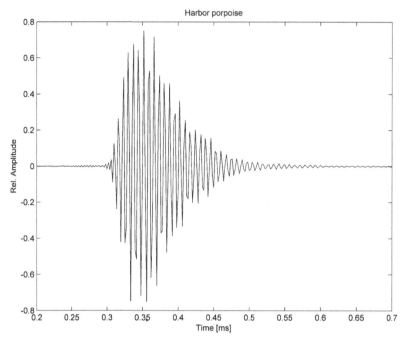

Figure 2.11   Harbour porpoise clicks (courtesy Peter T. Madsen).

Sperm whale clicks are, owing to their complexity, a challenge for optimal click detection. However, they do afford multiple opportunities for classification, tracking and identification.

2.3.6   Harbour porpoise

A further class of cetacean clicks is represented by the harbour porpoise. Figure 2.11 shows a typical harbour porpoise click; note that its duration is somewhat shorter than that of a Cuvier's beaked whale click but with much more oscillation, indicating a higher frequency of the sound wave. Whereas bottlenose dolphins and sperm whales emit short pulses with few cycles, harbour porpoises emit relatively long pulses, not necessarily long in time but in terms of oscillations, albeit with significant amplitude modulation. Harbour porpoise clicks are therefore considered as narrowband, or monochromatic clicks, whereas dolphin and sperm whale pulses are more broadband signals, as we will observe in a later chapter.

2.3.7   Inter-click interval

One special feature of echolocation clicks is that they are emitted in nearly regular sequences. These sequences of echolocation clicks, or click trains, are typically described by the inter-click interval (ICI), which quantifies the temporal separation of the clicks emitted from the same animal. The ICI is a statistical variable and may be characteristic of cetacean species. As a rule of thumb, the ICI of a sperm whale is around 0.5–2 s (Whitehead and Weilgart, 1990; Zimmer *et al.*, 2003; see also Figure 2.5); that of a beaked whale is about 0.3–0.4 s (Johnson *et al.*, 2004; Zimmer *et al.* 2005a; see Figure 2.1) and dolphins click faster with a typical ICI of 0.1 s or less (Madsen *et al.* 2004b; see Figure 2.3).

Very often the term ICI is replaced in the literature by the term IPI, especially when characterizing the click–pulse sequence of dolphins. There should, however, be no reason for confusion, as the only whale that emits multi-pulsed clicks is the sperm whale. Consequently, only for the sperm whale the term IPI means the time delay between the individual components of the clicks, whereas for all other whales and dolphins IPI may be used exchangeably with the term ICI.

2.4   Cetacean sounds in the frequency domain

Figures 2.2, 2.4, 2.7 and 2.11 show that cetacean clicks are in reality oscillating sound waves, which leads to the question of the value of the frequency $f(t)$ of these sound waves. There is a variety of methods to determine this oscillation frequency. Inspecting Figure 2.2 manually (e.g. by measuring the times of maximal amplitudes) results in a mean time between consecutive oscillations of 0.0248 ms, resulting in a signal frequency of 40.3 kHz, and for the harbour porpoise click (Figure 2.9) we obtain 0.0071 ms and therefore 140.4 kHz.

Matlab script to measure frequency by hand

```
% use 'ginput' to switch into input mode
% move cross-hair to local maxima
% and press the right button
% repat this for the six major oscillations (from left to right)
% terminate with pressing the 'Enter' key.
gg=ginput;
tm=gg(:,1);
dt=mean(diff(tm));
fr=1/dt
```

2.4.1   Spectral analysis

This manual method is not very precise and is therefore only appropriate for a fast initial inspection of single waveforms. A more standard way to obtain the frequency is to carry out a Fourier analysis and to obtain the most likely frequency from the Fourier spectrum. The Fourier analysis can be considered as a principal analytical tool in mathematics and physics (e.g. Papoulis, 1962).

Fourier transform

Formally, the Fourier analysis or Fourier transform is defined as an integral

$$F(\omega) = \int_{-\infty}^{\infty} P(t)\exp\{-i\omega t\}\,dt \tag{2.2}$$

where $P(t)$ is the sound pressure and ω is the circular frequency for which the Fourier transform is evaluated.

The Fourier spectrum is then the squared absolute value of the Fourier transform.

$$S(\omega) = |F(\omega)|^2 \tag{2.3}$$

Although the Fourier integral (Equation 2.2) maintains its importance in theoretical considerations, it is nowadays seldom used to estimate the spectrum (Equation 2.3) and then only when the integral may be evaluated in closed form. One such non-trivial case is that given for the Gabor pulse, as we shall see in the following example.

Example Consider the Gabor pulse of Equation 1.24, which can be expressed as

$$P(t) \equiv P_0 \exp\left\{-\frac{\lambda_0^2}{8\pi^2\sigma^2}(\omega_0 t - k_0 r)^2\right\}\cos\{\omega_0 t - k_0 r\} \tag{2.4}$$

Before estimating the Fourier transform it is convenient to introduce a new variable $x = \omega_0 t - k_0 r$ and therefore also $t = \dfrac{x + k_0 r}{\omega_0}$ and $dt = \dfrac{dx}{\omega_0}$. With these new variables the Fourier transform becomes

$$F(\omega) = P_0 \int_{-\infty}^{\infty} \left[\exp\left\{-\frac{x^2}{2k_0^2\sigma^2}\right\}\cos\{x\}\right]\exp\left\{-i\frac{\omega}{\omega_0}(x + k_0 r)\right\}\frac{dx}{\omega_0} \tag{2.5}$$

or, with $\cos\alpha = \dfrac{\exp(i\alpha) + \exp(-i\alpha)}{2}$

$$\begin{aligned}
F(\omega) &= \frac{P_0}{2\omega_0}\exp\left\{-i\frac{\omega}{\omega_0}k_0 r\right\}\int_{-\infty}^{\infty}\exp\left\{-\frac{x^2}{2k_0^2\sigma^2} - i\left(\frac{\omega}{\omega_0} - 1\right)x\right\}dx \\
&+ \frac{P_0}{2\omega_0}\exp\left\{-i\frac{\omega}{\omega_0}k_0 r\right\}\int_{-\infty}^{\infty}\exp\left\{-\frac{x^2}{2k_0^2\sigma^2} - i\left(\frac{\omega}{\omega_0} + 1\right)x\right\}dx
\end{aligned} \tag{2.6}$$

The integrals have now the form of

$$\int_{-\infty}^{\infty}\exp\{-ax^2 - ibx\}\,dx = \sqrt{\frac{\pi}{a}}\exp\left\{-\frac{b^2}{4a}\right\}$$

with $a = \frac{1}{2k_0^2\sigma^2}$ and $b = \frac{\omega}{\omega_0} \mp 1$, so that the Fourier transform (Equation 2.6) becomes

$$F(\omega) = \frac{P_0}{2\omega_0} \exp\left\{-i\frac{\omega}{\omega_0}k_0 r\right\} \sqrt{2\pi}k_0\sigma$$

$$\left[\exp\left\{-\frac{k_0^2\sigma^2}{2}\left(\frac{\omega}{\omega_0} - 1\right)^2\right\} + \exp\left\{-\frac{k_0^2\sigma^2}{2}\left(\frac{\omega}{\omega_0} + 1\right)^2\right\}\right]$$

or after further simplifications

$$F(\omega) = \frac{P_0\sqrt{2\pi}\sigma}{2}\frac{}{c}\exp\left\{-i\omega\frac{r}{c}\right\}$$

$$\left[\exp\left\{-\frac{\sigma^2}{2c^2}(\omega - \omega_0)^2\right\} + \exp\left\{-\frac{\sigma^2}{2c^2}(\omega + \omega_0)^2\right\}\right]$$

(2.7)

Inspecting the solution (Equation 2.7) we realize that the Fourier transform of the Gaussian pulse consists of the sum of two Gaussian functions, which become maximal for $\omega = \pm\omega_0$, that is for positive and negative values of the signal frequency ω_0.

Mathematical note

$$\int_{-\infty}^{\infty} \exp\{-ax^2 - ibx\}dx$$

$$= \int_{-\infty}^{\infty} \exp\left\{-a\left(x^2 + \frac{ib}{a}x + \left(\frac{ib}{2a}\right)^2 - \left(\frac{ib}{2a}\right)^2\right)\right\}dx$$

$$= \exp\left\{-\frac{b^2}{4a}\right\} \int_{-\infty}^{\infty} \exp\left\{-a\left(x + \frac{ib}{2a}\right)^2\right\}dx$$

$$= \exp\left\{-\frac{b^2}{4a}\right\}\sqrt{\frac{\pi}{a}}$$

Fast Fourier Transform

In the above example we obtained the Fourier transform of a Gaussian pulse by straightforward integration, which is only possible for specific examples. In practice, numerical methods are required to estimate the spectrum of a signal.

The standard numerical method to estimate the Fourier transform of a time series is the Fast Fourier Transform, or FFT (Brigham, 1974). The FFT is an efficient algorithm of the discrete Fourier transform (DFT) where the integral in Equation 2.2 is replaced by a finite sum

$$F_m = \frac{1}{f_S}\sum_{n=0}^{N-1} P_n \exp\{-i\omega_m n\}$$

(2.8)

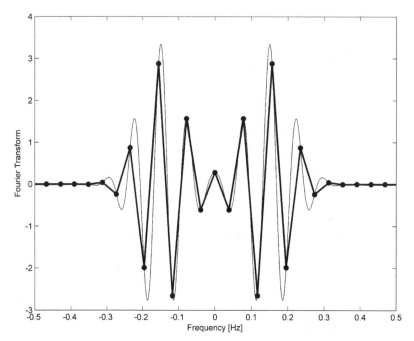

Figure 2.12   Real component of Fourier transform (thin line) and FFT (thick line).

where f_S is the sampling frequency of the time series P_n.

The DFT uses N discrete samples P_n, $n = 0, \ldots, N-1$, that is, the continuous function $P(t)$ is now sampled at discrete times. If we assume that P_n is sampled at time t_n then the sampling frequency is simply the reciprocal value of the time difference between two consecutive samples $\mathrm{d}t = t_n - t_{n-1}$, which in general is taken to be constant, that is:

$$f_S = \frac{1}{t_n - t_{n-1}}.$$

Although the DFT is defined without constraints on the frequency ω_m, practical implementations, for example the Fast Fourier Transform (FFT), replace the continuous frequency ω_m of the Fourier transform by discrete frequencies $\omega_m = 2\pi \frac{m}{M}$ where $m = 0, \cdots, M-1$. By the oscillating nature of the complex exponential term $\exp\{-\mathrm{i}\omega_m n\} = \exp\{-\mathrm{i}(\omega_m + 2\pi)n\}$, it is sufficient to limit the frequency between $\omega_m = 0$ and $\omega_m = 2\pi$, or equivalently to limit the number of discrete frequencies between $m = 0$ and $m = M-1$.

Figure 2.12 shows a direct comparison of the Fourier transform (Equation 2.2) and the FFT implementation of the DFT (Equation 2.8) of a Gaussian pulse. The thin line describes the closed form solution of the Fourier integral (Equation 2.2) and the dots give the FFT results. These FFT values in Figure 2.12 are connected by straight (thick) lines to support the visualization, but also to demonstrate that the Fourier transform coincides with the FFT values at the FFT frequencies. The common practice of interpolating linearly between FFT frequencies, as done in Figure 2.12, is in general no more than a visualization tool and does not result in a proper interpolated transform.

Matlab code

```
%Scr2_7
if ~exist('ifl'), ifl=1; end

%pulseFFT
r=20; % center location [km]
c=1.5; % sound speed [km/s]
lam=10; % signal wavelength [km]
sig=4; % Gaussian pulse width

% signal frequency [Hz]
fo=c/lam;
omo=2*pi*fo;

%sampling frequency [Hz];
fs=10;
%time vector for sampling pressure
t=0:(1/fs):25;
% sampled Gaussian pulse
Pt=exp(-1/2*((c*t-r)/sig).^2).* ...
    cos(2*pi*(c*t-r)/lam);

%frequency vector for Fourier transform
fa=-0.4:0.001:0.4;
% circular frequency
oma=2*pi*fa;
%Gaussian pulse Fourier transform
Fa=1/2*sqrt(2*pi)*sig/c*exp(-i*oma*r/c).* ...
    (exp(-sig^2/2*((oma-omo)/c).^2)+ ...
     exp(-sig^2/2*((oma+omo)/c).^2));

% number of FFT frequencies
if ifl==1
    nfft=256;
else
    nfft=512;
end
% call of FFT
Fct=fft(Pt,nfft)/fs;
% rotation to center zero frequency
Fc=fftshift(Fct);
% frequency vector for FFT
fc=((0:nfft-1)/nfft-0.5)*fs;

%display results
figure(1)
hp=plot(fa,real(Fa),'k-',fc,real(Fc),'k*-',fc, real(Fc),'ko');
set(hp(2),'linewidth',2)
xlim(0.5*[-1 1])
xlabel('Frequency [Hz]')
ylabel('Fourier Transform')
```

Comment on the call to the FFT

The simplest call to the FFT is y = fft (x), where x is a time series and y is the FFT output. If the length of the time series is N then the number of frequencies estimated by fft (x) is also N. The number of frequencies to be estimated may also be specified by the second input number y = fft (x, nfft) where it is customary to specify the number of frequencies as a power of 2 (e.g. nfft = 128, 256, 512, 1024, etc). The reason for this is that FFT implementations are computationally more efficient when the number of frequencies is a power of 2. The Matlab implementation of the FFT is further such that if nfft is less than the length of the time series vector x then the FFT uses only the first nfft samples and the rest is ignored. If the number of requested frequencies nfft is greater than the number of samples in the vector x then the FFT implementation extends to input vector x to a length of nfft by adding zero values.

Interpolating FFT

If we want to increase the frequency accuracy of the FFT to approach the Fourier integral we simply have to increase the number of frequencies, that is, to increase M (or nfft in the Matlab script). Figure 2.13 shows the result of the same Matlab script as used for Figure 2.12, with the exception that the number of frequencies to be estimated by the FFT is now increased from nfft = 256 to nfft = 512, i.e. twice as many frequencies and therefore a better approximation to the Fourier integral. In particular, the frequency at which the spectrum becomes maximal is now more accurately estimated.

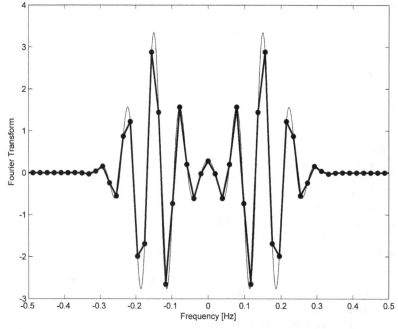

Figure 2.13   Interpolating FFT. Similar to Figure 2.12, but with nfft = 512.

Increasing the spectral accuracy is, however, not synonymous with increasing the spectral resolution, for which one needs to increase the number of samples in the dataset. It should be obvious that by adding zeros to the dataset, as done with the interpolating FFT, we do not add more information on the spectral content to obtain the dataset, and we should therefore not expect to obtain more detail in the spectrum.

A similar result could have been achieved by decreasing the sampling rate by a factor of 2, that is by replacing $fs = 10$ with $fs = 5$. Decreasing the sampling rate to increase the spectral fidelity is always possible if the spectral energy is expected to be at frequencies that are much lower than the sampling frequency. In the example of Figure 2.13 the maximal observed spectral content is below 0.4 Hz and therefore well below the sampling frequency of 10 Hz. The limit of this resampling is given when the sampling frequency corresponds to the maximal spectral extent including both positive and negative frequencies.

Nyquist criterion for sampling frequency

The above discussion on the spectral fidelity may be seen in a different light. A spectral analysis is only acceptable if the signal has zero, or negligible, energy in the region of and above half the sampling frequency, or otherwise the sampling frequency must be increased. This is equivalent to the *Nyquist criterion*, which requires that the sampling frequency must be higher than twice the maximum frequency present in the signal. If necessary, the data have to be lowpass filtered before being sampled to remove all frequency content above the Nyquist frequency (half the sampling frequency). All spectral energy that is found at higher frequencies than the half the sampling frequency will otherwise be folded (aliased) into the measured spectrum and contaminate the spectrum of interest. We will return in Chapter 10 to the constraints on selecting a proper sampling frequency in data acquisition systems.

2.4.2   Spectra of cetacean clicks

At this point we have the means to estimate the spectra (Equation 2.3) of the four selected cetacean clicks (Figures 2.2, 2.4, 2.7 and 2.11), which are given in Figure 2.14. It is custom to present only the positive frequencies in the spectra as for real valued signals the spectrum is symmetric around zero, i.e. the spectra are equal for positive and negative frequencies. The spectra in Figure 2.14 are normalized to the peak value to ease comparison of the different signals. We note that the spectra of the four clicks peak at different frequencies, that is, we obtain as the dominant frequency 39.75 kHz for the Cuvier's beaked whale click, 107.67 kHz for the bottlenose dolphin click, 8.63 kHz for the sperm whale pulse and 137.70 kHz for the harbour porpoise click.

Matlab code

To generate Figure 2.14 we first program a support function, which loads the data and carries out the FFT. As we want to carry out this functionality multiple times, it is convenient to offload this functionality into a separate Matlab function. After the support

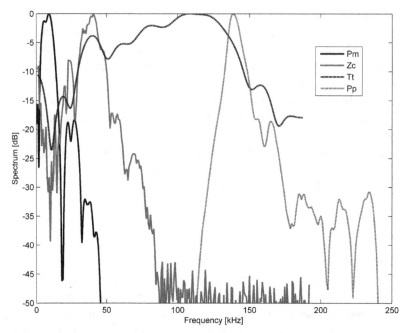

Figure 2.14 Spectra of cetacean echolocation clicks. (Pm: sperm whale; Zc: Cuvier's beaked whale; Tt: bottlenose dolphin; Pp: harbour porpoise)

function getSpectrum we find the script that loads the individual clicks and displays them in a single figure.

Support function

```
function [S,fc,x,fs]=getSpectrum(fname,nfft,xl)
%obtain sampling frequency
[x,fs]=wavread(fname,1);
kfs=fs/1000;
[x,fs]=wavread(fname,round(xl*kfs));
%
% call of FFT
Fc=fft(x,nfft)/fs*2;
S=10*log10(Fc.*conj(Fc));
% frequency vector for FFT
fc=(0:nfft-1)/nfft*kfs;
```

Script to generate Figure 2.14

```
%Scr2_8
% number of FFT frequencies
nfft=512;

fname1='../Pam_book_data/sw253a01_010719.wav';
sp1='Sperm whale'; xl1=10058+[0 1.5];
[S1,fc1,x1,fs1]=getSpectrum(fname1,nfft,xl1);
S1=S1-max(S1(2:end));
```

```
fname2='../Pam_book_data/zifioSelectScan.wav';
sp2='Cuvier''s beaked whale'; xl2=1942+[0 1];
[S2,fc2,x2,fs2]=getSpectrum(fname2,nfft,xl2);
S2=S2-max(S2(2:end));

fname3='../Pam_book_data/T0003217_scan_for_ ranging.wav';
sp3='Bottlenose Dolphin'; xl3=476.3+[0 0.1];
[S3,fc3,x3,fs3]=getSpectrum(fname3,nfft,xl3);
S3=S3-max(S3(2:end));

fname4='../Pam_book_data/Phocoena.wav';
sp4='Harbor porpoise'; xl4=[0.2 0.7];
[S4,fc4,x4,fs4]=getSpectrum(fname4,nfft,xl4);
S4=S4-max(S4(2:end));

ifr=2:nfft/2+1;
figure(1)
hp=plot(fc1(ifr),S1(ifr),'k-',...
        fc2(ifr),S2(ifr),'k-',...
        fc3(ifr),S3(ifr),'k-',...
        fc4(ifr),S4(ifr),'k-','linewidth',2);
set(hp(2),'color',0.5*[1 1 1])
set(hp(4),'color',0.5*[1 1 1])
ylim([-50 0])
xlabel('Frequency [kHz]')
ylabel('Spectrum [dB]')
legend('Pm','Zc','Tt','Pp',0);

fmax1=fc1(find(S1(ifr)==max(S1(ifr))));
fmax2=fc2(find(S2(ifr)==max(S2(ifr))));
fmax3=fc3(find(S3(ifr)==max(S3(ifr))));
fmax4=fc4(find(S4(ifr)==max(S4(ifr))));
res=[fmax1,fmax2,fmax3,fmax4]
```

2.5 Cetacean sounds in the time–frequency domain

The spectral analysis of cetacean sounds as presented in Section 2.4 implicitly assumes that the frequency of the signal is constant during the analysis period, or that we are not interested in an analysis of the time-dependence of the signal frequency function. Here we extend the analysis and ask for a concurrent time and frequency analysis.

2.5.1 Theoretical basis

Let us consider a time series or signal of the form

$$P(t) = A(t) \cos\{2\pi f(t)\, t\} \tag{2.9}$$

and let us estimate both the amplitude $A(t)$ and the frequency $f(t)$ as a function of time.

Short Time Fourier Transform (STFT)

The standard procedure for presenting time-varying signals is the spectrogram, which is the magnitude of the Short Time Fourier Transform (STFT) $X(\omega, \tau)$, defined as

$$X(\omega, \tau) = \int_{-\infty}^{\infty} P(t)w(t - \tau)\exp\{-i\omega t\}dt \qquad (2.10)$$

where $w(t - \tau)$ is a window function centred around time τ and $P(t)$ is the time series as defined in Equation 2.9.

As the name already indicates, the STFT is applied to a short snippet of the data that has been cut out of the time series by a window function $w(t - \tau)$. The reason for this approach is the assumption that, for signals with moderate variation in amplitude and frequency, short snippets may be characterized by signals with near-constant frequency.

Typical window functions are the Hann window, the Hamming window and the Gauss window, defined as follows

$$w_{\text{Hann}}(t - \tau) = \frac{1}{2}\left[1 + \cos(\pi\frac{t - \tau}{\tau})\right] \quad \text{for}|t - \tau| < \tau \qquad (2.11)$$

$$w_{\text{Hamming}}(t - \tau) = \frac{1}{2}\left[1.08 + 0.92\cos(\pi\frac{t - \tau}{\tau})\right] \quad \text{for}|t - \tau| < \tau \qquad (2.12)$$

$$w_{\text{Gauss}}(t - \tau) = \exp\left\{-\frac{1}{2}\left(\frac{t - \tau}{\sigma}\right)^2\right\} \qquad (2.13)$$

The Hann and Hamming windows are defined for $|t - \tau| < \tau$; the Gauss window is formally only limited by the length of the time series, but typically could be limited for $|t - \tau| < 3\sigma$, as for greater distances from the centre the window values, or weights, become very small.

Figure 2.15 demonstrates the spectrogram of the Cuvier's beaked whale click presented in Figure 2.2. Depending now on two dimensions, time and frequency, the spectrogram is conveniently presented as a grayscale coded image, where the intensity denotes the spectral level (expressed in dB values in Figure 2.15). In particular, Figure 2.15 shows that the Cuvier's beaked whale click is centred around time $t = 0$ and frequency $f = 40\,\text{kHz}$, covering a time span from at least -0.1 to $0.1\,\text{ms}$ and a frequency range from about 20 to over $60\,\text{kHz}$.

Figure 2.16 shows how the spectrogram is in practice constructed from the time series. The left side of this figure shows the complete time series and emphasizes the Hann windowed snippet to be used for the Fourier transform, the result of which is presented on the right column. The three rows show, from top to bottom, the window function applied to different times $t = -0.05, 0, 0.05\,\text{ms}$.

While already intuitive from Figure 2.15, the spectra of these three data snippets demonstrate that the frequency of the Cuvier's beaked whale click varies slowly and increases from $34\,\text{kHz}$ at $-0.05\,\text{ms}$ to $46\,\text{kHz}$ at $+0.05\,\text{ms}$. This moderate frequency

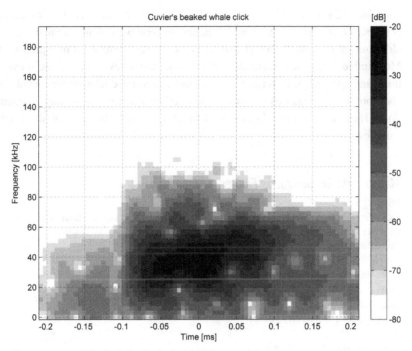

Figure 2.15 Spectrogram of Cuvier's beaked whale click.

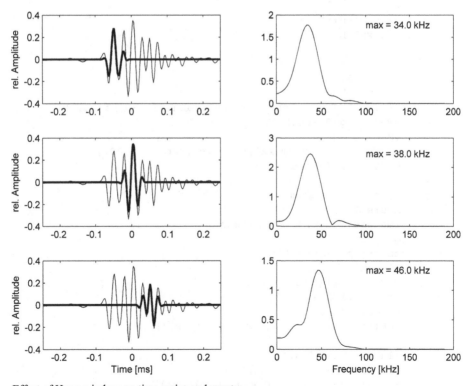

Figure 2.16 Effect of Hann window on time series and spectra.

variation (12 kHz in 0.1 ms) is typical of Cuvier's beaked whale echolocation clicks (Zimmer *et al.*, 2005a). Similar frequency sweeps were demonstrated for Blainville's (Johnson *et al.*, 2004, 2006) and Gervais' beaked whales (Gillespie *et al.*, 2009). Measurable frequency variations in echolocation clicks require multi-cycle oscillations and consequently do not exist in most dolphin clicks. Sperm whale pulses are also too short (measured in number of oscillations) to support any frequency variations. Harbour porpoise clicks would be long enough, but do not show a frequency variation.

Matlab code for generating Figures 2.15 and 2.16

```
%Scr2_9
[xx,fs] = wavread('../Pam_book_data/zifioSelect Scan.wav');
tt=(0:length(xx)-1)/fs;
fsk=fs/1000;
%
%extract single click
dts=0.25;
tso=1.94215;
ts=tso+[-1 1]*dts/1000;
%
tsel=tt>ts(1) & tt<ts(2);
ttl=(tt(tsel)-tso)*1000;
xx1=xx(tsel);
bias=mean(xx1);
xx1=xx1-bias;
%
% spectrogram
nfft=128;
%Matlab 7.5
%[S, F, T, P] = spectrogram(xx1,hann(32),30,nfft,fsk);
%M=10*log10(abs(P));
%Matlab 7.0
[B, F, T] = specgram(xx1,nfft,fsk,hann(32),30);
M=20*log10(abs(B));

%plot spectrogram
figure(1)
imagesc(T-dts, F, M); axis xy
grid on
colormap(1-gray(12))
cl=caxis;
caxis(-20+[-60 0])
%caxis(cl(2)+[-60 0])
hb=colorbar;
set(get(hb,'title'),'string','[dB]')

xlabel('Time [ms]')
ylabel('Frequency [kHz]')
title('Cuvier''s beaked whale click')
```

```
%apply window to time series
w1=[0*(1:62)';hann(32);0*(1:98)'];
z1=xx1.*w1;
w2=[0*(1:81)';hann(32);0*(1:79)'];
z2=xx1.*w2;
w3=[0*(1:100)';hann(32);0*(1:60)'];
z3=xx1.*w3;

% get FFT spectra
y1=abs(fft(z1));
y2=abs(fft(z2));
y3=abs(fft(z3));
% construct frequency vector
fr=(0:length(y1)-1)*fsk/length(y1);
ifr=fr<fsk/2;

% obtain frequency of spectral maxima
frmax1=fr(find(y1==max(y1),1,'first'));
frmax2=fr(find(y2==max(y2),1,'first'));
frmax3=fr(find(y3==max(y3),1,'first'));

%plot time series and spectra side by side
figure(2)
set(gcf,'position',[100, 100, 560, 600])
subplot(321)
hp=plot(tt1,xx1,'k',tt1,z1,'k'); xlim(tt1([1 end]))
set(hp(2),'linewidth',2)
ylabel('rel. Amplitude')
subplot(322)
plot(fr(ifr),y1(ifr),'k')
text(100,max(y1),sprintf('max = %.1f kHz',frmax1))
subplot(323)
hp=plot(tt1,xx1,'k',tt1,z2,'k'); xlim(tt1([1 end]))
set(hp(2),'linewidth',2)
ylabel('rel. Amplitude')
subplot(324)
plot(fr(ifr),y2(ifr),'k')
text(100,max(y2),sprintf('max = %.1f kHz',frmax2))
subplot(325)
hp=plot(tt1,xx1,'k',tt1,z3,'k'); xlim(tt1([1 end]))
set(hp(2),'linewidth',2)
xlabel('Time [ms]')
ylabel('rel. Amplitude')
subplot(326)
plot(fr(ifr),y3(ifr),'k')
text(100,max(y3),sprintf('max = %.1f kHz',frmax3))
xlabel('Frequency [kHz]')
```

To generate the spectrogram, a 32-point Hann window is embedded in a 192-sample data vector, with varying numbers of leading and trailing zeros.

Power spectral density
The use of window functions for the STFT leads to the question of their impact on the spectrum, or in other words, what the spectral values really mean.

From the definition of the Fourier transform (Equation 2.2) we note that the spectrum is an integral value and therefore describes some mean characteristics of the signal. To gain some insight we generate a plain sinusoidal signal with unit RMS intensity

$$x_n = A_0 \cos(2\pi f t_n) \tag{2.14}$$

where we select amplitude $A_0 = \sqrt{2}$, frequency $f = 50\,\text{kHz}$, and time t varying with sampling interval $dt = 1/384\,\text{ms}$ between 0 and $10 - dt\,\text{ms}$. We apply three window functions, one of which is a rectangular window where effectively all data samples are considered without changes. The other two window functions are Hann windows with variable length padded with zeros to remove the remaining data of the signal vector.

The spectra are now estimated according to the following equation

$$S_m = \frac{\left| \sum_{n=1}^{N} x_n w_n \exp\left\{ 2\pi i \frac{(n-1)(m-1)}{N} \right\} \right|^2}{N \sum_{n=1}^{N} w_n^2} \tag{2.15}$$

where w_n is the selected weighting function. That is, we normalize the weighted DFT with the length of the DFT multiplied by the sum of the window squared.

Figure 2.17 shows the result of this operation; note that only the rectangular window, which uses the unmodified time series, shows a spike at 50 kHz to 0 dB and negligible values elsewhere (the $-300\,\text{dB}$ in the figure are equivalent to a zero spectral level). The Hann windowed spectra show a broad peak around the signal frequency of 50 kHz and some residual spectral energy away from the signal frequency.

Figure 2.17 indicates that, although the location of the maximum of the spectrum (Equation 2.15) denotes the dominant frequency of the signal, the actual spectral values S_m are not the spectral signal power but the power spectral density of the signal. That is, by integrating/summing the spectral values S_m over all frequencies we obtain the mean, or RMS intensity, of the input signal x_n. For the example shown in Figure 2.17, this (relative) RMS intensity of the time series is 1 or 0 dB for all windows; the peak values of the spectra are 0 dB, $-16.6\,\text{dB}$, and $-22.7\,\text{dB}$ for the different window functions (Rect, Hann(32) and Hann(128)).

It is worthwhile to note that sharp peaks, as shown in Figure 2.17, only occur if the signal is strictly periodic, that is, if the phase at the end of the signal is equal to the initial phase of the signal or if the signal is composed of multiple complete oscillations of a single frequency. For example, a slightly changed frequency of 50.01 kHz would replace the sharp peak of Figure 2.17 by a spectral curve that smoothly increases from 0 dB to nearly 100 dB at 192 kHz.

From this discussion of Equation 2.15, we conclude that in order to obtain physically meaningful spectral values the FFT must be corrected for the influence of the window, at

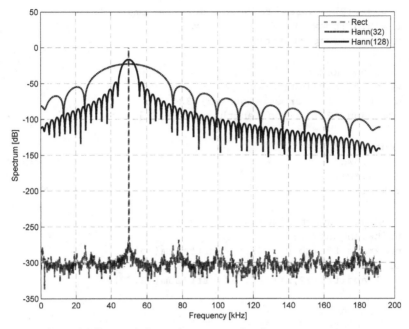

Figure 2.17 Comparison of different windows of spectrum estimation.

least as implemented in Matlab. However, very often only relative spectral values are desired, and for these cases this correction may be dropped.

In order to develop the spectrogram it was necessary to apply short windows to allow the visualization of time-varying signals. Shorter window functions, however, result in reduced frequency resolutions. If we define the frequency resolution as the width of the peak that is measured 3 dB below the peak (-3 dB bandwidth $\Delta f_{-3\mathrm{dB}}$) then by inspection of Figure 2.17 the spectral resolution of the Hann window may be approximated by

$$\Delta f_{-3\mathrm{dB}} \approx \frac{3}{2} \frac{f_S}{N_W} \tag{2.16}$$

where f_S is the sampling frequency and N_W is the length of the Hann window in samples.

The spectral resolution depends on the effective width of the window, or equivalently on the amount of non-zero data used for the Fourier analysis. This is in contrast to the accuracy of the spectrum, which may be increased by zero-padding the data as shown above when we discussed the interpolating FFT.

Matlab code to generate Figure 2.17

```
%Scr2_10
% simulate synthetic signal
```

```
Ao=sqrt(2); fs=384; fo=50.01; dt=1/fs; tmax=10;
tt=(0:dt:tmax-dt)';
xx=Ao*cos(2*pi*fo*tt);

%define windows
wo=ones(floor(length(xx)),1);
w1=[hann(32); zeros(length(xx)-32,1)];
w2=[hann(128); zeros(length(xx)-128,1)];

%cut time series
zo=xx.*wo;
z1=xx.*w1;
z2=xx.*w2;

% estimate spectral power
nfft=length(wo);
po=2*abs(fft(zo,nfft)).^2/(nfft*sum(wo.^2));
p1=2*abs(fft(z1,nfft)).^2/(nfft*sum(w1.^2));
p2=2*abs(fft(z2,nfft)).^2/(nfft*sum(w2.^2));

%plot results
fr=(0:nfft-1)*fs/nfft;
ifr=fr<fs/2;

figure(1)
plot(fr(ifr),10*log10(po(ifr)),'k:',...
     fr(ifr),10*log10(p1(ifr)),'k-',...
     fr(ifr),10*log10(p2(ifr)),'k','linewidth',2)
grid on
xlabel('Frequency [kHz]')
ylabel('Spectrum [dB]')
legend('Rect','Hann(32)','Hann(128)')
% obtain some numbers
% RMS value (using time series)
rms_x=sqrt(mean(xx.^2));

%integrate spectral power
rms_so=sqrt(sum(po(ifr)));
rms_s1=sqrt(sum(p1(ifr)));
rms_s2=sqrt(sum(p2(ifr)));

%peak level of spectral power
pko=max(10*log10(po(ifr)));
pk1=max(10*log10(p1(ifr)));
pk2=max(10*log10(p2(ifr)));

%display rms and peak values
rms_x
[rms_so rms_s1 rms_s2]
[pko pk1 pk2]
```

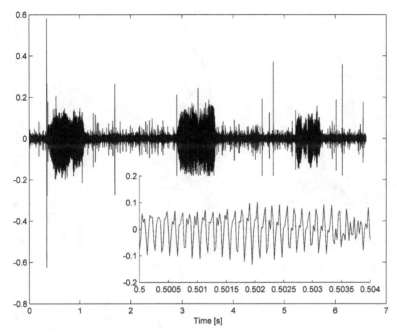

Figure 2.18 Time series of bottlenose dolphin whistles (courtesy G. Pavan).

2.5.2 Dolphin whistles

Although the time–frequency analysis of cetacean sound gives some interesting insight into beaked whale clicks, its main realm is the analysis of true tonal or pseudotonal signals, be they dolphin whistles, pulsed signals, or whale songs.

The presence of dolphins is in most cases announced by whistles. Whereas clicks are easily detected in the time domain, whistles are sometimes very hard to detect this way, if at all. Figure 2.18 shows such a time series where some ten clicks are easily detected but the three noise-like sounds need special investigation to be recognized as tonal signals, as done with the zoom window (insert in Figure 2.18).

Figure 2.19 shows the spectrogram of the same dataset with a 256-point Hann window, and 192-point overlap. As the data were sampled at 44 100 Hz this window covers a time span of 5.8 ms; according to Equation 2.16, this would result in a frequency resolution of 0.26 kHz, which seems appropriate to this type of data. Multiple frequency-modulated tonal whistles may be clearly noted.

Even though dolphin whistles have been known and studied for a long time, it seems impossible to establish a unique whistle characterization (see, for example, Esch *et al.*, 2009). Intuitively, however, there are some basic characteristics to whistles: length of signal, minimum and maximum frequency, start and stop frequency, interval between consecutive signals.

One analysis objective of bioacoustics is to describe dolphin whistles as accurately as is necessary to allow species classification or even identification of individuals and

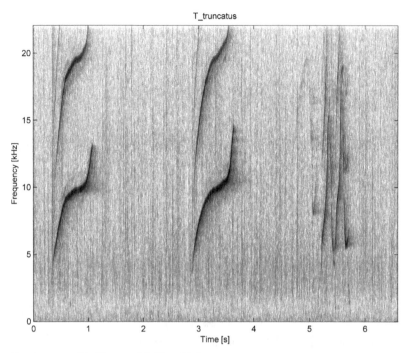

T_truncatus

Figure 2.19   Spectrogram of bottlenose dolphin whistles

different researchers follow different approaches. As research in whistle classification is ongoing, we fall back to a careful description of what we see in a spectrogram. In Figure 2.19, we note a tonal signal with variable frequency that is repeated in a very similar way, as shown around 0.7 and 3.3 s, but which may be also very different, as shown with the signal at 5.4 s. We may attribute the three signals to the same animal, as the frequency modulation of the first two whistles is similar, interfering signals are absent, and the signals have similar pressure amplitude and duration (Figure 2.18).

The two repetitive signals (0.7 and 3.3 s) are relatively simple upsweeps, showing, however, a second simultaneous signal at about twice the frequency. As these additional signals occur at multiple frequencies, they are frequently also called harmonics and the signal with the lowest frequency is then called the fundamental, or first harmonic signal. That these additional higher-frequency signals are really harmonics is demonstrated in Figure 2.20, which shows that the second higher-frequency signal is indeed the second harmonic as they coincide.

Harmonics are related with overtones, but they are counted differently, leading to some possible confusion. The first harmonic is the fundamental tone and the second harmonic is the first overtone, and so on.

The presence of harmonic signals is rather common in natural generation of tonal sounds and the variation in the intensity of these harmonics allows the recognition of individual speakers, at least between humans, where harmonics in speech are also called formants.

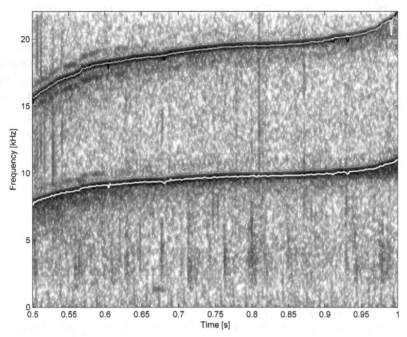

Figure 2.20   Zoom into Figure 2.19 with frequency traces. The white solid line is the fundamental signal; the white dashed line is the second harmonic.

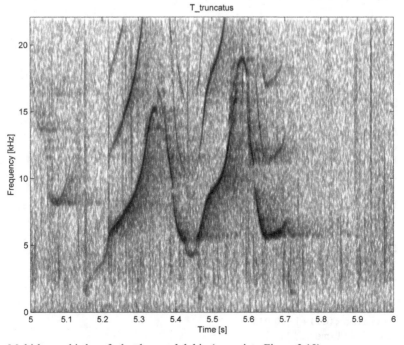

Figure 2.21   Multi-loop whistles of a bottlenose dolphin (zoom into Figure 2.19).

Dolphin whistles may be repetitive, as seen in Figure 2.19 for the first two whistles, but may also be very variable, as demonstrated in Figure 2.21, where the third signal from Figure 2.19 is shown in more detail. We recognize again the presence of harmonics and also that this particular signal is composed of up and down sweeps with much higher rates of frequency changes adding more difficulty to a global systematic whistle characterization scheme. It seems clear that in addition to the above-mentioned basic characterization we have to consider the rate by which the frequency changes, or how often the signal frequency varies between the extremes.

Matlab codes for Figures 2.18, 2.19 and 2.21

```
%Scr2_11
isel=1;

root='../pam_book_data/';

f=[];
ii=1; f(ii).name='T_t_short'; f(ii).title='T_truncatus';
f(ii).xl=[0 4];
ii=2; f(ii).name='S_coe_short'; f(ii).title='S_coeruleoalba';
f(2).xl=[0 1.5];

fname=f(isel).name;
tname=f(isel).title;
xl=f(isel).xl;

xx=[];
[xx,fs]=wavread([root fname]);
xx=xx(:,1);
tt=(0:length(xx)-1)/fs;

[B, F, T]=specgram(xx-mean(xx),1024,fs, hann(256),192);

figure(1)
clf
h1=axes;
ax=get(h1,'position');
line(tt,xx-mean(xx),'parent',h1,'color','k');
h2=axes('parent',gcf,'position',[0.37 ax(2) +0.06 0.5 0.3]);
line(tt,xx,'parent',h2,'color','k'),xlim(0.5 +[0 4]/1000)

figure(2)
imagesc(T, F/1000,20*log10(abs(B))), axis xy
colormap(1-gray)

cl=caxis;
caxis(max(cl)+[-60 0])
xlabel('Time [s]')
ylabel('Frequency [kHz]')
```

```
title(tname,'interpreter','none')

figure(3)
imagesc(T, F/1000,20*log10(abs(B))), axis xy
xlim([5 6])
colormap(1-gray)
cl=caxis;
caxis(max(cl)+[-60 0])
xlabel('Time [s]')
ylabel('Frequency [kHz]')
title(tname,'interpreter','none')

return
```

Script for Figure 2.20

```
%Scr2_12
root='../pam_book_data/';

fname='T_t_short'; tname='T_truncatus'; xl=[0 4];

xx=[];
[xx,fs]=wavread([root fname]);
xx=xx(:,1);
tt=(0:length(xx)-1)/fs;

[B, F, T]=specgram(xx-mean(xx),1024,fs, hann(256),192);

P=abs(B);
isel=find(T>0.5 & T<1.0);
if1=F<12000;
if2=F>12000;
i1=0*isel;
i2=0*isel;
for ii=1:length(isel)
    [d,i1(ii)]=max(P(if1,isel(ii)));
    [d,i2(ii)]=max(P(if2,isel(ii)));
end

fr1=F(i1)/1000;
fr2=12+F(i2)/1000;

figure(1)
hold off
imagesc(T, F/1000,20*log10(abs(B))), axis xy
xlim([0.5 1.0])
colormap(1-gray)
cl=caxis;
caxis(max(cl)+[-60 0])
xlabel('Time [s]')
ylabel('Frequency [kHz]')
```

```
hold on
plot(T(isel),fr1,'w-',T(isel),2*fr1,'k-',T(isel),fr2, ...
  'w--','linewidth',2)
hold off

return
```

The spectrogram in this script is estimated by the Matlab function

```
[B, F, T]=specgram(xx-mean(xx),1024,fs, hann(256),192);
```

This function is nowadays obsolete and should be replaced by

```
[S, F, T, P] = spectrogram(xx-mean(xx), hann(256),192,1024,fs);
```

similar to script Scr2_9. Apart from the order of the arguments, there is one big difference: the spectral power (in dB) is in the first script estimated as `20*log10(abs(B))`, and for the newer version it is `10*log10(abs(P))`.

2.5.3   Pulsed calls

We have seen so far that dolphins emit clicks and whistles, that is, pulse-type and tonal sounds. It seems logical that there should be something between these two categories; indeed, we have what are known as pulsed calls.

Figure 2.22 shows a sequence of pulsed calls of a long-finned pilot whale recorded in the Mediterranean Sea. These pulsed calls seem to be tonal signals that exhibit a series of harmonics, where, however, the first harmonic is not necessarily the dominant one, or the intensity of the higher-order harmonics may therefore not decrease with order number.

A detailed view (Figure 2.23) reveals that pulsed calls resemble more amplitude-modulated signals where a carrier frequency is modulated by a similar sinusoidal sound, generating the characteristic side-lobe structure seen in Figure 2.22.

Empirical mode decomposition (EMD)

To analyse this complex signal in more detail we do not use the Fourier analysis but apply the Huang empirical mode decomposition (EMD) to decompose the signal in a series of oscillating functions

$$x(t) = \sum_{n=1}^{N} C_n(t) + R_N(t) \tag{2.17}$$

where $C_n(t)$ are oscillating functions, also called IMF (intrinsic mode functions), which are characterized by a decreasing number of zero crossings with increasing index n, and $R_N(t)$ is in general a non-oscillating final trend (Huang et al., 1998; Flandrin et al., 2004).

From Figure 2.24, which shows the decomposition of the pulsed-call time series of Figure 2.23, we note that indeed the number of oscillations, and therefore zero crossings, decreases with increasing IMF number. The EMD method is a purely data-driven analysis method and I will go into a more detailed description of the algorithm by commenting on the Matlab code (see below).

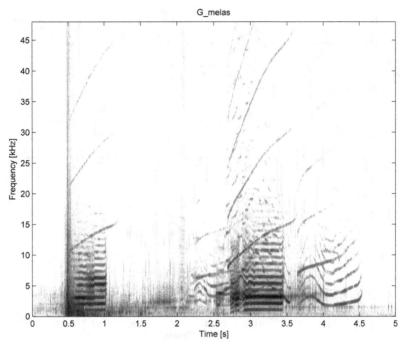

Figure 2.22   Pulsed call from long-finned pilot whales (courtesy P. Tyack).

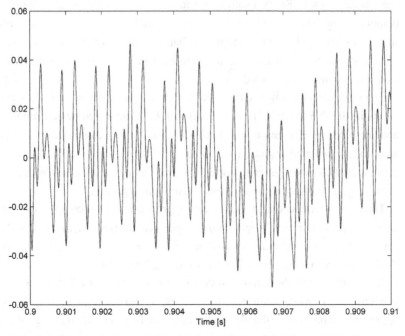

Figure 2.23   Pulsed-call time series, a zoom into the time series used for Figure 2.22.

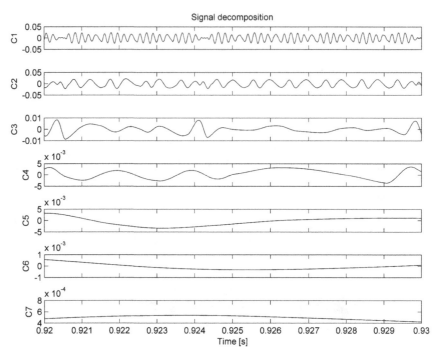

Figure 2.24 Signal decomposition of time series shown in Figure 2.23.

Comparing each IMF with the original time series (Figure 2.23) one observes that the basic time series IMFs are much easier to interpret than the original complex signals, demonstrating the power of this decomposition method. In particular, one notes that the first three IMFs (C1–C3) are simple oscillations with recurring features that may give some insight into the sound generation mechanism, or that may even be used to extract encoded information in pulsed calls.

The signal decomposition clearly shows that this particular pulsed call is composed of two units (0.9205 to 0.9243) and (0.9243 to 0.9299) separated by zero amplitudes in C1 and increased oscillations in C3. Each of these units is composed of nearly identical amplitude-modulated or phase-modulated signals as seen in C1 and C2, respectively. By combining these two dominant modes, we obtain a time series that is indeed very similar to the original pulsed-call time series, as shown in Figure 2.25.

The EMD is a rather out-of-mainstream approach to analyse complex time series and is aimed to provide more insight into the physical causes of waveforms (here sound generation) or the information content of the signal. However, the applications of EMD to cetacean sound are still limited (Adam, 2006a, b).

As the purpose of the EMD is to decompose a complex oscillating time series into a small number of more elementary time series, or IMFs, the question arises of how this approach compares with the classical Fourier series where periodic signals are decomposed into a series of trigonometric functions with constant amplitude:

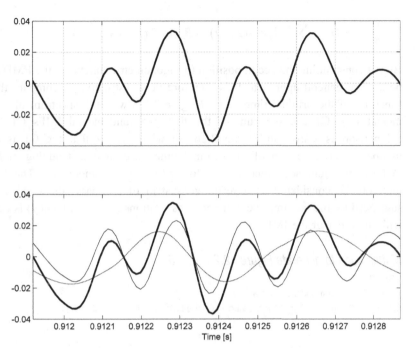

Figure 2.25   Dominant mode decomposition and synthesis of single pulse. Top panel: original time series, bottom panel: C1 as dashed line, C2 as thin solid, and C1+C2 as thick line.

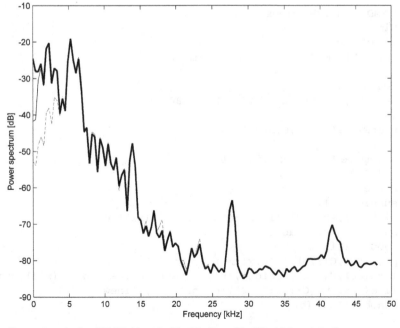

Figure 2.26   Spectral analysis of EMD (dotted: C1, thin line: C1+C2, thick: original).

$$x(t) = \sum_{n=1}^{N} (A_n \cos(2\pi f_n t) + B_n \sin(2\pi f_n t)) + R_N(t) \qquad (2.18)$$

By not constraining the decomposition to trigonometric functions, the EMD seems to be a more general approach. It is therefore very instructive to compare the power spectrum of the original time series (Figure 2.23) with the spectrum of the fundamental IMF C1 and the sum of the first two fundamental IMFs C1+C2. This comparison is shown in Figure 2.26, showing clearly that the C1 is nearly a highpass-filtered version of the original time series and that adding higher-order IMFs to the sum adds increasingly lower-frequency components. The EMD can therefore be considered as a recursive system of highpass filters, which is also suggested by the fact that the number of oscillations (mean frequency) is decreasing with the order of the IMF.

Matlab code to generate Figures 2.23 – 2.26

```
%Scr2_14
root='../pam_book_data/';
fname='G_melasTag3'; tname='G_melas'; xl=[0.9 0.95];
  xl1=[0.92 0.93];

[xx,fs]=wavread([root fname]);
tt=(0:length(xx)-1)'/fs;
xsel=tt>xl(1) & tt<xl(2);
xx1=xx(xsel,1);
tt1=tt(xsel);
%
% EMD signal decomposition
Nimf=10;
iimax=100;
[C, R, N] = EMD(tt1,xx1, Nimf,iimax);

figure(1)
set(gcf,'position',[100 100 560 680]);
ifg=1;
plotEMD(ifg,tt1,C,N,xl1)

% compare IMFs with original
xl3=0.91192+[0 0.00095];
figure(2)
subplot(211)
plot(tt1,xx1,'k','linewidth',2),xlim(xl3),grid
set(gca,'xticklabel',[])
subplot(212)
plot(tt1,C(:,1),'k'),xlim(xl3),grid
line(tt1,C(:,2),'color','k','linestyle','-'),xlim(xl3),grid
line(tt1,sum(C(:,1:2),2),'color','k','linewidth',2)
xlabel('Time [s]')

% spectral analysis of EMD
```

```
[Po, F]=psd(xx1,[],fs/1000);
P1=psd(C(:,1),[],fs/1000);
P2=psd(sum(C(:,1:2),2),[],fs/1000);

figure(3)
plot(F,10*log10(abs(Po)),'k','linewidth',2)
line(F,10*log10(abs(P1)),'color','k')
line(F,10*log10(abs(P2)),'color','k')
xlabel('Frequency [kHz]')
ylabel('Power spectrum [dB]')
```

Plotting support function

```
function plotEMD(ifg,tt,C,N,xl)
% plot IMFs
im=length(find(N>0));
figure(ifg)
for ii=1:im
    subplot(im*100+10+ii)
    plot(tt,C(:,ii),'k'), xlim(xl)
    ylabel(sprintf('C%d',ii))
    if ii<im, set(gca,'xticklabel',[]), end
    if ii==1, title('Signal decomposition'),end
end
xlabel('Time [s]')
```

EMD processing

```
function [C,R,N] = EMD(tt,xx,Nimf,iimax)
% [C,R,N]=function EMD(tt,xx,jmax,imax)
% EMD signal decomposition
%
% search for local maxima in time series
locMax = @(x) ...
    1+find(x(2:end-1)>x(3:end) & x(2:end-1) > =x(1:end-2));
%
% pre-allocate storage for residual R and IMFs C
R=zeros(length(xx),Nimf);
C=0*R;
% we use N to remember number of iterstions required oper IMF
N=zeros(1,Nimf);
%
% we sart with original time series
R(:,1)=xx;
for jj=1:Nimf-1 % upper limit must be one less than the number of IMFs
    %
    x1=R(:,jj); % get residual
    for ii=1:iimax % try max 20 times to find IMF
        xo=x1;              % data set to be worked on
        lmx=locMax(xo);     % find local maxima
        lmn=locMax(-xo);    % find local minima
        % check if we already got IMF
        if length(find(xo(lmn)>0))+ ...
```

```
    length(find(xo(lmx)<0))==0, break,end
    % construct envelope by interpolating extrema
    xmx=interp1(tt(lmx),xo(lmx), tt,'cubic','extrap');
    xmn=interp1(tt(lmn),xo(lmn),tt,' cubic','extrap');
    % Construct new data set by centring in between envelope
    xm=(xmx+xmn)/2;
    x1=xo-xm;
  end
  %Store number of iterations, IMS and residual
  N(jj)=ii;
  C(:,jj)=x1;
  R(:,jj+1)=R(:,jj)-x1;
  %
  % if first attempt yieded to IMF this should be the last one
  if ii==1, break,end
end
%
```

Comments on EMD code

The EMD estimates first an average, or mean, time series by forming the arithmetic mean of two envelope functions, one of which is connecting the local maxima and the other one is connecting the local minima of the time series. Subtracting this mean function from the original time series yields an intrinsic mode function (IMF) if the function is symmetric around zero amplitude, that is, if the amplitudes at the new local maxima are strictly positive and the amplitudes of the local minima are strictly negative. In cases where the modified time series is still asymmetric, an IMF has not been found and the procedure of constructing a new improved mean function continues. If an IMF has been found, it is not only stored but also subtracted from the original time series, resulting in a residual time series that may or may not be an oscillating function. In the case that this residual time series is a non-oscillating function, we have found a general trend and the EMD procedure is terminated, otherwise a new IMF will be sought that best describes the oscillating features of the residual time series. The application of the EMD algorithm is straightforward, except in the presence of significant levels of noise, where the algorithm has difficulty in finding a final trend. In fact, one feature of noise is that it may be considered as an infinite sum of deterministic, oscillating functions.

2.5.4   Buzzes and rapid pulse trains

Sequences of pulses or clicks are typical for echolocating toothed whales as presented earlier in this chapter. To qualify for echolocation these clicks or pulses must be spaced in time in such a way to allow the reception of prey echoes before emitting a new click or pulse. Very often toothed whales also emit very rapid pulse sequences (buzzes) with only limited echolocation capability, i.e. with an inter-pulse interval of the order of 1.5 ms or less. A typical example of such a buzz is shown in Figure 2.27 for a bottlenose dolphin, where the mean inter-pulse interval is 1.3 ms.

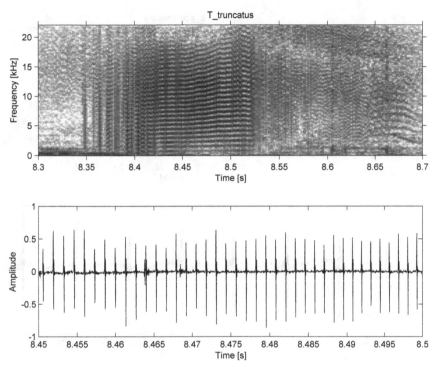

Figure 2.27   Buzz of a bottlenose dolphin (top panel: spectrogram, bottom panel: zoom into time series) (courtesy G. Pavan).

Eliminating standard echolocation application, one could consider these buzzes useful for investigation of nearby objects, or as part of social activities. Nevertheless, buzzes are worth considering separately, as Figure 2.27 shows clearly that what in a spectrogram appears as a multi-harmonic signal, is in reality a sequence of short pulses that are repeated with such a high repetition rate that multiple pulses fall within the spectrogram analysis window (see also Watkins, 1967). The spectrogram window in this case is a Hann window, which is 5.8 ms long (256 points at 44.1 kHz sampling frequency) covering therefore at least 4 pulses. Taking multiple pulses into a Fourier transform does result in a side-lobe-rich spectrum, as seen in the spectrogram of Figure 2.27. That these side-lobes are reasonably constant as a function of time is then an indication that the inter-pulse interval within the buzz is nearly constant.

Not only toothed whales but also baleen whales emit rapid pulse trains. Figure 2.28 shows a rapid pulse train by a minke whale. Here, the mean inter-pulse interval is 0.7 s (from 30 to 35 s), much longer than the inter-pulse interval of the bottlenose dolphin buzz and resembling more a jackhammer than the buzz of a bee.

As the spectrogram window (0.128 s) is much smaller than the inter-pulse interval, we should not be surprised that the spectrogram resolves the time of each pulse and does not show the multiple side-lobe structure as shown in Figure 2.27 for the bottlenose dolphin buzz.

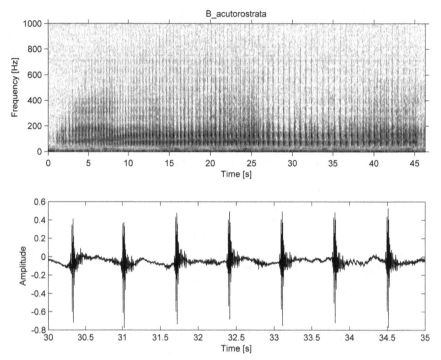

Figure 2.28   Rapid pulse train of a minke whale (top panel: spectrogram, bottom panel: partial time series) (courtesy D. Riesch).

2.5.5   Call trains

In contrast to toothed whales, baleen whales are in general not considered to employ echolocation for foraging, as they do not collect individual prey items but filter large quantities of water for small fish and zooplankton. However, baleen whales also emit trains of nearly identical calls. Figure 2.29 shows as an example a train of up-calls emitted by a north Atlantic right whale. The individual signal varies from 100 to 300 Hz and is repeated every 10 s or so. These right whale up-calls are considered to be contact calls (Parks and Tyack, 2005) and are therefore intensively used in passive acoustic monitoring of this highly endangered species. The signal-to-noise ratio (SNR) of this sample is very low; in the signal processing section (below) we will come back to this example.

As call trains could primarily be seen as a repetition of nearly identical calls, or signals, these call trains could indeed be considered as highly redundant sound emission without increased information content and therefore mainly suited to maintaining intra-species contact.

Other interesting and slightly different examples of call trains are emitted by blue and fin whales. They are not only the largest whales on earth but also the ones that emit sound with the lowest frequency. Both whales emit sounds around 20 Hz, that is, at frequencies that are known to travel the furthest due to very low sound absorption, as shown in Section 1.4.4.

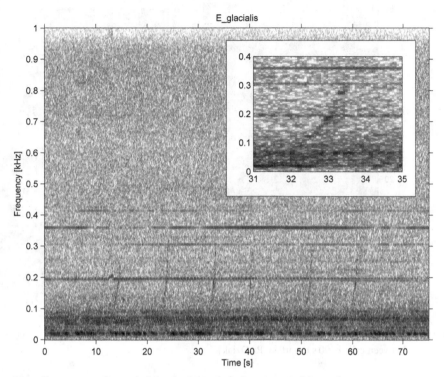

Figure 2.29   Up-call sequence of a north Atlantic right whale (courtesy A. Moscop).

Figure 2.30 shows some composite calls of Atlantic blue whales, which are rather long tonal or simple frequency-modulated sounds (*c.* 10 s for each component) repeated every 90 s or so. Blue whales emit these trains of very low-frequency calls in bouts of hours.

Fin whale calls are similar in frequency (also around 20 Hz); however, they are shorter (*c.* 1 s) and are repeated more frequently. In Figure 2.31, we see a fin whale call train with downsweep FM calls repeated on average every 17 s. The duration of a fin whale call train varies and may be hours.

In Figures 2.30 and 2.31, we note that these blue and fin whale call trains do not necessarily show a rigid inter-call interval but seem to have some rhythm. If we assume that these variations are not casual but intentional, then it seems clear that these variations, or if we prefer, the hidden code, is context-dependent and may vary with the whales' activity.

Typical suggestions for the purpose of this low-frequency sound are long-range communication and navigation, because these low-frequency signals are known to propagate over long distances with little absorption of the sound energy. The use of these call trains for communication implies that information is encoded in the signals, or in other words that the call train is modulated. In fact, we have different call units in the case of the blue whale, and also variations in the inter-call interval that would be suitable for carrying information.

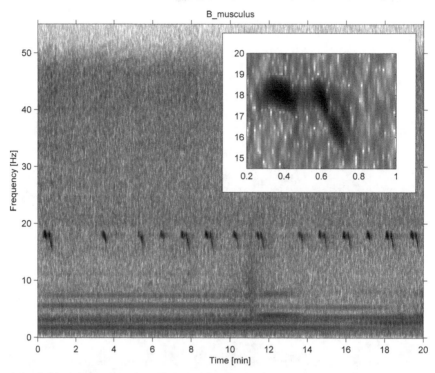

Figure 2.30   Atlantic blue whale composite calls (18 Hz for *c.* 9 s, and 19–16 Hz FM for *c.* 10 s), average interval between calls *c.* 90 s (courtesy R. Dziak).

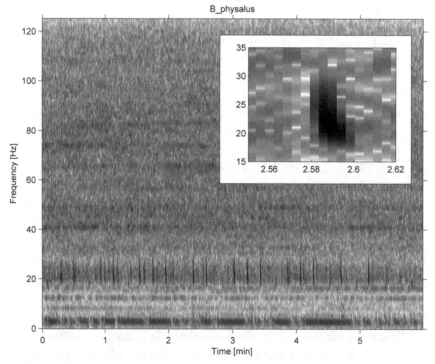

Figure 2.31   Call train of an Atlantic fin whale (courtesy R. Dziak).

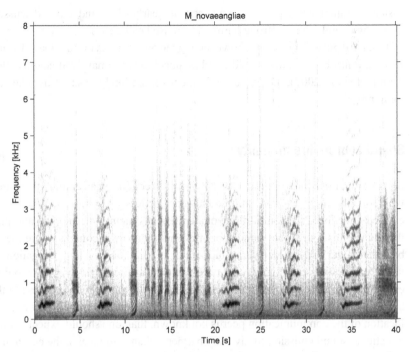

Figure 2.32   Cut of humpback whale song (courtesy P. Tyack).

The rather large inter-call interval, at least when compared with a rapid pulse train, as shown in Figure 2.28 for the minke whale, is also consistent with both functional suggestions (long-range communication or navigation), as both modes of operation require some time to listen to either a remote conspecific (communication) or an environmental echo (navigation).

2.5.6   Whale songs

If there were a song contest among whales, humpback whales would without doubt emerge victorious, at least if the jury were composed of *Homo sapiens*. As well as being aesthetically attractive, humpback whale songs have been and are still being studied extensively, mostly to understand why humpback whales sing. In fact only male humpbacks are known to sing and in general only when swimming alone, and songs are primarily performed in the winter breeding season (Payne *et al.*, 1983).

Humpback whales may sing continuously for hours and the song is typically composed of fewer than ten themes that are repeated in a particular order during the song (Tyack, 1999). A theme is made up of phrases or sequences lasting about 15 s. Figure 2.32 shows such phrases of a humpback song; note the variability of the components within such a humpback whale song phrase, which makes listening enjoyable, but may complicate the implementation of automated passive acoustic monitoring systems. Humpback songs are composed of nearly all sound types discussed so far (tonals with and without strong harmonics, pulsed calls, etc.).

Tonal and pulsed signals are, however, not the only sounds that are made by humpback whales; it has been demonstrated that humpback whales may emit echolocation clicks (Stimpert *et al.*, 2007), the purpose of which is more likely obstacle avoidance and not foraging.

2.6   Sound source directionality

Sound may or may not be emitted equally in all directions by an animal; there are reasons why animals might want to emit sound more or less directionally. Whether sound is emitted directionally or omni-directionally depends mainly on the type and purpose of the sound emission. If the sound is emitted in a monostatic context where the animal is both the sender and the receiver of the sound, i.e. echolocation and navigation, then one might expect the sound energy to be emitted in a directional way to avoid wasteful emission of sound energy in directions that do not contribute to the success of the actual acoustic activity. If, on the contrary, sound is intended for communication, and if the location of the communication peer is not known, than one should expect a more omni-directional sound emission to give any partner a chance to receive the communication.

The directionality of a sound source depends in general on the relation of the wavelength of the emitted sound to the dimension of the sound projector: highly directional sound emissions require large sound projectors. In general, one can say that sound generated by a single organ, be it the larynx or the phonic lips, is predominantly omni-directional when the dimension of the sound-producing organ is small compared with the wavelength of the generated sound. For most toothed whales, or odontocetes, the sound that is produced by the phonic lips first passes the melon, or the combination of spermaceti organ and junk in the case of sperm whales, before entering the water (Norris and Harvey, 1974; Cranford *et al.*, 1996). The purpose of the melon is to focus the sound into the direction of interest, so that the final emitted sound becomes directional.

Compared with echolocation clicks, whistles are relatively low-frequency sound emissions, having fundamental wavelengths that are comparable to the physical dimensions of the sound projectors of dolphins. One might expect that a 2 m bottlenose dolphin would have a melon size of less than 20 cm, so that the ratio of melon size to wavelength becomes less than one for frequencies below 7.5 kHz.

As sound waves must interfere in a constructive and a destructive way to generate directionality, a high ratio of melon size to wavelength is required for significant directionality. Broadband signal emissions, however, have a variable directivity: signal energy emitted at higher frequencies is more directive than energy emitted at lower frequencies (see e.g. Lammers and Au, 2003). Interestingly, most whistles vary over a considerable frequency range, and frequently have harmonics, resulting in a significant variation of melon size to wavelength ratio and giving a temporal variation in directionality.

As spectrograms of pulsed sounds also show frequency components at higher frequencies, one could speculate that these higher-frequency components are also part of a
mixed directionality of the pulsed call (see e.g. Miller, 2002). In this context, it is not of
relevance whether the pulsed call is the result of amplitude modulation or due to an
integration window that is too long. In both cases, energy emitted at higher frequencies is
more directive than energy emitted at lower frequencies.

2.6.1 Physical acoustics of sound source

Concentrating on the sound beam of odontocetes, it is convenient to model the final
sound projector (dolphin melon, or sperm whale junk) as a plane circular piston, a well
analysed and (for most applications) accepted sound projector model. This model should
be considered as a first-order, single-parameter approximation, as the precise description
of real sound radiation patterns requires more than one parameter.

The far field sound radiation from a plane circular piston is given in complex notation
by

$$P(r, \vartheta, t) = \frac{P_0}{r} \exp\{i(\omega t - kr)\} \left(\frac{2 J_1(ka \sin \vartheta)}{ka \sin \vartheta} \right) \tag{2.19}$$

where a is the radius of the circular piston, r is the distance from the piston ($r \gg a$, i.e. far-
field), ϑ is the off-axis angle, that is the angle between the sound axis (normal to the disk
plane) and the direction to the observer, and $J_1(x)$ is defined by

$$J_1(z) = \frac{z}{\pi} \int_0^{\pi} \cos\{z \cos \varphi\} \sin^2 \varphi \, d\varphi \tag{2.20}$$

$J_1(x)$ is also known as a first-order Bessel function of the first kind.

The directivity pattern of a sound source is the sound intensity relative to the on-axis
sound intensity

$$D(\vartheta) = \left| \frac{P(r, \vartheta, t)}{P(r, 0, t)} \right|^2 = \left(\frac{2 J_1(ka \sin \vartheta)}{ka \sin \vartheta} \right)^2 \tag{2.21}$$

From Equation 2.21 we note that the directivity pattern of a circular piston depends, due
to rotational symmetry, only on the off-axis angle ϑ.

The directivity index (DI) describes the intensity gain achieved by a narrowband or
monochromatic directional sound source relative to an omni-directional sound source
and may be defined by

$$DI = 10 \log \left(\frac{4\pi}{\Psi} \right) \tag{2.22}$$

where Ψ is the equivalent solid angle of the sound beam, which is estimated for a planar
circular piston and for $ka \gg 1$ as

$$\Psi = 2\pi \int_0^\pi D(\vartheta) \sin \vartheta \mathrm{d}\vartheta \approx \frac{4\pi}{(ka)^2} \tag{2.23}$$

and therefore the directivity index of a plane circular piston becomes

$$DI = 20 \log(ka) \quad \text{for } ka \gg 1 \tag{2.24}$$

In general, the directivity pattern, as given by Equation 2.21, has a forward–backward ambiguity, as $D(\vartheta) = D(\pi - \vartheta)$. In reality, only the forward-oriented sound beam is of interest, which is equivalent to modelling the sound projector as a circular piston in an infinite baffle, eliminating all backward-oriented energy and side effects of the finite piston.

The beam pattern and the directivity index presented here are frequency-dependent and are therefore only valid for strictly narrowband sound emissions. Broadband sound emissions show a variable directivity, where sound energy emitted at higher frequencies is more directive than energy emitted at lower frequencies.

Au *et al.* (1999) give an empirical equation for the directivity index that is based on measurements on harbour porpoise, bottlenose dolphin, pseudo-orca, and beluga whale:

$$DI = 28.89 - 1.04\left(\frac{d}{\lambda}\right) + 0.04\left(\frac{d}{\lambda}\right)^2 \tag{2.25}$$

where d is the diameter of the head and λ is the wavelength at peak frequency.

Comparing Equations 2.25 and 2.24, one notes that Equation 2.24 increases logarithmically with the sound aperture, whereas Equation 2.24 is a quadratic function. Therefore, the shapes of these two expressions seem to be in contrast to each other, but this is mainly due to the use of an extremely high DI of the beluga whale ($DI = 32.6$ dB) forcing a positive quadratic component; a reduction by 3 dB of the DI of the beluga whale would allow an easy fit to Equation 2.24. Apart from these details, the two formulas generate similar values for the selected species if one assumes that the radius of the circular piston a (used in Equation 2.24) is about 16% of the diameter of the animal's head d (used in Equation 2.25). In other words, the effective diameter of the radiating surface is one third of the diameter of the head; this, however, is to be expected from the anatomy of dolphin heads.

2.7   Cetacean source levels

To generate sound requires energy; it therefore makes sense to analyse bio-acoustic sound levels in terms of the energy that is required to generate this sound.

The on-axis RMS source level SL_0, expressed in dB, is obtained from the total radiated sound energy E_A, measured in Ws, by

$$SL_0 = 171 + 10 \log(E_A) - 10 \log(T_S) + DI \tag{2.26}$$

where

Table 2.1 Typical cetacean source characteristics

Species	Signal energy	Signal length	Directivity index	Source level
Calls				
Blue whale (*Bm*)	1000 Ws	20 s	0 dB	188 dB
Fin whale (*Bp*)	40 Ws	1 s	0 dB	187 dB
Humpback whale (*Mn*)	4 Ws	1 s	0 dB	177 dB
Minke whale (*Ba*)	1 Ws	0.5 s	0 dB	174 dB
Clicks				
Sperm whale (*Pm*)	1 Ws	120 μs	27 dB	237 dB
Cuvier's beated whale (*Zc*)	1 mWs	200 μs	25 dB	203 dB
Common bottlenose dolphin (*Tt*)	1 mWs	25 μs	26 dB	213 dB
Porpoise (*Pp*)	1 μWs	100 μs	22 dB	173 dB
Whistles				
Common bottlenose dolphin (*Tt*)	10 mWs	1s	0 dB	151 dB

The source levels are given as RMS values in dB re 1 μPa at 1 m. All values should be considered as indicative.

E_A is the total acoustic energy of the whale [Ws] (see also Equation 1.48)
T_S is the effective duration of the transmitted signal [s]
DI is the directivity index of the sound source [dB]

The value of 171 is equivalent to $-10\log\left(\frac{4\pi}{\rho c}10^{-12}\right)$, a constant required to convert the omnidirectional sound intensity to sound pressure ($\rho = 1000$ kg/m^3, $c = 1500$ m/s).

Table 2.1 gives an overview of the source characteristics that we can expect for cetacean sounds. As a rule of thumb one can say that the total energy for a cetacean signal (single call, click, whistle, etc.) increases with the size of the animal. The largest whales (blue whales) emit calls with about 1000 Ws and small harbour porpoises generate clicks with about 1 μWs acoustic energy per click, or call.

For the bottlenose dolphin, we find two numbers, 1 mWs for clicks and 10 mWs for whistles. As dolphins usually have a moderate click rates, say 10 clicks/s, the emitted click energy integrated over a time that is equivalent to the duration of whistles is more or less comparable to the whistle energy.

Table 2.1 is not at all representative, and it is the merit of the increasing interest in passive acoustic research that more and more publications present quantitative cetacean source descriptions. The reader is invited to scan the list of references and further reading for publications on sound descriptions of baleen and toothed whales.

2.8   Order Cetacea

To conclude the presentation of cetacean sounds, it seems appropriate to present all cetaceans in a more formal way. The classification of cetaceans follows closely Dale W. Rice, *Marine Mammals of the World: Systematics and Distribution* (Rice, 1998), which has become a standard taxonomy reference in the field. However, the presentation of super- and subfamilies is ignored.

In addition to the vernacular and scientific names, for most species a list of publications is given; however, this may not be exhaustive and should be considered as an introduction to the species-relevant literature. For some species the available information is abundant and (in the case of sperm whales) may even be considered as biased towards easy study subjects, but for other species (especially some beaked whales) the published information is still very scarce.

2.8.1   Suborder Mysticeti: baleen whales

Family Balaenidae: right whales and bowhead whale

- Bowhead whale, *Balaena mysticetus*
  Blackwell *et al.*, 2007; Clark, 1989; Clark and Ellison, 2000; Clark and Johnson, 1984; Clark *et al.*, 1986; Cosens and Blouw, 2003; Cummings and Holliday, 1987; George *et al.*, 2004; Heide-Jørgensen *et al.*, 2003, 2006, 2007; Ljungblad *et al.*, 1980, 1986, 1982; Raftery and Zeh, 1998; Richardson *et al.*, 1995b; Stafford *et al.*, 2008; Würsig *et al.*, 1985
- North Atlantic right whale, *Eubalaena glacialis*
  Baumgartner and Mate, 2003; Gillespie, 2004; Kraus *et al.*, 1986; Matthews *et al.*, 2001; Murison and Gaskin, 1989; Parks and Tyack, 2005, 2006; Urazghildiiev and Clark, 2007b
- North Pacific right whale, *Eubalaena japonica*
  Brownell *et al.*, 2001; McDonald and Moore, 2002; Mellinger *et al.*, 2004b; Shelden *et al.*, 2005
- Southern right whale, *Eubalaena australis*
  Clark, 1982, 1983; Payne and Payne, 1971

Family Neobalaenidae: pygmy right whale

- Pygmy right whale, *Caperea marginata*
  Dawbin and Cato, 1992; Kemper, 2002

Family Balaenopteridae: rorquals

- Common minke whale, *Balaenoptera acutorostrata*
  Gedamke *et al.*, 2001
- Antarctic minke whale, *Balaenoptera bonaerensis*
- Sei whale, *Balaenoptera borealis*
  Baumgartner and Fratantoni, 2008
- Bryde's whale, *Balaenoptera brydei*
- Eden's whale, *Balaenoptera edeni*
- Blue whale, *Balaenoptera musculus*
  Alling *et al.*, 1991; Brueggeman *et al.*, 1985; Clark and Fristrup, 1997; Cummings and Thompson, 1971; Edds, 1982; Fiedler *et al.*, 1998; Mate *et al.*, 1999; McDonald *et al.*, 1995; Mellinger and Clark, 2003; Reeves *et al.*, 2004; Rivers, 1997; Širović *et al.*,

2004, 2007; Stafford, 2003; Stafford *et al.*, 1998, 1999a, 2001, 2004; Teranishi *et al.*, 1997; Thode *et al.*, 2000; Thompson *et al.*, 1996; Yochem and Leatherwood, 1985

- Fin whale, *Balaenoptera physalus*
  Charif *et al.* 2002; Clark and Fristrup, 1997; Clark *et al.*, 2002; Croll *et al.*, 2002; McDonald and Fox, 1999; McDonald *et al.*, 1995; Monestiez *et al.*, 2006; Panigada *et al.*, 2008; Rebull *et al.*, 2006; Schevill *et al.*, 1964; Širović *et al.*, 2004, 2007; Thompson, 1992; Watkins, 1981; Watkins *et al.*, 1987
- Humpback whale, *Megaptera novaeangliae*
  Au *et al.*, 2006; Baker *et al.*, 1985; Baraff *et al.*, 1991; Cato, 1991; Cato *et al.*, 2001; Cerchio *et al.*, 2001; Chabot, 1988; Charif *et al.*, 2001; Clapham and Mattila, 1990; Clark and Clapham, 2004; Dawbin, 1966; Dolphin, 1987; Frankel *et al.*, 1995; Gabriele and Frankel, 2003; Helweg *et al.*, 1990; Martin *et al.*, 1984; Mate *et al.*, 1998; McSweeney *et al.*, 1989; Norris *et al.*, 1999; Payne and Guinee, 1983; Payne and McVay, 1971; Payne *et al.*, 1983; Reeves *et al.*, 2004; Silber, 1986; Stimpert *et al.*, 2007; Thompson *et al.*, 1986; Weinrich *et al.*, 1997; Winn *et al.*, 1981

Family Eschrichtiidae

- Grey whale, *Eschrichtius robustus*
  Crane and Lashkari, 1996; Cummings *et al.*, 1968; Fish *et al.*, 1974; Gardner and Chávez-Rosales, 2000; Moore and Ljungblad, 1984; Stafford *et al.*, 2007b

2.8.2   Suborder Odontoceti: toothed whales

Family Delphinidae: dolphins

- Commerson's dolphin, *Cephalorhynchus commersonii*
  Kyhn *et al.*, 2010
- Chilean dolphin, *Cephalorhynchus eutropia*
  Ribeiro *et al.*, 2007
- Heaviside's dolphin, *Cephalorhynchus heavisidii*
  Watkins *et al.*, 1977
- Hector's dolphin, *Cephalorhynchus hectori*
  Dawson, 1991; Dawson and Thorpe, 1990
- Long-beaked common dolphin, *Delphinus capensis*
  Bernal *et al.*, 2003
- Short-beaked common dolphin, *Delphinus delphis*
  Goold, 2009
- Arabian common dolphin, *Delphinus tropicalis*
  Smeek *et al.*, 1996
- Pygmy killer whale, *Feresa attenuata*
  Madsen *et al.*, 2004a
- Short-finned pilot whale, *Globicephala macrorhynchus*
  Nores and Pérez, 1988

- Long-finned pilot whale, *Globicephala melas*
  Baird *et al.*, 2002; Cañadas and Sagarminaga, 2000; Heide-Jørgensen *et al.*, 2002
- Risso's dolphin, *Grampus griseus*
  Madsen *et al.*, 2004b; Soldevilla *et al.*, 2008
- Fraser's dolphin, *Lagenodelphis hosei*
  Perrin *et al.*, 1973
- Atlantic white-sided dolphin, *Lagenorhynchus acutus*
  Selzer and Payne, 1988
- White-beaked dolphin, *Lagenorhynchus albirostris*
  Rasmussen *et al.*, 2002, 2004, 2006
- Peale's dolphin, *Lagenorhynchus australis*
  Kyhn *et al.*, 2010
- Hourglass dolphin, *Lagenorhynchus cruciger*
  Kasamatsu and Joyce, 1995
- Pacific white-sided dolphin, *Lagenorhynchus obliquidens*
  Soldevilla *et al.*, 2008, 2010
- Dusky dolphin, *Lagenorhynchus obscurus*
  Würsig and Würsig, 1980
- Northern right whale dolphin, *Lissodelphis borealis*
  Jefferson *et al.*, 1994; Rankin *et al.*, 2007
- Southern right whale dolphin, *Lissodelphis peronii*
  Jefferson *et al.*, 1994
- Irrawaddy dolphin, *Orcaella brevirostris*
  van Parijs *et al.*, 2000
- Australian snubfin dolphin, *Orcaella heinsohni*
  Parra *et al.*, 2006
- Killer whale, *Orcinus orca*
  Au *et al.*, 2004; Deecke *et al.*, 2005, 1999; Ford and Fisher, 1982, Gaetz *et al.*, 1993; Miller, 2002; Simon *et al.*, 2007
- Melon-headed whale, *Peponocephala electra*
  Frankel and Yin, 2010
- False killer whale, *Pseudorca crassidens*
  Au *et al.*, 1995; Madsen *et al.*, 2004b
- Tucuxi, *Sotalia fluviatilis*
  Santos *et al.*, 2000
- Costero, *Sotalia guianensis*
  Rossi-Santos *et al.*, 2007
- Pacific humpback dolphin, *Sousa chinensis*
  van Parijs *et al.*, 2002
- Indian humpback dolphin, *Sousa plumbea*
- Atlantic humpback dolphin, *Sousa teuszii*
  van Waerebeek *et al.*, 2004
- Pantropical spotted dolphin, *Stenella attenuata*
  Baird *et al.*, 2001; Schotten *et al.*, 2003

- Clymene dolphin, *Stenella clymene*
  Fertl *et al.*, 2003
- Striped dolphin, *Stenella coeruleoalba*
  Gordon *et al.*, 2000; Panigada *et al.*, 2008
- Atlantic spotted dolphin, *Stenella frontalis*
  Au and Herzing, 2003; Herzing, 1996
- Spinner dolphin, *Stenella longirostris*
  Bazúan-Durán and Au, 2004; Lammers and Au, 2003; Lammers *et al.*, 2004, 2006;
  Schotten *et al.*, 2003
- Rough-toothed dolphin, *Steno bredanensis*
  Watkins *et al.*, 1987
- Indian ocean bottlenose dolphin, *Tursiops aduncus*
  Hawkins and Gartside, 2009, 2010
- Common bottlenose dolphin, *Tursiops truncatus*
  Akamatsu *et al.*, 1998; Au *et al.*, 1974, 1986, 1982; Buck and Tyack, 1993; Caldwell
  et al., 1990; Esch *et al.*, 2009; Freitag and Tyack, 1993; Herzing, 1996; Janik, 2000;
  Janik *et al.*, 1994; Murchison, 1980; Norris *et al.*, 1961; Renaud and Popper, 1975;
  Sayigh *et al.*, 2007; Simar *et al.*, 2010

Family Monodontidae

- Beluga, *Delphinapterus leucas*
  Au *et al.*, 1985, 1987; Belikov and Bel'kovich, 2003; Erbe and Farmer, 1998; Fish and
  Mowbray, 1962; Karlsen *et al.*, 2002; Schevill and Lawrence, 1949; Sjare and Smith,
  1986; van Parijs *et al.*, 2003
- Narwhal, *Monodon monoceros*
  Ford and Fisher, 1978; Miller *et al.*, 1995; Møhl *et al.*, 1990; Shapiro, 2006; Watkins
  et al., 1971

Family Phocoenidae: porpoises

- Finless porpoise, *Neophocaena phocaenoides*
  Akamatsu *et al.*, 1998; Wang *et al.*, 2005
- Spectacled porpoise, *Phocoena dioptrica*
- Harbour porpoise, *Phocoena phocoena*
  Amudmin, 1991; Au *et al.*, 1999; Carlström, 2005; Forney *et al.*, 1991; Gillespie and
  Chappell, 2002; Goodson and Sturtivant, 1996; Hansen *et al.*, 2008; Møhl and
  Andersen, 1973; Verfuss *et al.*, 2005
- Vaquita, *Phocoena sinus*
  Silber, 1991
- Burmeister's porpoise, *Phocoena spinipinnis*
- Dall's porpoise, *Phocoenoides dalli*
  Evans and Awbrey, 1984

Family Physeteridae: sperm whales

- Sperm whale, *Physeter catodon* (syn. *P. macrocephalus*)
  Adler-Fenchel, 1980; Amano and Yoshioka, 2003; Antunes *et al.*, 2010; Backus and

Schevill, 1966; Barlow and Taylor, 2005; Bedholm and Møhl, 2006; Clarke *et al.*, 1993; Cranford, 1999; Cranford *et al.*, 1996; Douglas *et al.*, 2005; Drouot *et al.*, 2004a, b; Frantzis and Alexiadou, 2008; Gillespie, 1997; Goold and Jones, 1995; Gordon, 1987; Gordon and Steiner, 1992; Gordon *et al.*, 2000; Hastie *et al.*, 2003; Jaquet and Whitehead, 1999; Jaquet *et al.*, 2001, 2003; Leaper *et al.*, 1992; Madsen, 2002a, b; Madsen *et al.*, 2002; Mellinger *et al.*, 2004a; Miller *et al.*, 2004a, b; Møhl, 2001; Møhl *et al.*, 2000, 2002, 2003; Moore *et al.*, 1993; Mullins *et al.*, 1988; Norris and Harvey, 1972; Nosal and Frazer, 2006, 2007; Papastavrou *et al.*, 1989; Pavan *et al.*, 2000; Rendell and Whitehead, 2004; Schulz *et al.*, 2009; Teloni, 2005; Teloni *et al.*, 2005, 2007; Thode, 2004; Tiemann *et al.*, 2006, 2007; van der Schaar *et al.*, 2009; Wahlberg, 2002; Watkins, 1980; Watkins and Daher, 2004; Watkins and Moore, 1982; Watkins and Schevill, 1977; Watkins *et al.*, 1993, 1999, 2002; Watwood *et al.*, 2006; Weilgart and Whitehead, 1997; Whitehead and Weilgart, 1990, 1991; Whitehead *et al.*, 1989; Zimmer *et al.*, 2003, 2005b, c

Family Kogiidae

- Pygmy sperm whale, *Kogia breviceps*
  Marten, 2000; Madsen *et al.*, 2005a
- Dwarf sperm whale, *Kogia sima*
  Dunphy-Daly *et al.*, 2008

Family Ziphidae: beaked whales

Barlow *et al.*, 2006; Tyack *et al.*, 2006

- Arnoux's beaked whale, *Berardius arnouxii*
  Hobson and Martin, 1996; Rogers, 1999
- Baird's beaked whale, *Berardius bairdii*
  Dawson *et al.*, 1998
- Northern bottlenose whale, *Hyperoodon ampullatus*
  Gowans *et al.*, 2000b, 2001; Hooker and Baird, 1999a; Hooker and Whitehead, 2002; Whitehead and Wimmer, 2005; Whitehead *et al.*, 1997; Wimmer and Whitehead, 2004
- Southern bottlenose whale, *Hyperoodon planifrons*
- Indo-Pacific beaked whale, *Indopacetus pacificus*
  Anderson *et al.*, 2006
- Sowerby's beaked whale, *Mesoplodon bidens*
  Hooker and Baird, 1999b
- Andrews' beaked whale, *Mesoplodon bowdoini*
- Hubbs' beaked whale, *Mesoplodon carlhubbsi*
  Marten, 2000
- Blainville's beaked whale, *Mesoplodon densirostris*
  Johnson *et al.*, 2004, 2006; Madsen *et al.*, 2005b; Marques *et al.*, 2009; Rankin and Barlow, 2007; Schorr *et al.*, 2009; Ward *et al.*, 2008
- Gervais' beaked whale, *Mesoplodon europaeus*
  Gillespie *et al.*, 2009

- Ginkgo-toothed beaked whale, *Mesoplodon ginkgodens*
- Gray's beaked whale, *Mesoplodon grayi*
- Hector's beaked whale, *Mesoplodon hectori*
- Layard's beaked whale, *Mesoplodon layardii*
- True's beaked whale, *Mesoplodon mirus*
- Perrin's beaked whale, *Mesoplodon perrini*
- Pygmy beaked whale, *Mesoplodon peruvianus*
- Stejneger's beaked whale, *Mesoplodon stejnegeri*
- Spade-toothed whale, *Mesoplodon traversii*
- Tasman beaked whale, *Tasmacetus shepherdi*
- Cuvier's beaked whale, *Ziphius cavirostris*
  Frantzis *et al.*, 2002; Johnson *et al.*, 2004, 2006; Moulins *et al.*, 2006; Zimmer *et al.*, 2005a, 2008

Family Iniidae

- Amazon river dolphin, *Inia geoffrensis*
  May-Collado and Wartzok, 2007; Podos *et al.*, 2002

Family Lipotidae

- Baiji (extinct?), *Lipotes vexillifer*
  Akamatsu *et al.*, 1998; Wang *et al.*, 2006

Family Pontoporiidae

- Franciscana, *Pontoporia blainvillei*
  Secchi *et al.*, 2002

Family Platanistidae

- Ganges river dolphin, *Platanista gangetica*
  Sinha and Sharma, 2003
- Indus river dolphin, *Platanista minor*

3 Sonar equations

This chapter presents and discusses all parts that add up to the sonar equation, which is the workhorse of sonar design and performance analysis. Here the focus remains on the passive sonar equation, which is further adapted for detecting cetacean sounds.

Starting with the definition of what constitutes passive acoustic detection, the introduction of the signal-to-noise ratio and the detection threshold, the remainder of this chapter discusses the different components of the sonar equation:

Source level
Off-axis attenuation
Sound propagation
Noise level
Array gain of the receiver
Processing gain of the receiver

Discussing sound propagation will be a major part of this chapter and will cover simplified geometric models, but will also introduce reference models from the acoustic modelling community, especially the Bellhop Gaussian ray-trace model.

As the noise level is another key element of the passive sonar equation, its sources and levels will be discussed in more detail.

3.1 Passive sonar equations

The sonar equations describe in simple terms the conditions under which sonar systems succeed in remotely detecting signals. They serve two important practical functions: performance prediction of existing sonar systems, and design support of new sonar implementations. The sonar equations come in different flavours depending on whether they are intended for active or passive sonar usage: the passive sonar equation is the relevant form for PAM.

There is a variety of ways to formulate the passive sonar equations. Common to all of them is that they describe the conditions under which the sonar system is fulfilling its task. To detect a signal (cetacean sound) a decision will be made, by a human observer or an automatic detector, as to whether a signal is present or not. If no signal is present the sonar operator will only listen to what is called background noise, that is, the operator or the PAM system has no indication on the presence of a signal. On the other side, the sonar operator will decide on the presence of a signal if he or she notes a variation in the background noise that can be attributed with some confidence to the presence of the signal. That is, sonar operators analyse the background noise and detect (decide on the presence of) signals if what they hear differs in amplitude, energy or statistics from the background noise.

To avoid subjective decisions, the basic sonar equation is used to determine a critical or minimal signal-to-noise ratio that is required for the sonar implementation to call for positive signal detections. To detect a signal, we relate a minimal signal-to-noise ratio *SNR* to a fixed and well-defined threshold *TH*

$$SNR_{\min}(R, \vartheta) = TH \tag{3.1}$$

If the actual *SNR* is above this threshold, then the sonar system will indicate the presence of a signal; otherwise the system will decide that a signal is absent.

The signal-to-noise ratio is typically expressed in dB; for passive acoustic monitoring of cetaceans it may be described as follows

$$SNR(R, \vartheta) = ASL(\vartheta) - TL(R) - NL + AG + PG \tag{3.2}$$

where

$ASL(\vartheta)$ is the apparent source level of the sound source [dB // 1 µPa @ 1 m] where the receiver is not necessarily located in line with the sound axis (i.e. it may be off axis, where ϑ is the off-axis angle)

$TL(R)$ is the range dependent transmission loss (*R* range [m])

NL is the background, or ambient, noise level masking the signal

AG is the gain of the receiving hydrophone array

PG is the gain of the processing system

The *SNR*, as defined in Equation 3.2, describes the signal–to-noise ratio at the moment of the decision, say in the headphone of an operator who is listening to the underwater hydrophones. For this reason, Equation 3.2 includes a term that describes the range-dependent transmission loss $TL(R)$. It also includes the potential gains due to directional sensitivity of multiple hydrophones (receiving array) AG and also due to any signal specific data processing PG, which is typically used to reduce unwanted noise contamination of the signal.

3.2 Apparent source level

The apparent source level is a typical quantity when listening for cetacean sounds, and may be considered as composed of two components: one describing the maximum source level, the direction of which is also called the acoustic axis, the other describing the reduction of sound intensity as a function of off-axis angle

$$ASL(\vartheta) = SL_0 - DL(\vartheta) \tag{3.3}$$

where

SL_0 is the on-axis source level

$DL(\vartheta)$ is the directional loss of the source as a function of off-axis angle.

If we assume that the sound source is rotationally invariant for cetaceans, then the apparent source level depends only on the off-axis angle, which is the angle between the acoustic axis (the direction where most of the acoustic energy goes) and the direction of the receiver. The directional loss of the sound source is given by the source directivity pattern (e.g. Equation 2.21) expressed in dB and its negative value is then called off-axis attenuation.

Broadband off-axis attenuation of Gaussian bio-sonar pulses

Off-axis attenuation is especially relevant for the echolocation of toothed whales emitting bio-sonar pulses. These pulses are in general very short and therefore broadband and are emitted from a small, nearly circular area on the head of the animal. If we approximate these bio-sonar pulses by Gaussian pulses emitted from a circular piston and use Equations 2.7 and 2.21, we obtain, for positive frequencies, the following frequency-dependent directivity pattern

$$D(\omega, \vartheta) = \exp\left\{-\frac{\sigma^2}{c^2}(\omega - \omega_0)^2\right\}\left(\frac{2J_1(ka\sin\vartheta)}{ka\sin\vartheta}\right)^2 \tag{3.4}$$

We see that for a Gaussian bio-sonar pulse the directivity pattern is weighted with a Gaussian shape weighting function.

Noting that $c = \lambda f = \dfrac{\omega}{k}$ and defining $\overline{ka} = k_0 a$ and $\beta = \dfrac{a}{\sigma}$ then we may express the spectral directivity pattern in terms of ka

$$D(ka, \vartheta) = \exp\left\{-\left(\frac{ka - \overline{ka}}{\beta}\right)^2\right\}\left(\frac{2J_1(ka\sin\vartheta)}{ka\sin\vartheta}\right)^2 \tag{3.5}$$

The broadband directivity pattern is then obtained by integrating the spectral directivity pattern over all frequencies, or equivalently over all ka values

$$D_{bb}(\vartheta) = \frac{\int_{-\infty}^{\infty}\exp\left\{-\left(\frac{ka - \overline{ka}}{\beta}\right)^2\right\}\left(\frac{2J_1(ka\sin\vartheta)}{ka\sin\vartheta}\right)\,\mathrm{d}(ka)}{\int_{-\infty}^{\infty}\exp\left\{-\left(\frac{ka - \overline{ka}}{\beta}\right)^2\right\}\mathrm{d}(ka)} \tag{3.6}$$

In practice the integration may be replaced by a summation ranging from -3β to $+3\beta$.

Figure 3.1 shows a comparison of narrowband and broadband off-axis attenuation.

The off-axis loss as given in Equation 3.6 is, in addition to the off-axis angle, a function of the centre value of ka: $\overline{ka} = k_0 a$ and the width β of the spectral distribution of the pulse measured in terms of ka. It is therefore intuitive to characterize also the broadband off-axis attenuation in terms of the mean value of ka, or via Equation 2.24 by a nominal narrowband directivity index. It makes further sense to relate the width of the Gaussian weighting function to the mean ka value of the bio-sonar pulse.

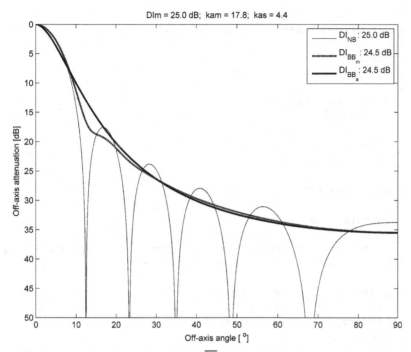

Figure 3.1   Off-axis attenuation of broadband piston ($\overline{ka} = 17.8, \beta = 4.4$). The narrowband off-axis attenuation is shown in grey; the broadband off-axis loss is shown as a dashed black line; the thick solid line describes the approximation to the broadband off-axis attenuation.

In cases where the estimation of the off-axis attenuation is too cumbersome, an approximation may be used

$$D_{bx}(\vartheta) = C_1 \frac{(C_2 \sin \vartheta)^2}{1 + |C_2 \sin \vartheta| + (C_2 \sin \vartheta)^2} \qquad (3.7)$$

where the two constants C_1 and C_2 are given by

$$\begin{aligned} C_1 &= 47 \\ C_2 &= 0.218 ka \end{aligned} \qquad (3.8)$$

As ka may be replaced by the directivity index (Equation 2.24), one could consider the two parameters given in Equation 3.8 as applicable to most short echolocation pulses, depending only on the directivity index.

This approximation is shown in Figure 3.1 as a thick, solid curve.

Matlab code

```
%Scr3_1
DIa=25;
%
kam=10^(DIa/20);
kas=kam/4;
```

```
%
ka=(kam+(-floor(3*kas):ceil(3*kas)))';
%
dth=0.1;
th=0:dth:90;
%
aa=ka*sin(th/180*pi);
bb0=piston(aa);

wwo=exp(-((ka-kam)/kas).^2);
ww=wwo*ones(size(th));

bbm=mean(ww.*bb0.^2)./mean(ww);

ibm=find(ka==kam);

DI1=2*sum(sin(th/180*pi))/sum(bb0(ibm,:).^2.*sin(th/180*pi));
DIm=2*sum(sin(th/180*pi))/sum(bbm.*sin(th/180*pi));

% approximation of beam pattern
cx=0.218*kam*sin(th*pi/180);
DLx= 47*cx.^2./(1+abs(cx)+cx.^2);
bpx=10.^(-DLx/10);
%
% convert to dB
DL0=-20*log10(abs(bb0(ibm,:)));
DLb=-10*log10(abs(bbm));
DIx=2*sum(sin(th/180*pi))/sum(bpx.*sin(th/180*pi));

figure(1)
plot(th,DL0,'k')
line(th,DLb,'color','k','linewidth', 2,'linestyle','-')
line(th,DLx,'color','k','linewidth',2)
ylim([0 50]);
set(gca,'ydir','rev')
xlabel('Off-axis angle [^o]')
ylabel('Off-axis attenuation [dB]')

title(sprintf('DIm = %.1f dB; kam = %.1f; kas = %.1f', ...
        DIa,kam,kas))

legend(['DI_{NB_}: ' sprintf('%.1f dB',10*log10(DI1))], ...
        ['DI_{BB_m}: ' sprintf('%.1f dB',10*log10(DIm))], ...
        ['DI_{BB_a}: ' sprintf('%.1f dB',10*log10(DIx))]);
```

3.3   Sound propagation

When sound propagates through any medium, its intensity is decreasing with time, or equivalently as a function of distance from the sound source. As one may expect, this propagation loss depends heavily on geometry, sound speed profile and frequency of the sound wave.

From the solution to the spherical wave equation (Equations 1.20 or 1.51), we know that the sound intensity of a spherical wave varies in inverse proportion to the range squared

$$\frac{I(r,t)}{I(1,t)} = r^{-2} \tag{3.9}$$

The decrease in sound intensity as a function of range is commonly expressed as transmission loss TL, which is measured in dB and defined by

$$TL(r) = 10\log\left(\frac{I(1,t)}{I(r,t)}\right) \tag{3.10}$$

Equation 3.10 defines the transmission loss due to spatial distribution of sound energy. In the case of spherical spreading this results in the well-known spherical spreading law: $TL(r) = 20\log(r)$.

In general, the transmission loss is augmented by the absorption as described in Section 1.4.4 so that the transmission loss is estimated according to

$$TL(r) = 10\log\left(\frac{I(1,t)}{I(r,t)}\right) + \alpha r \tag{3.11}$$

which for spherical spreading becomes

$$TL(r) = 20\log(r) + \alpha r \tag{3.12}$$

3.3.1   Transmission loss modelling

It has already been mentioned that sound propagation in real oceans might rarely be considered as spherical, because spherical spreading requires constant sound speed and no limiting boundaries. In most cases, we have variable sound speed profiles and interactions with sea boundaries (surface and bottom).

Although it is possible to generate a sequence of cases where the sound propagation is modelled more rigorously or even analytically, and although each of these special cases provides physical insight into acoustic propagation problems, real acoustic propagation modelling is rarely done analytically but by using more or less sophisticated acoustic propagation models.

As with all physical models, the results of acoustic models must be understood as an approximation to the solution. That is, acoustic propagation models are developed to provide an approximate solution of the wave equation. In this sense the transmission loss based on the spherical spreading law (Equation 3.12) is also a physical model with only two parameters (range r and absorption coefficient α), which is valid as long the physical conditions controlling the validity are satisfied. More complex models will require more input parameters to generate more realistic results, but as usual, the quality of the output will be governed by the quality of the input. Complex physical models have in recent

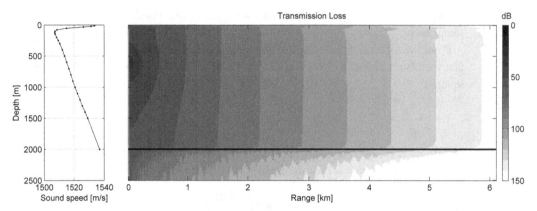

Figure 3.2   Transmission loss estimation using Bellhop. The thick horizontal line indicates the bottom at 2000 m.

years become the realm of numerical, i.e. computer-based, approaches to solving physical problems.

There are a variety of computer models for sound propagation; these however, were written for specific acoustic problems. In recent years, the Bellhop program (Porter and Bucker, 1987) has evolved as some sort of favourite in the bioacoustics community. It is freely available on the Internet and is computationally relatively fast. Bellhop is a ray-based model and as such appropriate for high sound frequencies: the term 'high' means that the wavelength of the sound is small compared with the available water space and geometry of interest. As a rule of thumb, one can assume that ray-based models may be appropriate if the water depth is much larger than the wavelength of sound. For very low frequencies or very shallow waters, sound propagation is better described by programs that are more specific to the physics of waveguides (e.g. using mode propagation).

Ray-based and normal-mode-based programs are available on the Internet from the Ocean Acoustics library (http://oalib.hlsresearch.com/), which contains a suite of acoustic modelling software for general use.

A typical transmission loss plot is shown in Figure 3.2. It shows on the left the sound speed profile and on the right a greyscale image of the modelled transmission loss. The selected frequency is 40 kHz and therefore very well suited for ray-based modelling. This plot shows that close to the position of the source (range 0 km and depth 600 m) the transmission loss is small (by definition TL is 0 dB at 1 m distance from the source). As expected from the spherical spreading law, TL increases fast to 40 dB at 100 m distance. Although the 40 db and 50 dB steps are reasonably circular, indicating spherical spreading, we note that beginning with the 60 dB step the TL curvature flattens out, becoming increasingly depth-independent.

Matlab code
The Matlab code to generate Figure 3.2 is twofold, one encapsulating the interface protocol of Bellhop and one general script calling the interface function and plotting the results.

Bellhop interface

```
function [pressure,Pos]=doBellhop(freq,sv,BOT,sd, rd,bty,ang)
% ready for TL estimation
% clean up temp files
warning off
delete ENVFIL
delete BTYFIL
delete LOGFIL
delete SHDFIL
warning on

% prepare input files for Bellhop
fid=fopen('ENVFIL','w');
% TITLE
    fprintf(fid,'''PAM_WMXZ''\n');
% FREQ (Hz)
    fprintf(fid,'%f,\n',freq);
% NMEDIA
    fprintf(fid,'1,\n');
% C-linear, Vacuum, db/lambda, Thorpe
    fprintf(fid,'''CVWT'',\n');
% ignored, ignored, DEPTH of bottom (m)
    fprintf(fid,'0 0.0 %f \n',sv(end,1));
    for jj=1:size(sv,1)
        fprintf(fid,'%f %f /\n',sv(jj,1),sv(jj,2));
    end
%0.0 is bottom roughness in m
    fprintf(fid,'''A*'' 0.0\n');
    fprintf(fid,'%f %f 0.0 %f %f 0 \n', ...
        BOT.depth,BOT.pSpeed,BOT.dens,BOT.pAtt);
% the following lines are specific to Bellhop
% NSD
    fprintf(fid,'%d\n',length(sd));
% SD(1:NSD) (m)
    fprintf(fid,'%f %f /\n',sd([1 end]));
% NRD
    fprintf(fid,'%d\n',length(rd));
% RD(1:NRD) (m)
    fprintf(fid,'%f %f /\n',rd([1 end]));
% NR,
    fprintf(fid,'%d\n',size(bty,1));
% R(1:NR) (km)
    fprintf(fid,'%f %f /\n',bty([1 end],1));
% ''R/C/I/S''
    fprintf(fid,'''I''\n'); %incoherent
% NBEAMS
    fprintf(fid,'%d\n',length(ang));
% ALPHA1,2 (degrees)
    fprintf(fid,'%f %f /\n',ang([1 end]));
% STEP (m), ZBOX (m), RBOX (km)
    fprintf(fid,'0.0 %f %f,\n', ...
```

```
         1.01*sv(end,1),1.01*bty(end,1));
fclose(fid);

%%%%%%%%%%%%% Bathy
fid = fopen( 'BTYFIL', 'w' );
fprintf(fid, '''%c'' \n', 'C');
    fprintf(fid,'%d\n',size(bty,1));
    for ii=1:size(bty,1)
        fprintf(fid,'%f %f\n',bty(ii,1),bty(ii,2));
    end
fclose(fid);

%execute Bellhop
tic
!bellhop <ENVFIL >LOGFIL
toc

%read TL
[titleText, plottype, freq, atten, Pos, pressure ]=...
read_shd_bin( 'SHDFIL');
```

The Bellhop interface function prepares some input files that are required by the Bellhop program, executes the Bellhop program and finally reads the transmission loss results, which are then passed back as an output of the function.

The script to estimate transmission loss, as shown in Figure 3.2, by calling doBellhop is given next. It first calls the function getSoundSpeed to obtain the sound speed for a given month and geographic location before invoking the Bellhop interface.

```
%Scr3_2
freq=40000; rmax=6.1;
%freq=15000; rmax=30.1;
%freq=3000; rmax=100.1;
%source depth
sd=600;

%plot parameters
ncol=15; %number of colors
dBs = 10; %dB/step
%cmap=flipud(jet(ncol));
cmap=gray(ncol);

%receiver depth [m]
rd=0:10:2500;

% sound speed profile
month=8;
lat0=41;
lng0=5;
%

[sv,T,S]=getSoundSpeed(month,lat0,lng0);
```

```
BOT.depth=2000;
BOT.dens=2.4;
BOT.pSpeed=3000;
BOT.pAtt=0.1; %dB/lambda

% extend sv if necessary
if rd(end)>BOT.depth
    sv(end+1,1)=BOT.depth;
    sv(end,2)=sv(end-1,2)+(sv(end,1)-sv(end-1,1))/64.1;
end

%describe bottom
rng=linspace(0,rmax,501)';
bty=[rng, 2000+0*rng];
%
ang=-89:89;
[pressure,Pos]=doBellhop(freq,sv,BOT,sd,rd,bty, ang);

pressure(pressure==0)=1e-38;
RL=-20*log10(abs(squeeze(pressure)));

save(sprintf('TL%.0f',freq),'pressure','Pos');

figure(1)
clf
set(gcf,'PaperOrientation','landscape');
set(gcf,'PaperPositionMode','auto');
set(gcf,'position',[100 300 860 330]);

%sv profile -------------------------
ax1=axes('units','pixel','position',[60,40,100, 250]);
plot(sv(:,2),sv(:,1),'k.-'), axis ij,grid on
xlabel('Sound speed [m/s]')
ylabel('Depth [m]')
ylim(rd([1 end]))
%RL image ------------------------
ax2=axes('units','pixel','position',[200,40,610,250]);
hold off
hi=imagesc(Pos.r.range/1000,Pos.r.depth,RL);
caxis( [0 ncol*dBs] )
set(gca,'yticklabel',[])
title('Transmission Loss','interpreter','none')
colormap(cmap)
xlabel('Range [km]')
%
hc=colorbar;
set(hc,'units','pixel','position',[820 40 10 250])
set(get(hc,'title'),'string',' dB')
set(hc,'ydir','reverse')
%
hold on
```

```
plot(bty(:,1),bty(:,2),'k','linewidth',2)
hold off
```

As indicated above numerical propagation models tend to be complex and require an increased amount of input. Relevant input parameters for Bellhop transmission loss modelling are given next.

General data
Frequency: `freq` in Hz
Source depth: `sd` in m

Environmental data
Sound speed profile: `sv` as vector: `sv(:,1)` depth in m, `sv(:,2)` sound speed in m/s
Bathymetry: `bty` as vector: `bty(:,1)` range in km, `bty(:,2)` depth in m
Bottom characteristics (pressure-wave sound speed and attenuation)

Modelling parameter
Number of range steps, min and max range values
Number of receiver depth steps, min and max receiver depth
Number of beams, min and max beam angle

Notes on parameter selection
To keep computation time reasonable it is advisable to keep the number of beams, ranges and depth values as small as possible: 250 depth, 500 range and 180 beam values seem to be reasonable for modern desktop computers. In cases of doubt, one should start with reduced numbers of values and slowly increase the computational parameters to reach the desired resolution.

3.3.2   Alternative geometric spreading laws

Equation 3.12 gives the transmission loss of a spherical wave in an iso-velocity and infinite medium:

$$TL(r) = 20\log(r) + ar \tag{3.13}$$

If the medium is limited between two horizontal perfectly reflecting planes and the source is a line source between these two planes then the spreading is cylindrical and the transmission loss is estimated by

$$TL(r) = 10\log(r) + ar \tag{3.14}$$

Figure 3.3 compares the two geometric spreading laws with a result from the Bellhop modelling exercise carried out earlier (Figure 3.2). One may observe that the cylindrical spreading law never adequately describes the modelled transmission loss. The spherical spreading law consistently overestimates the modelled *TL*.

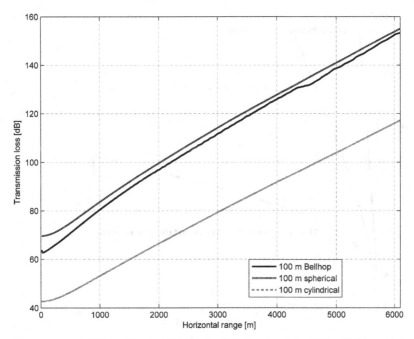

Figure 3.3   Comparison of different transmission loss models for a receiver depth of 100 m and a source depth
of 600 m. Solid lines, Bellhop numerical model; dashed lines, spherical spreading model
(Equation 3.12); dotted lines, cylindrical spreading model.

Even if it is generally accepted that iso-velocity spreading laws hardly describe reality,
there seems to be a need to characterize the transmission loss with the simplicity of
geometric spreading laws. One possible extension to the standard geometric spreading
laws is to assume spherical spreading up to a critical range and then to switch into a
modified spreading law

$$TL(r) = 20\log(\min(r, r_0)) + \gamma \log\left(\max(1, \frac{r}{r_0})\right) + \alpha r \tag{3.15}$$

where r_0 is the transition range separating the spherical regime from the modified
transmission loss regime, which is characterized by γ.

Figure 3.4 presents Bellhop TL modelling results for three frequencies 3, 15, and 40
kHz. The modelled data represent the TL for a source depth of 600 m and a receiver depth
of 100 m. In addition to the Bellhop model results (grey lines) Figure 3.4 also shows
(dashed lines) a fit of Equation 3.15 to the individual Bellhop curves. The formulas
describing the modified spreading laws are given within the figure. One notes that all
three cases result in different spreading law coefficients and consequently no single
modified spreading law may be defined, but Equation 3.15 may be used to approximate
the result of the Bellhop model where a more compact transmission loss formula is
desired.

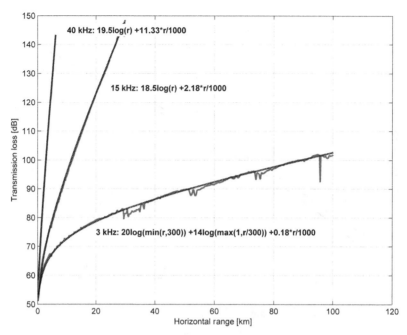

Figure 3.4   Comparison of different geometric spreading law approximations to numerically modelled transmission loss.

Figure 3.4 demonstrates that Equation 3.15 is reasonable to fit the modelled TL for long distances. How far this approach may be used to describe the results of a numerical TL model with a simplified equation must be assessed on a case-by-case basis.

3.4   Noise level

The sonar equation relates the received signal level to the background noise level that is masking the detection of the signal. The concept of noise is therefore more related to its effect than to its acoustic description. As stated in Chapter 1 we differentiate interferences from noise, and consider as noise only sound that may be described by a truly random process, either because it originated as a random process, or because it is constructed by random addition of a large number of otherwise non-random signals.

The causes of noise may then be grouped into two categories:

- *Ambient noise* originates outside the system and is therefore independent of the PAM system and its operation. Typical natural causes are wind, waves, rain and animals; typical anthropogenic causes are shipping and industrial activities.
- *Self-noise* is generated by the PAM system itself and includes radiated noise of the PAM vessel, flow noise, and thermal noise from the electronics.

3.4.1   Spectral noise level

Noise, as a random process, cannot be described by deterministic functions as we did for signals (e.g. Gaussian pulse: Equation 1.24). Noise may, however, be described by the mean intensity

$$\bar{I} = \frac{1}{\rho_0 c} \frac{1}{T} \int_0^T x^2(t) \mathrm{d}t \tag{3.16}$$

where $x(t)$ is the noise time series and T is the integration time, which is selected at an appropriate length to guarantee convergence of the intensity.

By definition, noise is broadband; in order to characterize noise, one has to quantify the bandwidth for which the noise level is measured. In practice, noisy time series $x(t)$ are bandpass filtered with a filter of bandwidth Δf around a centre frequency f_m, resulting in a new random time series $y(t)$.

The spectral density of this new noisy time series is then obtained by estimating the mean intensity and dividing the result by the bandwidth so that the spectral noise level in dB becomes:

$$NL0(f) = 10 \log \left(\frac{1}{T \Delta f} \int_0^T y^2(t) \mathrm{d}t \right) \tag{3.17}$$

The spectral noise level is measured in dB re 1 μPa^2/Hz; it is frequency-dependent and allows a general spectral description of noise. As the spectral noise level is estimated by means of a bandpass filter, it is also important to note the bandwidth of the filter, especially when the filters overlap in frequency. Even if the noise spectrum is given in terms per Hz, noise filters are typically much wider than 1 Hz, and hence narrow noise filters generate more spectral details than wide ones.

3.4.2   Noise masking effect

To obtain the masking noise NL in a broadband receiver, the spectral noise level $NL0$ is typically multiplied by the bandwidth, or in dB terms

$$NL = NL0(f_m) + 10 \log(B) \tag{3.18}$$

where f_m is the centre frequency of the receiver and B is the receiver bandwidth measured in Hz.

Equation 3.18 treats the spectral noise level as a constant within the receiver bandwidth. This, however, is only an approximation for the case where the analysis bandwidth used to estimate the spectral noise level is identical with the receiver bandwidth B.

In cases where the receiver is extreme broadband it is necessary to estimate the masking noise level by directly integrating the noise spectrum over the frequency band of interest B.

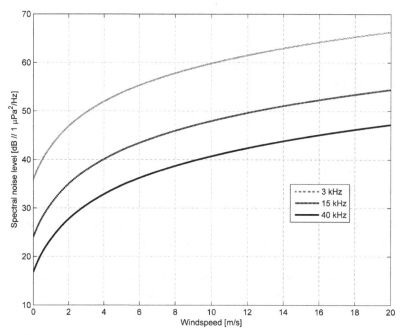

Figure 3.5   Surface noise spectral levels.

$$NL = 10 \log \left(\int_B 10^{\frac{NL0(f)}{10}} \mathrm{d}f \right) \tag{3.19}$$

Obviously, for a constant noise spectrum, Equation 3.19 translates to Equation 3.18.

3.4.3   Surface noise

The surface noise in ice-free oceans is mostly wind-driven and depends, in addition to the frequency, on the wind speed. To describe the spectral ambient noise level between 1 and 100 kHz (in dB re 1 μPa^2/Hz), the following fit to the Knudsen noise model (Knudsen, *et al.*, 1948; Lurton, 2002) may be used

$$NL_{Surf\_0}(f) = 44 + 23 \log\{v + 1\} - 17 \log\{\max(1, f)\} \tag{3.20}$$

where v is the wind speed in m/s and f is the frequency in kHz.

Figure 3.5 shows, for three different frequencies, the surface noise spectral level as function of wind speed.

3.4.4   Attenuation of surface noise with depth

Surface noise, as the name indicates, is generated at the water surface. The question arises, as to how this noise contribution varies with the depth of a receiver. There are two effects occurring when the receiver moves away from a noisy surface. First, one should expect that the contribution of individual noise sources decreases due to propagation loss, and second, one should expect that the number of surface noise sources increases that are within the same range of the receiver and therefore add together. Combining both effects, one obtains the following expression for the depth effect of surface-generated noise (Lurton, 2002)

$$NL_{\text{Surf\_0}}(d,f) = NL_{\text{Surf\_0}}(f) - a_f d - 10\log\left(1 + \frac{a_f d}{8.686}\right) \tag{3.21}$$

where d is the receiver depth, measured in m, and a_f is the absorption coefficient, measured in dB/m.

The depth-dependent reduction of surface noise becomes significant at higher frequencies. So the surface-generated noise at a frequency of 40 kHz and at a receiver depth of 1000 m is at least 10 dB lower than it is at 1 m water depth.

3.4.5   Shipping noise

Noise from (mainly remote) shipping is the main cause of ambient noise at low frequencies, that is, between some tens of Hz and 1 kHz. Ships carry a variety of noisy machinery (engines, generators, winches, etc.), the sound of which is transmitted via the hull into the sea. Furthermore, rotating propellers may create characteristic broadband cavitation noise, the level of which increases with the speed of rotation of the propeller and therefore with the speed of the vessel.

Shipping noise intensity is very variable, depends strongly on the distance of the ships, and is not easy to quantify, but for remote shipping, the following empirical model (Wenz, 1962; Lurton, 2002) may be used for an indicative order of magnitude of dB level:

$$NL_{\text{Ship\_0}}(f) = 60 + 10\mu - 20\log\{\max(0.1, f)\} \tag{3.22}$$

$\mu = 0, 1, 2, 3$ for (quiet, low, medium, heavy) ship traffic, and frequency f is given in kHz. For frequencies below 100 Hz, shipping noise may be assumed to be frequency-independent.

3.4.6   Turbulence noise

At very low frequencies, between 1 Hz and 10 Hz, ambient noise is dominated by noise originating from oceanic turbulence. It falls off very rapidly with frequency and may be modelled according to (Wenz, 1962)

$$NL_{\text{Turb}} = 17 - 30\log(f) \tag{3.23}$$

where the frequency f is given in kHz.

3.4.7   Thermal noise

At very high frequencies, say in excess of 100 kHz, ambient noise is again dominated by oceanic effects, but this time by the random molecular motion in the sea. This thermal noise may be modelled for an ideal hydrophone according to (Lurton, 2002)

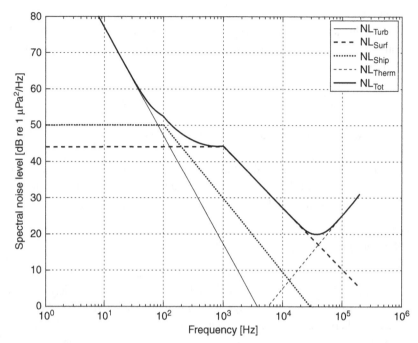

Figure 3.6   Ambient noise spectra for sea state zero, quiet ship traffic, and hydrophone depth of 100 m.

$$NL_{\text{Therm}} = -15 + 20\log(f) \tag{3.24}$$

where the frequency f is given in kHz.

3.4.8   Overall ambient noise spectrum

Combining all ambient noise contributions, we obtain the overall ambient noise spectrum level, which will obviously depend on sea state, ship traffic and hydrophone depth. It is, however, instructive to visualize the modelled minimum ambient noise level, which is a good measure when assessing the performance of a PAM system. If we wanted to detect very faint signals, that is, distant signals with low signal-to-noise ratio, then an ideal PAM system should be designed to be able to measure and quantify this minimal ambient noise.

Figure 3.6 shows the minimal ambient noise spectral level for a hydrophone depth of 100 m, sea state zero, and quiet ship traffic. This is not necessarily the lowest ambient noise level one may observe in some remote locations of the ocean, as all noise contributions are determined empirically and at places that, to some extent, should be considered as noisy. For example, Equation 3.21 indicates that increasing the hydrophone depth will decrease the contribution of surface-generated noise. Oceanic turbulence noise may also be shielded by ocean bottom topography. On the other hand, some places may never be as quiet as is modelled here. Although the above models are useful for performance assessment, they are no substitute for real ambient noise measurements *in situ*.

Matlab code

```
%Scr3_6
f=(1:200000)';
fkz=f/1000;
D=0; %0:3 %quiet shipping
w=0; %m/s % no wind
%
T=13;
S=38;
Z=100;
c=1500;
pH=7.8;
aa=FrancoisGarrison(fkz,T,S,Z,c,pH);
dcorr=aa*Z/1000;

NL1=17-30*log10(fkz);
NL2=44+23*log10(w+1)-17*log10(max(1,fkz));
NL3=30+10*D-20*log10(max(0.1,fkz));
NL4=-15+20*log10(fkz);

NL3d=NL3-dcorr-10*log10(1+dcorr/8.686);

NL=10*log10(10.^(NL1/10)+10.^(NL2/10)+ ...
             10.^(NL3d/10)+10.^(NL4/10));

figure(1)
hp=plot(f,[NL1 NL2 NL3d NL4 NL],'k','linewidth',1);
set(hp(1),'linestyle','-')
set(hp(2),'linestyle',':')
set(hp(3),'linestyle','--')
set(hp(4),'linestyle','--')
set(hp([2 3 5]),'linewidth',2)
ylim([0 80])
grid on
ylabel('Spectral noise level [dB re 1 \muPa^2/Hz]')
xlabel('Frequency [Hz]')
set(gca,'xscale','log','xminorgrid','off')
legend('NL_{Turb}','NL_{Surf}','NL_{Ship}', ...
       'NL_{Therm}','NL_{Tot}')
return
```

3.5   Array gain

PAM systems use one or more hydrophones to intercept the sound in the sea. Hydrophones are typically selected to have the same sensitivity in all directions, that is, they are omni-directional. This is in general desirable for PAM applications, as the search for cetaceans implicitly suggests that the direction to the cetaceans is *a priori* not known. Using multiple hydrophones, however, allows us in certain circumstances to focus the sensitivity of the receiving array towards predefined directions and to reduce environmental noise that is arriving from unwanted directions, resulting in an array gain AG. This AG is defined as an improvement of the signal-to-noise ratio due to the array when compared with an omni-directional hydrophone. Consequently, the estimation of the array gain requires knowledge of the directionality of the array and the anisotropy of the ambient noise field.

3.5.1   Array gain estimation

To obtain a value for AG it is convenient to consider the following approach by Urick (1983), which uses statistical properties of signal and noise. Consider a linear array of n hydrophones with equal sensitivity and a distant source abeam (90°) of the array. Let us now sum the n different hydrophones with and without signal. In the presence of a sufficiently strong signal we obtain for the array output

$$S_n(t) = \sum_{i=1}^{n} s_i(t) \tag{3.25}$$

where $s_i(t)$ are the outputs generated by the hydrophones in the presence of noise-free signal.

In the absence of a signal we obtain

$$N_n(t) = \sum_{i=1}^{n} n_i(t) \tag{3.26}$$

where $n_i(t)$ are the outputs generated by the hydrophones in absence of a signal.

The average signal-to-noise ratio is then given by

$$SNR_n = \frac{\overline{(S_n(t))^2}}{\overline{(N_n(t))^2}} = \frac{\overline{\left(\sum_{i=1}^{n} s_i(t)\right)^2}}{\overline{\left(\sum_{i=1}^{n} n_i(t)\right)^2}} \tag{3.27}$$

Expanding the sum squared and taking the time average on the individual elements (pair of hydrophones) one obtains a cross-correlation of the hydrophone measurements, be it signal or noise.

Using the definition of the cross-correlation coefficient ρ_{ij}

$$\rho_{ij} = \frac{\overline{x_i(t)x_j(t)}}{\sqrt{\overline{x_i^2(t)x_j^2(t)}}} \tag{3.28}$$

the average signal-to-noise ratio becomes with $\overline{s_i^2(t)} = \overline{s^2}$ and $\overline{n_i^2(t)} = \overline{n^2}$

$$SNR_n = \frac{\overline{s^2}}{\overline{n^2}} \frac{\sum_i \sum_j (\rho_S)_{ij}}{\sum_i \sum_j (\rho_N)_{ij}} = SNR_1 \frac{\sum_i \sum_j (\rho_S)_{ij}}{\sum_i \sum_j (\rho_N)_{ij}} \tag{3.29}$$

and the array gain becomes

$$AG = 10 \log\left(\frac{SNR_n}{SNR_1}\right) = 10 \log\left(\frac{\sum_i \sum_j (\rho_S)_{ij}}{\sum_i \sum_j (\rho_N)_{ij}}\right) \tag{3.30}$$

The array gain depends therefore on the cross-correlation coefficients between all pairs of hydrophones, for both the signal and the noise. The cross-correlation coefficient for the signal tends to stay, if not close to unity, higher than the cross-correlation coefficient for noise, resulting in a positive array gain.

A simple example assumes complete coherence of the signal over the array and completely uncorrelated noise measurements by the different hydrophones. With these assumptions we obtain for the array gain

$$AG = 10 \log\left(\frac{n^2}{n}\right) = 10 \log(n) \tag{3.31}$$

However, noise is rarely completely uncorrelated for broadband receiving systems. With partially coherent noise the array gain degrades and becomes

$$AG = 10 \log\left(\frac{n}{1 + (n-1)\rho_n}\right) \tag{3.32}$$

where ρ_n is the cross-correlation coefficient of noise between the hydrophones, which are now assumed to be equidistant. If the hydrophones are very close together in relation to the wavelength, then the noise cross-correlation coefficient ρ_n approaches unity and AG approaches zero dB.

3.6   Processing gain

Similar to the array gain, which describes how array processing reduces the impact of ambient noise on the performance of the PAM system, the signal processing gain describes the improvement of the signal-to-noise ratio due to signal processing that occurs after the reception of the sound by one or more hydrophones. Consequently, the signal processing gain PG depends on the detailed implementation of the processing system and the suitability of the processing implementation for detecting signals against masking noise. A more detailed description must therefore follow when describing the signal processing of cetacean signals.

3.7   Detection threshold

The other quantity in the initial sonar equation (Equation 3.1) is the detection threshold TH, which by definition is the minimum signal-to-noise ratio required to detect a signal. Original interpretations of the sonar equation refer the detection threshold to the SNR at the output of the hydrophone or array, and consider the processing gain a part of the detection threshold. However, as signal processing is becoming more important in PAM, we treat the processing gain separately and consider the detection threshold as the signal-to-noise ratio at the final visual or aural presentation of the sound, or more generally, at the decision level. An additional advantage with this approach is that there is no need for processing-dependent detection thresholds and the detection threshold depends only on the noise statistics at the receiver output, which may easily be measured and determined.

For a given detection threshold $TH = \log(th)$, our detections (decisions for signal presence) fall into two classes: true detections due to the presence of signals and false detection due to excessive noise levels. A common method is therefore to relate the detection threshold to the number of false detections, or false alarms, by limiting the probability of encountering false alarms $P_{FA}(th)$

$$P_{FA}(th) = \int_{th}^{\infty} w_{Noise}(y)\,dy \tag{3.33}$$

where $w_{Noise}(y)$ is the probability density function (PDF) of the noise.

In the absence of any knowledge of the noise statistics, it is convenient to assume a Rayleigh distribution for the noise PDF (see also Section 1.5.2)

$$w_{Noise}(y) = \frac{y}{\sigma_0^2} \exp\left\{-\frac{y^2}{2\sigma_0^2}\right\} \tag{3.34}$$

where σ_0^2 is the noise variance, which in the context of the sonar equation becomes effectively 1, as the sonar equation is formulated in terms of signal-to-noise, i.e. we replace $\frac{y}{\sigma_0}$ with the signal-to-noise ratio.

Integrating Equation 3.33 according to Equation 3.34, we obtain for the probability of a false alarm

$$P_{FA}(th) = \exp\left\{-\frac{th^2}{2}\right\} \tag{3.35}$$

resulting in a straightforward expression for the threshold as a function of the false alarm probability

$$th = \sqrt{-2\ln(P_{FA})} \tag{3.36}$$

Equation 3.36 is widely used even if real ocean noise hardly follows a true Rayleigh distribution. If real noise statistics are available then it is preferable to estimate the desired threshold by means of Equation 3.33, using a measured noise PDF.

Part II

Signal processing (designing the tools)

This part discusses the algorithms that are needed to detect, classify, locate and track (DCLT) cetaceans. Chapter 4 first presents and discusses techniques that are commonly used to detect signals in passive sonar applications, providing the necessary information for the subsequent signal processing chain. Chapter 5 introduces the concept of classification as a means of reducing erroneous detections, interferences and unrelated detections. The ability to localize animals is addressed in Chapter 6, in which I discuss well-known techniques for passive range estimation: multi-hydrophone ranging, triangulation, multi-path ranging and beam-forming. Tracking is the last component in the DCLT signal processing toolset and is presented together with localization methods in Chapter 6. Considering PAM, tracking plays an important role if one not only tries to localize one or more acoustically active cetaceans, but also tries to monitor their behaviour, which requires tracking, or the continuous estimation of individual animal locations.

4 Detection methods

This chapter presents and discusses techniques that are commonly used to detect signals in passive sonar applications. Detection algorithms generate, from acoustic raw data, the information that is required by the subsequent steps in the PAM data processing chain.

Starting with the detection concept as presented in Chapter 3 to introduce the sonar equation, this chapter discusses first the standard 'threshold' detector, which is implemented in different flavours, and introduces then the 'sequential probability ratio detector', which is a non-linear recursive detector that has found increasing interest in sonar systems research.

Strongly related to the detection process is the selection of a proper method for reducing unrelated noise content by spectral filtering. This volume concentrates on major techniques found in the relevant literature: spectral equalization, bandpass filtering and matched filtering.

All detection methods are tested with real data and the performance differences are discussed. Parallel to the detection performance this chapter will address the issue of false alarms and discuss the Receiver Operation Characteristics (ROC) curves of detectors.

4.1   Detection of echolocation clicks

A PAM application may receive a time series similar to the one depicted in Figure 4.1, where we see 3 seconds' worth of data of an hour-long reception of underwater sound received on a single hydrophone. The figure shows a sequence of short transients embedded in a continuous noise floor as measured by the hydrophone.

Inspecting this figure one realizes that, although the different transients are clearly visible, it is not obvious how many signals may be found within this time series. Starting with the strongest transient one may count at least 16 transients that could qualify as a signal, but the weakest transients may easily represent also noise fluctuations. There is therefore a need for a quantitative method to establish what is a transient of interest and should be considered as a signal and which potential signal is only due to noise fluctuations and consequently of no interest to the PAM application.

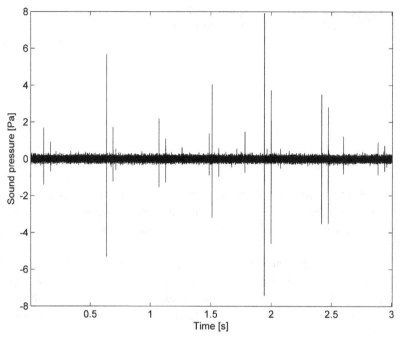

Figure 4.1   Typical time series of echolocating beaked whales.

Matlab code to generate Figure 4.1

```
%Scr4_1
% load data
[xx,fs] = wavread('../Pam_book_data/zifioSelectScan.wav');
%
% construct time vector
tt = (1:length(xx))'/fs;
%
% subtruct mean to remove constant offset
ss = xx - mean(xx);
%
%calibrate time series (convert to Pascal)
cal=10^(27/20); %unit is equiv 27 dB re 1 Pa
ss=cal*ss;
%
figure(1)
plot(tt,ss,'k')
set(gca,'fontsize',12)
xlabel('Time [s]')
ylabel('Sound pressure [Pa]')
xlim(tt([1 end]))
```

4.1.1   Threshold detector

In Section 1.5 we described signals as sound messages that are of interest to our PAM application and which we wanted to receive with our PAM system. Inspecting Figure 4.1,

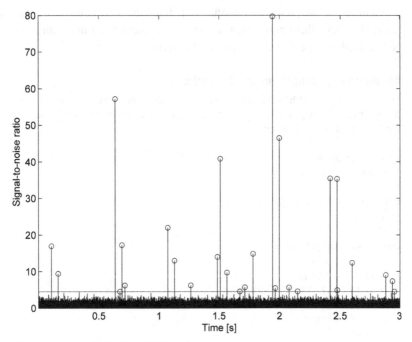

Figure 4.2 Detections of multiple signals (circles).

we expect that the recorded dataset contains some short signals (visible transients) embedded in noise. We consider a transient to be a signal if it satisfies some signal detection criteria and decide for a signal when its peak amplitude exceeds significantly the noise values; that is, we relate the signal to the background noise and denote an observed transient as a signal if its peak signal-to-noise ratio (*SNR*) exceeds a threshold. All transients below the threshold are considered noise and all transients above the threshold are called signals.

Definition of detection

A transient is considered a detected signal if its signal-to-noise ratio *SNR* exceeds a threshold *TH*

$$SNR > TH \qquad (4.1)$$

This relation (here formulated as an inequality) not only constitutes a general definition but also contains the basic recipe for the detection process: in order to detect a signal one should first estimate the noise at the time of the transient, estimate the signal-to-noise ratio, and then compare it with a threshold, the value of which is determined independently.

Figure 4.2 shows the result of the detection process using the data in Figure 4.1. For the detection process a threshold of *th* = 4.5, equivalent to *TH* = 13 dB, was chosen, resulting in 27 detections. The signal-to-noise ratio is presented in Figure 4.2 as a ratio and not as a dB level, to facilitate the comparison with previous figures, but this should not be a problem as the inequality in Equation 4.1, holds in both linear and logarithmic (dB) scales. To obtain this figure it was necessary to implement a simple click detector, which is described in the comments to the Matlab script.

Matlab script of simple threshold detector

```
function [jth,ith]=do_SimpleDetection0 (snr,tt,th, dt_max)
% function [jth,ith]=do_SimpleDetection0 (snr,th, dt_max)
% simple click detector
%
% thresholding
ith=find(snr>th) ;
%
% detection pruning
jth=ith;
jj=1;
for ii=2:length(ith)
    ki=ith(ii) ;
    kj=ith(jj) ;
    if tt(ki)-tt(kj)>dt_max
        jj=ii;
        continue
    else
        if snr(ki)>snr(kj) % actual snr value is larger
                           % than old one
            jth(jj)=0;      % remove old one
            jj=ii;          % take the actual as old
        else
            jth(ii)=0;      % remove actual one
        end
    end
end
%
%reduce detection vector
jth(jth==0)=[] ;
```

Comments on the Matlab function

To implement a simple detector we first threshold the *SNR* and then prune the threshold crossings by requiring that detections not only exceed the threshold but are also local maxima within a running window of 2 ms. dt_max=2ms, as defined in script Scr4_1. In other words, within a 2 ms window we only keep the threshold-crossing with the maximum signal-to-noise ratio. This method defines a minimum distance between two consecutive detections, which becomes the length of the pruning window. The size of the pruning window is context-dependent and must be determined together with the detection threshold.

Matlab script for running the detector and plotting the results

```
%Scr4_2
%if required (un)comment next line
Scr4_1
```

```
%
%noise estimate over complete time series
nn=sqrt(mean(ss.^2));
%
% estimate signal-to-noise ratio
snr = abs(ss)/nn;
%
% determine threshold
th=4.5;
% detection pruning window
dt_max=2/1000; %2 ms
%
% simple click detector
jth=do_SimpleDetection0(snr,tt,th,dt_max);

% plot all
figure(1)
plot(tt,snr,'k', tt(jth),snr(jth),'ko', ...
     tt([1 end]),th*[1 1],'k--')
xlim(tt([1 end]))
set(gca,'fontsize',12)
xlabel('Time [s]')
ylabel('Signal-to-noise ratio')
```

Comments on the script

The script executes after the previous script that generated Figure 4.1 and starts by obtaining a noise estimate, taking the root mean square (RMS) value of the data. We use the whole dataset, including the signals, to estimate the noise, as the signals are very short transients and we are willing to overestimate the noise level somewhat.

To estimate the signal-to-noise ratio we simply divide the absolute value of the de-meaned dataset by the noise estimate. In this example the threshold is guessed by visual inspection of the *SNR* plot. Later, we will use a method to determine an improved selection of the threshold. The Matlab script estimates the signal-to-noise ratio as a pure ratio and therefore requires the threshold to be given, not in dB, but in terms of a required ratio.

4.1.2 Signal amplitudes, signal power and signal envelope

We note from the Matlab script that we took the absolute value of the signal before estimating the signal-to-noise ratio. Although this procedure is intuitive, it is not necessarily the best approach.

From Equation 2.1 we realize that the absolute pressure value of a general signal is given by

$$|P(t)| = \sqrt{P(t)^2} = \frac{A(t)}{\sqrt{2}}\sqrt{1+\cos\{2\omega t\}} \tag{4.2}$$

that is, the absolute value (pressure) is a function, which varies between $A(t)$ and zero, resulting in multiple threshold crossings and therefore multiple detections for the same signal. We removed these multiple detections by pruning the detections and allowing only the detection with the highest SNR to survive.

However, it is the amplitude function, or envelope $A(t)$ in Equation 4.2, in which we are really interested for our description of the signal and consequently for the estimate of the signal-to-noise ratio. It is therefore necessary to separate the amplitude function from the purely oscillating part, which in Equation 2.1 is the cosine function, the absolute value of which resulted in Equation 4.2 in the additional $\sqrt{\frac{1+\cos\{2\omega t\}}{2}}$ term.

Analytic signal form

If we were able to replace the real valued signal description by a complex one

$$P_C(t) = A(t)\exp\{-i\omega t\} \tag{4.3}$$

then taking the absolute value would yield only the desired amplitude function

$$A(t) = |P_C(t)| \tag{4.4}$$

because the absolute value of a purely complex number is unity $|\exp\{-i\omega t\}| = 1$.

There exists a simple method to convert a real valued signal into a complex representation of the form of Equation 4.3. This method is based on the observation that the Fourier transforms of real valued functions have negative frequencies whereas the Fourier transform of the complex extended description of the signal exhibits only positive frequencies. To obtain what is also called an analytic extension, or an analytic form of the signal, it is sufficient to carry out the following operation

$$P_C(t) = FFT^{-1}(W_C \times (FFTP(t))) \tag{4.5}$$

where W_C is a weighting function that modifies the spectrum of the real valued signal $P(t)$ by setting the negative frequencies to zero, so that the inverse Fourier FFT^{-1} transform results in a complex signal description $P_C(t)$. That is, in Matlab terms we must choose $W_C(1) = 1$, $W_C(n) = 2$, for $n = 2 \ldots$ nfft/2 (positive frequencies), and $W_C(n) = 0$, for $n =$ nfft/2+1 $\ldots$ nfft (negative frequencies).

Figure 4.3 presents the amplitude or envelope function in which the signal is taken from the time series presented in Figure 4.1 and the envelope is obtained as absolute value of the analytic time series.

Hilbert transform

The above method of Equation 4.5 is suited for short time series that can be Fourier transformed with a single operation. For very long time series, or continuous processing, transitional edge effects occur between consecutive datasets, making it hard to generate an analytic signal. An alternative method to Equation 4.5 is based on the observation that for an analytic, complex valued time series, the imaginary part is 90° phase shifted relative to the real part. A digital filter that accomplishes this task and that is suited for continuous processing is known as the Hilbert transform. The analytic time series is then

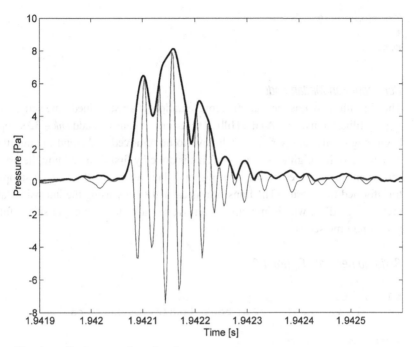

Figure 4.3   Signal amplitude or envelope function.

constructed by keeping the original time series as the real part and considering the Hilbert-transformed time series as the imaginary part of the analytical time series.

The Matlab code shows a practical implementation of this Hilbert transform (Stearns and David, 1988), which, however, differs from the ideal filter in three ways: the impulse response is finite, the filter is causal, and we apply an additional window function to improve the gain characteristics of the practical implementation. The causality of the filter delays the filter response by half the window length, a fact that must be considered when constructing the analytical signal.

Matlab code for estimating the complex valued analytic time series

```
function z=doAnalytic(x,nfft)
%
% define impulse for Hilbert transform
nk=32;
k=(1:nk)';
h=zeros(2*nk+1,1);
h(nk+1+k)=(2/pi)*(k-2*floor(k/2))./k;
h(nk+1-k)=-h(nk+1+k);
h=h.*hamming(2*nk+1);
%
% use FFT filt to process long data and correct for delay
y=fftfilt(h,[x;zeros(nk,size(x,2))]); y(1:nk,:)=[];
%
```

```
% construct complex signal
z=x+i*y;
return
```

Comments on Matlab code

The algorithm obtains the analytic representation of a real valued time series by estimating its Hilbert transform. As the Hilbert transform is causal, it adds $nk = 32$ samples at the beginning of the time series, which have to be removed at the end to align the Hilbert transform to the original time series. This is done by first zero padding the original time series with nk samples before filtering, and then removing the first nk samples in the transformed time series. The transform is carried out by using the intrinsic Matlab filter method `fftfilt`, which implements an overlap and add technique suited for filtering very long time series.

Script to generate Figure 4.3

```
%Scr4_3
% load data
[xx,fs] = wavread('../Pam_book_data/zifioSelect Scan.wav');
%
% construct time vector
tt = (1:length(xx))'/fs;
%
% subtract mean to remove constant offset
ss = xx - mean(xx);
%
%calibrate time series
cal=10^(27/20); %unit is equiv 147 dB re 1 uPa
ss=cal*ss;
%
% envelope
yy_a=doAnalytic(ss,512);
%
xl=1.9421+[-2 5]*1e-4;
%
figure(1)
clf
hp=plot(tt,ss,'k',tt,abs(yy_a),'k');
set(hp(2),'linewidth',2)
set(gca,'fontsize',12)
xlabel('Time [s]')
ylabel('Pressure [Pa]')
xlim(xl)
```

4.1.3   Signal extraction

Characterizing the detection of a signal by the event time and by the value of the maximum SNR constitutes the minimal information one should expect from a signal

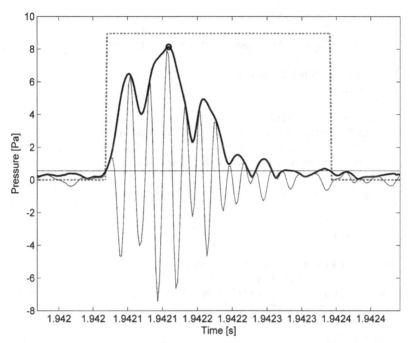

Figure 4.4   Extracted signal in pressure scale. The dotted line marks the signal extent for extraction.
The indicated threshold (dashed line) was transferred for visualization purposes from an SNR scale
to a pressure scale.

detector. As we have seen in the description of cetacean sounds, the assignment of signals
to different cetacean species may be complicated, and it therefore seems appropriate
also to extract the time series of the different detected signals from the data stream for
potential use in a later classification step. To extract the signals we have to modify the
detector to provide the start and stop times of the detected signal. A simple approach to
obtaining start and stop times is to find all threshold crossings that are within the
detection window around the previously found detection event. To extract the time series
it is then useful to add some data samples before and after the detected signal boundaries.

To complete the description of the detected signal it is appropriate to add to the detector
output all necessary information to allow a complete reconstruction of the detection
process, which also includes, for example, the noise level.

Figure 4.4 shows an extracted signal, the detection of which is based on the envelope or
amplitude function of the signal pressure. The envelope-based SNR is preferable to an
absolute pressure-based SNR, as the envelope-based detection tends to detect the maximum
extent of the signal. This is because the pressure envelope is by definition never less than the
absolute pressure, but may on occasion exceed the values given by the absolute pressure.

Matlab code of the refined detection algorithm for estimation of start and stop time

```
function [jth,i1,i2]=do_Simple Detection1(snr,tt,th,dt_max)

% function [jth,i1,i2]=do_SimpleDetection(snr,th, dt_max)
```

```
% simple click detector
%
[jth,ith]=do_SimpleDetection0(snr,tt,th,dt_max);

% estimate signal length
j1=0*jth;
j2=0*jth;
for ii=1:length(jth)
    %find all threshold crossings around detection
    thc=find(abs(tt(jth(ii))-tt(ith))<=dt_max);
    % store first and last threshold crossing
    j1(ii)=min(thc);
    j2(ii)=max(thc);
end
i1=ith(j1);
i2=ith(j2);

%eliminate overlapping detections
iov=find(i1(2:end)-i2(1:end-1)<0);
imx=snr(jth(iov))>snr(jth(iov+1));
iov(imx)=iov(imx)+1;
jth(iov)=[];
i1(iov)=[];
i2(iov)=[];
```

Matlab code to extract the signal time series and to plot a single detection

```
%Scr4_4
%
Scr4_3
%
nn_a=sqrt(mean(abs(yy_a).^2));
snr_a=abs(yy_a)/nn_a;
%
% determine threshold
th=4.5;
% detection pruning window
dt_max=2/1000; %2 ms
%
% simple click detector
%
[jth,i1,i2]=do_SimpleDetection1(snr_a,tt,th, dt_max);
%
%extract signals
di=ceil(fs*1e-4);
Det=[];
for ii=1:length(jth)
    Det(ii).tdet=tt(i1(ii)-di);
    Det(ii).idet=[jth(ii),i1(ii),i2(ii)];
    Det(ii).odet=di;
    Det(ii).ss=ss((i1(ii)-di):(i2(ii)+di));
```

```
    Det(ii).nn=nn_a;
end
%
% select single detection
ii=14;
xx1=Det(ii).ss;
tt1=Det(ii).tdet+(0:(length(xx1)-1))'/fs;
%
% construct envelope
yy=doAnalytic(xx1,512);
A=abs(yy);
nn_a=Det(ii).nn;
%
% mark detection
D=0*tt1;
ipk=Det(ii).odet+1+(Det(ii).idet(1)-Det(ii).idet(2));
idt=Det(ii).odet+1+(0:Det(ii).idet(3)-Det(ii).idet(2));
D(idt)=A(ipk)*1.1;
%
% plot selected detection
figure(1)
hp=plot(tt1,xx1,'k',tt1,A,'k',...
        tt1(ipk),A(ipk),'ko',...
        tt1,D,'k:',...
        tt1([1 end]),th*nn_a*[1 1],'k-');
xlim(tt1([1 end]))
%set(hp(1),'color',0.5*[1 1 1])
set(hp(2:4),'linewidth',2)
set(gca,'fontsize',12)
xlabel('Time [s]')
ylabel('Pressure [Pa]')
```

4.1.4 Pre-processing raw data

Based on the suggestion of the sonar equation, we implemented and analysed a simple threshold detector at the beginning of this chapter. Although the instructions of the sonar equations are straightforward and rather intuitive, the first example has already demonstrated difficulties with the implementation. In particular, it turned out that what is expected to be a single signal, or detection event, may result in multiple threshold crossings of the signal-to-noise ratio (i.e. multiple possible detections). All threshold crossings that are associated with a single detection event are highly correlated and may be reduced with proper processing. A first method pruned the number of threshold crossings within a running window, the length of which was chosen to reflect the expected signal length. The actual signal length was then defined by the time between the first and last threshold crossing around the event time where the highest signal-to-noise ratio occurred.

The signal detection described so far used the instantaneous absolute pressure or the pressure envelope as basis of the SNR estimation. This is appropriate when the threshold

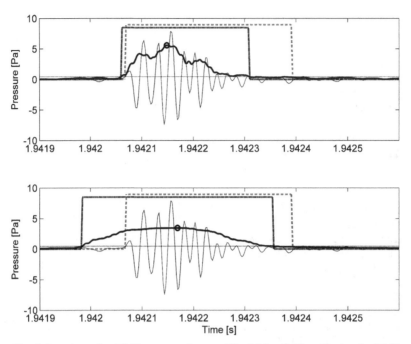

Figure 4.5   Signal detection using RMS pressure detector. The thick solid line denotes the RMS pressure, the thick dashed solid line the result of the detector, and the dotted line the envelope detector result from Figure 4.4. The RMS estimation of the top panel is based on a 25 μs window (10 samples) integration and the bottom panel in based on a 200 μs window (78 samples).

crossing of a single pressure value constitutes a valid signal detection, which is hardly the case for real and noisy datasets. We expect that real signals are of finite length covering more than one pressure measurement. The question arises of whether one could avoid these spurious threshold crossings that are too short to stem from real signals and therefore improve the detection performance by using for the estimation of the signal-to-noise ratio an average of the pressure measurements instead of the instantaneous pressure amplitudes. The assumption is that, as signals are more correlated than noise, any averaging will reduce the noise, will therefore increase the signal-to-noise ratio, and will consequently improve the detection performance.

In the case discussed so far, the sampling frequency is 384 kHz, about six times the highest frequency of 60 kHz, and the signal is long enough to result in over 70 samples, giving the opportunity to implement noise averaging by means of a RMS pressure detector. From Chapter 1 we know that the mean intensity is estimated by averaging the sound pressure squared over some time window, which may either be a cycle of the pressure wave or over the whole signal.

The RMS pressure at a time t is then estimated by

$$P_{RMS}(t) = \sqrt{\frac{1}{T} \int_{t-T/2}^{t+T/2} P^2(\tau) \mathrm{d}\tau} \qquad (4.6)$$

whereby the (rectangular) window of length T is centred around desired time t.

Figure 4.5 shows the impact of the averaging, the now smoothed RMS pressure function, on the resulting detection. The results differ from the envelope-based detector, but whether there is a true improvement in detection performance cannot be told from this figure but requires a more global analysis, which will be carried out later in this chapter.

We note also from Figure 4.5 that the longer the integration window the smoother the estimated RMS pressure becomes. The start position of the signal then occurs earlier than with the envelope threshold detector, as the PRM pressure estimate already 'sees' the signal well before the signal onset (half window length). The end position of the signal also tends to be earlier than with the envelope threshold detector, as the smoothing property of the window suppresses weak trailing signal oscillations that are present in the dataset.

In recent years, some publications have promoted what is called the Taeger–Kaiser (TK) algorithm, claiming that it is a measure of the proper sound wave energy (e.g. Kaiser, 1990; Kandia and Stylianou, 2006).

Let the sound pressure be described by

$$P(t) = P_0 \cos(\omega t) \tag{4.7}$$

we form the product

$$P(t - dt)P(t + dt) = P_0^2 \cos(\omega(t - dt)) \cos(\omega t(t + dt)) \tag{4.8}$$

Using the trigonometric identity

$$\cos(\alpha - \beta)\cos(\alpha + \beta) = \cos^2(\alpha) - \sin^2(\beta) \tag{4.9}$$

we obtain

$$P(t - dt)P(t + dt) = P_0^2 \left[\cos^2(\omega t) - \sin^2(\omega dt)\right]$$

or

$$P(t - dt)P(t + dt) = P^2(t) - P_0^2 \sin^2(\omega dt) \tag{4.10}$$

that is

$$P_0^2 \sin^2(\omega dt) = P^2(t) - P(t - dt)P(t + dt) \tag{4.11}$$

which for small ωdt becomes

$$P_0^2(\omega dt)^2 = P^2(t) - P(t - dt)P(t + dt) \tag{4.12}$$

We note that the left-hand side of Equation 4.11 is proportional to the maximal pressure squared modified by a term that depends only on the ratio of the signal frequency f_0 to the sample frequency f_s, as $\omega dt = 2\pi \frac{f_0}{f_s}$.

Noting that the mean sound wave intensity (or single cycle sound energy) is related to the physical amplitude s_0 of the sound wave by

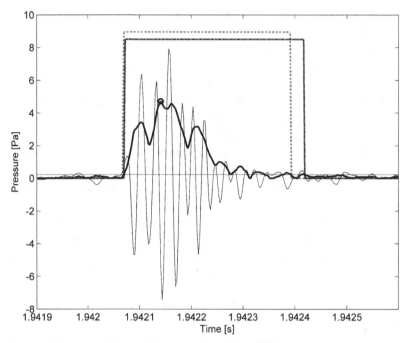

Figure 4.6   Taeger–Kaiser-based threshold detector. The thick solid line shows the Taeger–Kaiser amplitude estimate (Equation 4.14). The dashed line shows the result of the detector, and the dotted line is the envelope detector result from Figure 4.4.

$$\bar{I} = (\rho_0 c)\frac{\omega^2 s_0^2}{2} \tag{4.13}$$

it was suggested by Kaiser (1990) that Equation 4.12 could be used to estimate the mean intensity or energy (Equation 4.13). Unfortunately, the mean acoustic intensity is proportional to P_0^2 and not to $(\omega P_0)^2$. If we were to measure the particle displacement $s(t)$ of the sound field and not the sound pressure $P(t)$, then an algorithm similar to Equation 4.12, but now applied to the displacement $s(t)$, would measure the sound intensity up to a factor. In fact, Kaiser (1990) presented the algorithm for a simple mechanical oscillator, where motion amplitude is observed directly.

Nevertheless, because the left-hand side of Equation 4.11 is constant for any given frequency, it could be interesting to use the right-hand side of Equation 4.11 to define a frequency-weighted pressure amplitude estimate P_{TK}

$$P_{TK}(t) = \sqrt{\max(0, P^2(t) - P(t - dt)P(t + dt))} \tag{4.14}$$

To avoid accidental imaginary solutions, the argument of the square root in Equation 4.14 is restricted to non-negative values.

Figure 4.6 shows for the TK method the frequency-weighted pressure amplitude estimate (thick line) and the result of the threshold detector (dashed line) in comparison with the envelope threshold detector (dotted line). As to be expected, the P_{TK} estimate seems to be a scaled version of the envelope function shown in Figure 4.4. The detected signal has a similar start time but a different estimate for the end time.

4.1.5   Sequential probability ratio detector (Page test)

An alternative implementation to the simple threshold detector discussed so far is a sequential detector. This detector implements what is called a sequential probability ratio test (Wald, 1947). It does not necessarily decide for every available sample on the presence or absence of a signal, as is done by the threshold detector, but the algorithm may defer the decision until after the reception of a new data point. This procedure is repeated until a desired decision (signal present, or signal absent) is made. For a sequential detector one could also envision an algorithm that does not implement a predefined simple threshold, but changes its detection strategy as the number of indecisive data points increases.

A particular implementation of a sequential probability ratio test is the Page test (Page, 1954), where the a test variable $S(t)$ is estimated by

$$S(t) = S(t_1) + \int_{t_1}^{t} (snr^2(x) - b)\mathrm{d}x \tag{4.15}$$

with b being a predetermined bias and snr is the signal-to-noise ratio in linear scale. At each time step t the following decision rule is applied

$$S(t) = \begin{cases} > S_{TH} : & \text{target since } t_1 \\ < 0 \quad : & \text{no target since } t_1 \end{cases} \tag{4.16}$$

The integration (Equation 4.15) continues until a decision can be made. After each decision, the integration is reinitiated $t_1 = t$ and the new integration constant $S(t_1)$ is set according to

$$S(t_1) = \begin{cases} S_{TH} : & \text{target at } t_1 \\ 0 \quad : & \text{no target at } t_1 \end{cases} \tag{4.17}$$

The Page test generates a sparse sequence of signal and noise detections, filled with data points that are neither signal nor noise. As the results of a Page test contain samples where no decision has been made, it is necessary to decide *a posteriori* on these data points. It seems clear that undecided samples between similar decisions, be they signal or noise, are to be considered as signal or noise, respectively. In cases where the decisions are changing from one type to another, noise to signal, or signal to noise, we recall that the exact start and stop time of the signal is not known but lies within the two different decisions. To maximize the signal duration it is convenient to consider all undecided samples prior to and following the last signal decision as part of the signal presence.

Figure 4.7 shows the output of the page test detector applied to the envelope estimation of the signal. Interesting to note are the multiple positive decisions before and after the true signal (dash–dotted line), indicating a somewhat too low threshold and the risk of

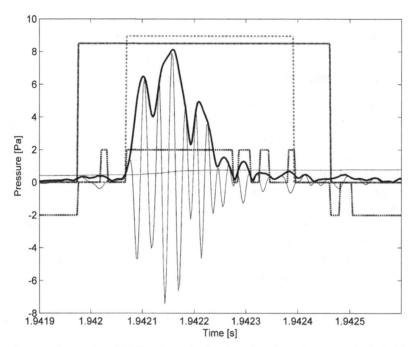

Figure 4.7   Page test detector. The thick line shows the signal envelope from Figure 4.3. The dashed line shows the
result of the detector, and the dotted line is the result from the envelope threshold detector. The dash–
dotted line shows the intermediate decisions before final reassignments of the undecided samples.

spurious or false detections. Tuning threshold and bias may improve the detection
performance.

Compared to the simple threshold detector, which decides on the signal to noise ratio,
the Page test accumulates a negatively biased squared signal-to-noise power ratio, which
for a constant noise estimate would be equivalent to accumulating the energy of the
signal. Only if this pseudo-energy exceeds a threshold does the Page test decide for the
presence of a signal. The method of choosing the threshold should therefore reflect
the minimum energy requirement for a good signal. In addition to the threshold, the Page
test requires the definition of a bias, which should ensure that time series without signals
do not lead to frequent signal decisions in presence of only noise. Any value higher than
the mean of snr^2 should be fine: a higher value results in fewer false alarms. Typical
values for the bias between 5 and 6 were found to be adequate as an initial guess. The
combination of threshold and bias also determines the time the detector needs to settle
after the presence of a signal before deciding again for noise. Overall, although the
selection of threshold and pruning window is straightforward for the threshold detector,
the proper choice for bias and threshold is more complex for the Page test detector and
requires some experience.

Matlab code of the Page test detector

```
function [dd,snr2,ss,nn,dd1]=doPagetest(xx,bias, Thr,aa0,ss,nn)
dd=0*xx;
```

```
xx2=xx.^2;
aa=aa0;
snr2=0*xx;
for ii=1:length(dd)
    nn=nn*(1-aa)+xx2(ii)*aa;
    snr2(ii)=xx2(ii)/nn;
    ss=ss+(snr2(ii)-bias);
    if ss<0
        ss=0;
        aa=aa0;
        dd(ii)=-1;
    end
    if ss>Thr
        ss=Thr;
        aa=aa0/100;
        dd(ii)=1;
    end
end
dd1=dd;
%decide a posteriori for undecided samples
%decide first for signal
for ii=length(dd)-1:-1:1
    if dd(ii)==0 && dd(ii+1)>0 , dd(ii)=dd(ii+1); end
end

for ii=2:length(dd)
    if dd(ii)==0 && dd(ii-1)>0 , dd(ii)=dd(ii-1); end
end
%decide for noise in all other cases
for ii=1:length(dd)
    if dd(ii)<0, dd(ii)=0; end
end
```

Comments on the Matlab code

To simplify the algorithm we work in terms of power, that is we first square the input vector xx. Bias and threshold are therefore related to the squared SNR values.

While looping through all data

- estimate exponential averaged noise value
- estimate SNR squared
- get a test function by accumulating the negatively biased squared SNR
- if the test function is below zero, clip to zero and decide for noise
- if the test function is above threshold, clip to threshold and decide for signal

After looping through all data the samples where no decision has been made are adjusted to detection around positive decisions and to noise otherwise.

4.1.6   Detection performance

So far, we have implemented a variety of click detectors and it is now necessary to address the question of which one of the methods may be the best one. For this, we first have to define a metric to measure the performance.

We expect the detector to detect signals if and only if they are present. From Figure 4.2 we see that signals have variable strength: some are strong and easily detected, but others are weak and may be mistaken for noise. The performance of the detector is mainly controlled by the threshold *TH* and here starts the dilemma. With a high threshold value, we will avoid false detections, but detect only some of the signals and miss the weaker ones; with a very low threshold, we will detect all signals but will also make same false detections, i.e. our detector will be confused by noise.

The metric of the detector performance will therefore depend on two types of result, one counting the correct signal detections and one counting the wrong detections or false alarms. For this, we have to know which of our detections correspond to true signals. However, from the information we have extracted so far, we cannot know which of the detections are correct and which are wrong. For this, we either have to classify the detection correctly, or have to know the true situation. Anticipating the results of a future classification step, we assume that we already know the event times of the correct signals. With this knowledge, we are now able to count the correct detections for a varying threshold value.

If we denote by N_T the total number of detections, with N_S the total number of signals present in the dataset, and with N_C the number of correctly detected signals, then the probability of (correct) detection P_{det} and the probability of false alarm P_{FA} may be approximated by the relative fractions

$$P_{\text{det}} = \frac{N_C}{N_S} \tag{4.18}$$

and

$$P_{\text{FA}} = \frac{N_T - N_C}{N_T} \tag{4.19}$$

Varying the threshold, we will obtain a series of total detections, correct detections and false alarms allowing us to estimate the probability of detection and probability of false alarm as a function of selected threshold. Figure 4.8 shows the detection probability and the false alarm probability as a function of threshold varying from *th* = 2 to *th* = 30.

There are some observations that could be made: With increasing threshold, the false alarm probability decreases faster than the detection probability. For thresholds *th* ≥ 15 all methods are free of false alarms, but with variable detection probability; and all techniques detect all signals for *th* ≤ 3 but with high probability of false alarms.

A single method seems to perform better than the others and that is the Taeger–Kaiser method. Its P_{FA} decreases to zero at a *th* ≥ 10, well before most of the other methods. This result is also shown in Figure 4.9, which compares directly P_{FA} and P_{det} in what is called a Receiver Operation Characteristics (ROC) curve. To interpret a ROC curve we should

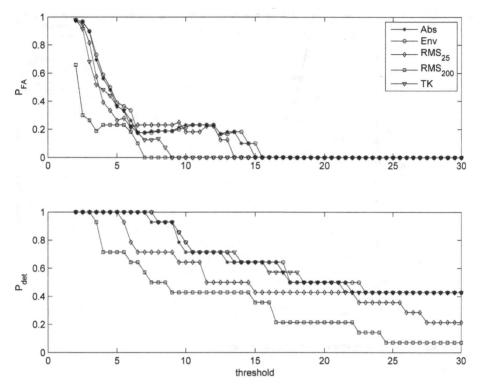

Figure 4.8   Probability of false alarm (top panel) and probability of detection (bottom panel) as a function of threshold.

ask ourselves what constitutes a good PAM system in terms of P_{FA} and P_{det}. It should be clear that we want a system that has a high probability of detection P_{det} and at the same time a low probability of false alarm P_{FA}, that is we want systems for which the ROC curve passes as close as possible through the upper left corner ($P_{det} = 1$ and $P_{FA} = 0$).

In Figure 4.9, we note that the TK method achieves highest P_{det} for lowest P_{FA}; all other techniques produce more false alarms for high P_{det}, and the 25 µs RMS estimate performs worst as it achieves the lowest P_{det} for low P_{FA}.

In Figure 4.8, only pure threshold detectors are compared with each other, as the Page test detector is based on a different threshold criterion, but there is no reason not to generate a ROC curve for the Page test also. From Figure 4.9 we may deduce that the performance of the Page test detector is similar to the performance of the majority of the detectors, with a tendency to be somewhat worse than the equivalent envelope threshold detector. However, the direct comparison between the two methods (threshold and Page test detector) is not easy as the different parameters involved are not compatible and the actual choice is not the result of an optimization process. In fact, the curves presented are in reality only a representation of a family of curves and some small adjustments in the bias (Page test) and the running window length (threshold detector) may slightly change the ROC curves.

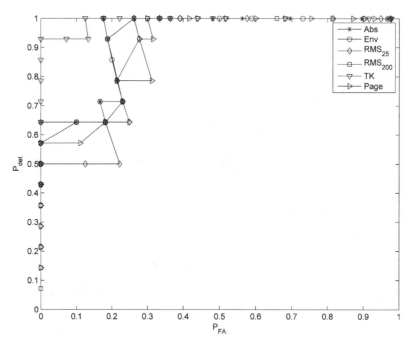

Figure 4.9 Receiver operation characteristics curve.

The better performance of the TK method is mainly due to the implicit spectral weighting that acts at low frequencies as a slowly varying highpass filter. It is therefore appropriate to investigate the influence of data filters that selectively weight the spectral content of the data.

4.2 Data filter

In the previous section, we have seen that the quality of a detector depends strongly on its capability to detect week signals against noise. In fact, if we consider for the moment only threshold detectors, we note that the detector *per se* is always the same function, but we modified the signal-to-noise ratio estimate that is used by the detector. We should expect that if we were to increase the signal-to-noise ratio for the same signal, then the detector should perform better.

The best way to increase the signal-to-noise ratio is to decrease the impact of the noise. Figure 4.10 shows the spectral density of the noise in the absence of any beaked whale clicks. We know from the description of Cuvier's beaked whale clicks that the dominant frequency for their echolocation clicks is around 40 kHz. From Figure 4.10 we note that the noise estimate at 40 kHz is about 30 dB below the low-frequency noise level, say around 10 kHz. This means that low-frequency noise is significantly masking weak beaked whale clicks. There are two methods to reduce this masking impact of the

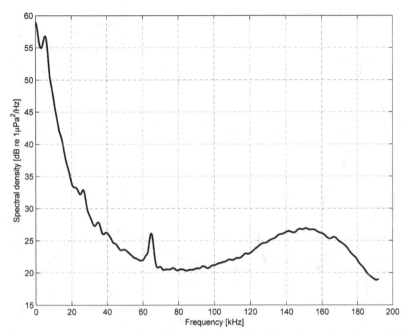

Figure 4.10 Noise spectrum based on selected times from beaked whale echolocation dataset (0.25–0.6 s in Figure 4.1).

noise. The first one is to flatten out the noise, that is, to equalize the background noise; the second one is to suppress completely unwanted noise spectra.

4.2.1   Noise equalization filter

Noise equalization is a process that transforms a frequency-dependent, or coloured, noise spectrum (e.g. Figure 4.10) into a nearly flat, or white, power spectral density. To simplify the discussion it is convenient to address first the low-frequency noise in Figure 4.10. To equalize the low-frequency noise spectrum we have to attenuate the lower frequencies according to their spectral level. From our discussions of the sonar equation, we know that surface noise decreases by about 17 dB per frequency decade and ship noise decreases by 20 dB per frequency decade. This is equivalent to a spectral variation of 5–6 dB per octave, i.e. by doubling the frequency the noise spectrum will be reduced by 5–6 dB.

What we are looking for is a slowly increasing highpass filter that keeps the data untouched for frequencies above 80 kHz, where the spectral minimum is found in Figure 4.10. To minimize the number of calculations that are necessary for filtering the data, we opt for what is called an infinite impulse response (IIR) filter, which for order 2 may be defined by

$$y_n = \frac{b_1 x_n + b_2 x_{n-1} + b_3 x_{n-2}}{a_2 y_{n-1} + a_3 y_{n-2}} \tag{4.20}$$

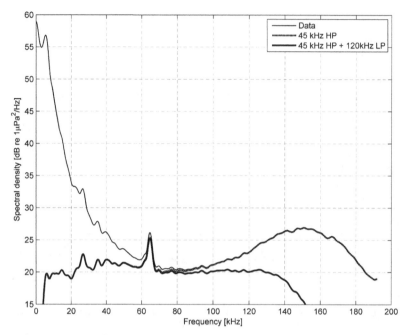

Figure 4.11   Noise equalization filters. The thin curve is original noise spectrum; the thick curve is the final filtered noise spectrum. For frequencies above 80 kHz the original spectrum (thin line) coincides with the highpass-filtered spectrum (dashed line).

where the x_n denote the data to be filtered (filter input) and the y_n the filtered data (filter output) (e.g. Hamming, 1977). IIR filters are recursive and require appropriate design of the filter coefficients a_n, b_n to avoid instabilities. When using IIR filters, it is therefore always appropriate to inspect the filter performance to ensure that the filtered data show the desired characteristics. Here we use Matlab functions to generate and visualize filter coefficients.

Figure 4.11 shows two spectra that are the result of a highpass filter, and a combination of a highpass and lowpass filter, overlaid on the measured noise spectrum. The filter coefficients were generated with the Matlab function `butter` as shown in the Matlab code.

From Figure 4.11 we conclude that for the present dataset a sequence of two Butterworth IIR filters equalized very well the noise from below 10 kHz to over 130 kHz. The peak at 65 kHz is a continuous spectral interference (weak tone) that cannot be removed with this approach, but if this tone turns out to be a problem for our signal detectors, further processing may be required to remove this spectral line.

Equalizing the noise, however, signifies also that the spectral component of the signal is modified. In our case, the Cuvier's beaked whale clicks are varying from, say 20 to 60 kHz and by normalizing the noise, we suppress the lower frequencies of the signal spectrum by about 12 dB with respect to the upper frequencies. Even if the signal-to-noise ratio is improved by reducing the influence of strong low-frequency noise, the same filtering may affect the detector in its overall performance. If, for example, the length of the signal is significantly affected by the lower-frequency components of the signal, then attenuating the lower spectral content may influence the length estimation of the signal. Although this may or may not be important for the decision of signal presence, such spectral distortions may significantly influence any further spectrum-based classification processes. Classification algorithms should therefore be constructed in such a way that they reflect any data filtering in earlier signal processing steps.

Matlab code snippet to equalize data

```
%equalization
%Butterworth filter of fixed order
[b1,a1] = butter(2, 2*45000/fs, 'high');
[b2,a2] = butter(2, 2*120000/fs, 'low');
yy1=filter(b1,a1,ss);
yy2=filter(b2,a2,yy1);
```

Comment on the Matlab code

To generate the filter, the function requires as input the value of the 3dB cut-off frequency, which is the frequency where the attenuation should be 3 dB. The original noise curve suggests 45 kHz as the 3 dB cut-off frequency for the highpass and 120 kHz for the lowpass. To pass the filter cut-off frequencies to the filter functions they are to be expressed as a fraction of half the sampling frequency.

4.2.2   Bandpass filter

Comparing the noise alone spectrum (Figure 4.10) and the signal spectrum (Figure 2.14) one could suggest that in order to improve the signal-to-noise ratio one should simply suppress all noise outside the spectrum of interest. This operation is best done with a bandpass filter that is tuned to the spectral width of the signal. As we again have to filter a large quantity of data, the first choice falls once more on a Butterworth IIR filter. In order to achieve better off-signal attenuation we approach the filter design slightly differently and define a filter description in terms of cut-off frequencies and stop band attenuation and let the computer determine the filter order.

Figure 4.12 shows the bandpass-filtered noise. The filter was specified by a passband, which varied from 20 to 60 kHz, where all frequencies pass without attenuation. The filter was further designed to suppress by 60 dB all frequencies outside the range 10–100 kHz.

Applying a bandpass filter maintains the spectral power within the passband window, but requires *a priori* knowledge of the spectral content of the signal of interest. If the chosen passband cut-off frequencies are too narrow then spectral signal information is more suppressed than, for example, with the noise equalization filter. If they are too wide, especially at the lower-frequency side, then noise suppression and therefore the increase in SNR may not be as great as expected.

Figure 4.12 shows, in addition to the broadband filter, a narrowband filter, for which the passband was selected to be 37–43 kHz, that is, just around the expected spectral peak of the Cuvier's beaked whale signal.

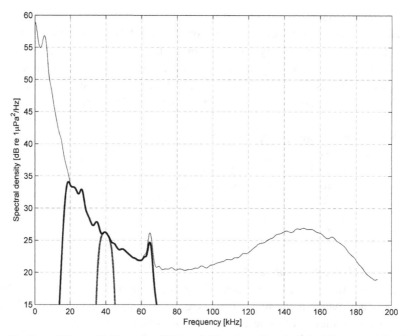

Figure 4.12   Bandpass filter applied to noise. Thick solid line, broadband; dashed line, narrowband filter.

Matlab code
Support function

```
function y=BP_filter(x,df1,df2)
%band-pass Butterworth filter
% df1 id the relative pass band
% df2 is the relative stop band
[n,Wn] = buttord(df1,df2,0.1,60);
[b,a]=butter(n,Wn);
gd=grpdelay(b,a);
nk=round(max(gd)/2);
u=filter(b,a,[x; zeros(nk,size(x,2))]); u(1:nk,:)=[];
y=doAnalytic(u,512);
```

Comment on the Matlab code
The filter cut-off frequencies are again given as a fraction of half the sampling frequency. The function shifts the filter output to coincide in time with the original dataset. A call to doAnalytic transforms the filter into a complex valued analytic time series.

Matlab code for calling the bandpass filter
```
xx=BP_filter(ss,df1,df2);
nn=sqrt(mean(abs(xx).^2));
snr = abs(xx)/nn;
```

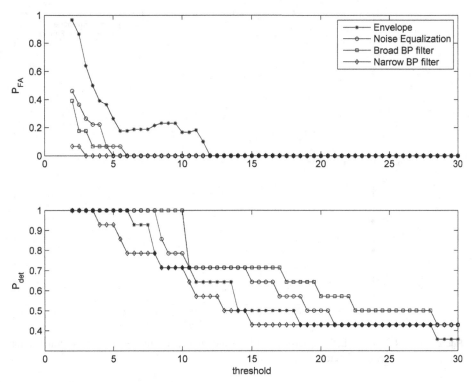

Figure 4.13   Detection performance of noise reduction filters.

where the frequencies are defined as follows

Broad bandpass

```
df1=[20 60]/fs*2000; %pass-band
df2=[10 100]/fs*2000; %stop-band
```

Narrow bandpass

```
df1=[37 43]/fs*2000; %pass-band
df2=[30 50]/fs*2000; %stop-band
```

4.2.3   Filter performance

Figure 4.13 compares the detection performance only in terms of P_{det} and P_{FA}, as the ROC curves of the filtered methods are the same. From the figure we may note that the minimal threshold for which no false alarms occurred is 12 (21.6 dB) for the original envelope detector, 6 (15.6 dB) for the noise equalization filter, 5 (14 dB) for the wide bandpass filter, and 3 (9.5 dB) for the narrow bandpass filter. The required threshold to detect the weakest signal is 6 (15.6 dB) for the original envelope detector, 8 (18.1 dB) for the noise equalization filter, 10 (20 dB) for the wide bandpass filter, and 3.5 (10.9 dB) for the narrow bandpass filter.

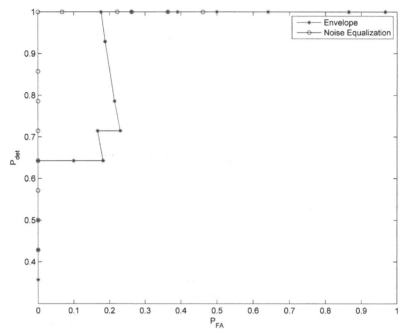

Figure 4.14   ROC curve before (star-marked line) and after (circle-marked line) noise equalization filter for envelope detector.

For the original envelope detector the strongest false alarm is 6 dB above the weakest signal and a trade-off between P_{FA} and P_{det} is necessary for the threshold selection. For all other modified detectors (equalization, broad and narrow bandpass filters) the strongest false alarm is now well below the weakest signal by some 2.5 dB for the noise equalization filter, 6 dB for the wide bandpass filter and 1.3 dB for the narrow bandpass filter, allowing the selection of an optimal threshold detecting all signals without false alarms. From this result we may deduce that the wide bandpass filter is the best filter as it allows the detection of signals that are up to 6 dB weaker than the ones that are present in the dataset, without increasing the false alarms.

Figure 4.14 shows the improvement of the detector performance due to noise equalization in terms of a ROC curve. The improved detector is perfect insofar as it achieved 100% detection probability for 0% false alarm probability. Of course, this result is specific to the analysed dataset and in cases where weaker signals are present then the both detector ROC curves will shift away from the 100% P_{det} – 0% P_{FA} corner. Only the case of the equalization filter is shown as the ROC curves of all other filters coincide with the equalization filter, i.e. they are equally perfect.

4.3   Detection of FM signals with pulse compression (matched filter)

In the previous section, we learned that with a bandpass filter we may increase the signal-to-noise ratio of the signal, improving the detector performance. Now we explore whether there exists an optimal filter that not only improves but also maximizes the signal-to-noise ratio. As we may expect, the optimal or best filter will depend on the specific signal to be detected and it is therefore only appropriate to assume that we know the signal that we wanted to detect.

4.3.1   Matched filter theory

The matched filter is a standard signal processing method for radar and sonar processing where the signal is known by definition. This matched filter is an optimal filter for a known signal. The following theory follows the presentation in Winkler (1977).

Given a known signal $s(t)$ with energy

$$E_0 = \int_0^T |s(t)|^2 \mathrm{d}t \tag{4.21}$$

the output at time T of a linear filter with the impulse response $h(t)$ is in general given by the convolution of the time series $x(\tau)$ and filter impulse response $h(\tau)$

$$y_X(T) = \int_0^T x(\tau)h(T-\tau)\mathrm{d}\tau \tag{4.22}$$

Applying a filter to a signal, we obtain a result that in general carries less energy than the signal energy, but which may reach the signal energy if we choose the filter to be proportional to the time-reversed signal, that is, we match the filter to the signal

$$h(t) = ks(T-t) \tag{4.23}$$

as we then obtain

$$y_S(T) = \int_0^T s(\tau)h(T-\tau)\mathrm{d}\tau = k\int_0^T s(\tau)s(\tau)\mathrm{d}\tau = kE_0 \tag{4.24}$$

As the filter affects not only signals but also noise (e.g. between signals) the question arises of how the filter processes noise.

It is convenient to assume the noise $n(t)$ to be a random process with zero mean and autocorrelation

$$R_{NN}(t,\tau) = E\{n(t)n(\tau)\} \tag{4.25}$$

where $E\{.\}$ denotes the mean or expectation operator.

The expected noise power at the filter output is then written as

$$E\left\{|y_N(T)|^2\right\} = \int_0^T \left(\int_0^T R_{NN}(t,\tau)h(T-\tau)\mathrm{d}\tau\right)h(T-t)\mathrm{d}t \tag{4.26}$$

It is now common to assume that the noise is white, that is, its spectrum density is constant for all frequencies and the autocorrelation function is zero for all $t \neq \tau$, so that we obtain

$$E\left\{|y_N(T)|^2\right\} = k^2 R_{NN}(0) \int_0^T s^2(T-t)\,dt \qquad (4.27)$$

If we define the signal-to-noise ratio at the filter output at time T as

$$\left(\frac{S}{N}\right)_{\max}^2 = \frac{|y_S(T)|^2}{E\left\{|y_N(T)|^2\right\}} \qquad (4.28)$$

then we obtain

$$\left(\frac{S}{N}\right)_{\max} = \sqrt{\frac{E_0}{R_{NN}(0)}} \qquad (4.29)$$

Relating the signal energy E_0 to the mean signal intensity I_0 and noting that the input noise N_0 is given by $N_0 = B R_{NN}(0)$ where B is the bandwidth of the signal, then we obtain

$$\left(\frac{S}{N}\right)_{\max} = \sqrt{BT}\left(\sqrt{\frac{I_0}{N_0}}\right) = \sqrt{BT}\left(\frac{S}{N}\right)_{IN} \qquad (4.30)$$

The processing gain of the matched filter, which is defined for intensity ratios, is therefore

$$PG_{MF} = 10\log(BT) \qquad (4.31)$$

The matched filter improves the signal-to-noise ratio by replacing the mean signal intensity with the signal energy and the broadband noise by the spectral noise value. The greater the time–bandwidth product BT, the greater the processing gain. Even if real noise is seldom white noise, as assumed for Equation 4.27, the processing gain as given in Equation 4.31 remains a good approximation of what may be obtained with a matched filter, as long as the signal is well known.

4.3.2   Matched filtering of right whale up-calls

In this section we take a sequence of very low-level northern right whale up-calls and try to filter the signals with a matched filter. Figure 4.15 repeats Figure 2.29 and shows a spectrogram of about 75 seconds of data containing six barely visible up-calls from a northern right whale, which are marked with downward arrows.

Constructing the replica

In order to apply a matched filter we have to construct a replica, which is a signal waveform that may be used as the input into our matched filter. There are multiple options, one of which is to use a very high-quality (noise free) example of a real signal;

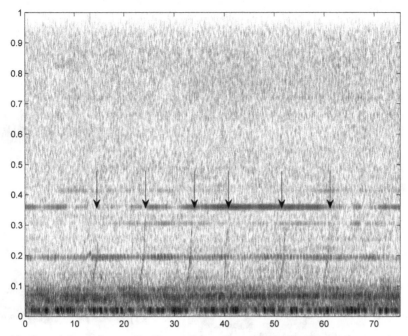

Figure 4.15 Northern right whale up-calls. Six weak up-calls are marked by arrows.

another is to construct an artificial signal that reflects the frequency function of the real signal as precise as possible. The latter case is of importance where high-quality signals are not available.

Figure 4.16 shows a direct comparison of the data before and after the matched filter. Here I opted for a synthetic reconstruction of the right whale up-call. In particular, I generated a cubic frequency function, which closely follows the visualized up-call, as can be seen in the lower left panel of Figure 4.16.

To generate an up-call replica we define the replica as a complex valued signal

$$R(t) = W(t) \exp\{-2\pi i f(t)t\} \qquad (4.32)$$

where $W(t)$ is a amplitude weighting function and the frequency function $f(t)$ is constructed from a set of numbers $q_1, \ldots, q_4$ via the following formula:

$$f(t) = \frac{q_1}{4}t^3 + \frac{q_2}{3}t^2 + \frac{q_3}{2}t + q_4 \qquad (4.33)$$

The order of the coefficients q_n has been chosen to coincide with the output of the polyfit Matlab function, which may be used to perform a polynomial fit of manually obtained frequency read outs.

In Figure 4.16 one notes further that the signal is not at all visible in the time series, which is dominated by low-frequency sounds (< 100 Hz) (upper left panel in Figure 4.16), but after matched filtering there is a strong peak at the beginning of the up-call (upper right panel, Figure 4.16). This is also a specific characteristic of the matched filter; as the matched filter time-compresses the signal into a sharp peak, the

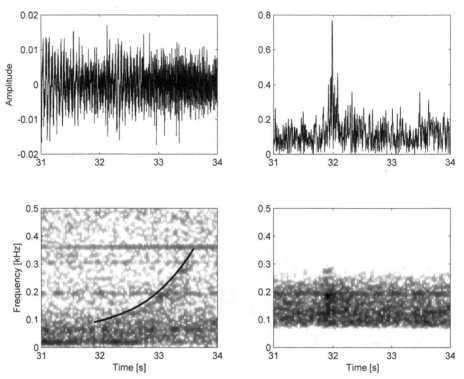

Figure 4.16 Single right whale up-call before (left) and after (right) pulse compression. Black line indicates the frequency function of the replica (shifted up by 0.02 kHz to keep the right whale up-call visible).

location of this peak is of importance. Because of the way in which the matched filter is constructed, the peak is placed at the beginning of the signal, but in some applications it may be more practical to have the peak at the end of the signal (see comment on Matlab code).

Comparing the two spectrograms in Figure 4.16, we can also note that the matched filter transformed a long broadband signal into a short broadband pulse. The bandwidth of the replica is maintained, but the signal is compressed in time and its resulting length is reduced.

Figure 4.17 finally compares the original time series with the matched filter output. While the original time series shows some low-frequency noise-related spikes, the matched filter output clearly shows the six signals marked in Figure 4.15. Figure 4.17 shows three additional peaks (c. 18 s, c. 47 s, c. 63 s), that are not visible in Figure 4.15; it is difficult to affirm that these additional peaks are also right whale up-calls, but they may be very weak ones.

It is therefore fair to say that after applying a matched filter it should be easier for a detector to decide on signal presence, even if the synthetic replica is not a perfect match. However, owing to possible mismatch of replica and signal, it may be difficult to achieve

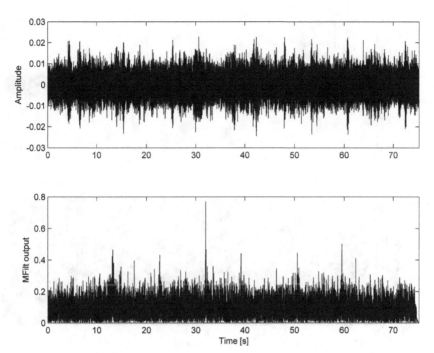

Figure 4.17   Matched filter results.

the estimated processing gain of 25 dB ($BT = 340$ for $T = 1.7$ s and $B = 200$ Hz) for the
selected example of a real right whale up-call.

Matlab code for replica generation

```
% generating a synthetic right whale up call
qq=[0.025 0.02 0.05 0.07]; % based on kHz spectrogram
tr=31.9:0.1:33.6;

dt=1/fs;
tt_r=(0:dt:(tr(end)-tr(1)))';
ff_r=polyval(qq.*[1/4 1/3 1/2 1],tt_r);
nw=length(tt_r);
R = hann(nw).*exp(-2*pi*i*ff_r.*tt_r*1000);
```

Matlab code for matched filter

```
%using fftfilt to match filter the data
y=fftfilt(flipud(R),[ss;zeros(floor(nw),1)]); y(1:floor(nw))=[];
```

Comments on the matched filter code

If we use `fftfilt` to estimate the matched filter, then we have to time-inverse the replica
according to Equation 4.23. This is done for a column vector by `flipud(R)`. In cases
where we wanted the peak at the end of the signal then the match filter simplifies to

```
y=fftfilt(flipud(R),ss);
```

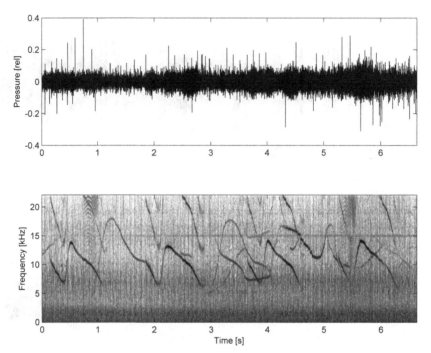

Figure 4.18   Example of multiple whistles of striped dolphins (courtesy G. Pavan).

4.4   Detection of dolphin whistles

Although the detection of short stereotyped FM calls may be considered as a special application of a time domain transient detector where a matched filter transforms the FM call into a short transient, this may not be so easy when we are confronted with dolphin whistles.

Dolphin whistles are extremely variable in both duration and frequency coverage. A simple matched filter approach may only be suitable if a specific whistle is searched for inside a large dataset, but in general PAM applications, whistles of potential interest appear with such a variety as to challenge classical matched filter approaches.

Although it is easy to state that all signals may be described by taking the real part of some complex valued function

$$S(t) = A(t) \exp\{-2\pi i f(t) t\} \tag{4.34}$$

it is in general not trivial to reconstruct from the measurements both the amplitude and the frequency function. As a matter of simplification, whistles are therefore only described by the frequency function $f(t)$ and the development of a whistle detectors will most likely be based on the spectrogram.

As example, let us consider Figure 4.18, where we see the time series and the spectrogram of multiple striped dolphin whistles. First, we realize that the dolphins tend to mix clicks and whistles, so that signal processing becomes challenging. As

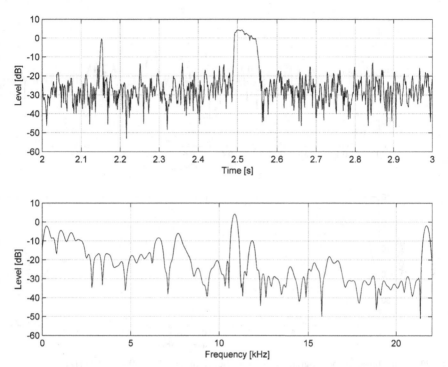

Figure 4.19   Cuts through spectrograms. Top panel: time series for fixed frequency (10.85 kHz). Bottom panel: spectrum for fixed time (2.5 s).

expected, clicks are easier to see in the time series, whereas whistles are more obvious in the spectrogram display.

It seems intuitive to conclude that whistle detection is more a two-dimensional signal processing problem than a simple one-dimensional one, which we discussed for the detection of echolocation clicks. However, there are no reasons not to consider whistle detection in the context of the sonar equation, that is, to assess the signal-to noise ratio and decide for signal presence if the SNR exceeds a predefined threshold.

Figure 4.19 shows a horizontal and a vertical cut through the spectrogram crossing at the time–frequency 'co-ordinates' of 2.5 s and 10.85 kHz, which was chosen intentionally as a whistle passes through this point. In fact, the time series shows a peak at 2.5 s and the spectrum shows for the same time a peak at 10.85 kHz. Whistles are now easily detectable as broad peaks both in the time series cut and in the spectral cut.

If we recall that spectrograms are based on STFT with short windows we recognize without difficulty that the time series cut may be considered as a very narrow bandpass-filtered dataset. The width of the spectral peak of a tonal signal is determined by the length and shape of the STFT windowing function as shown in Figure 2.17. Inspecting Figure 4.19, one may find the presence of at least three tonal peaks at 10.86, 11.84 and 21.75 kHz. The third peak is most likely the second harmonic of the first peak, as its frequency is doubled.

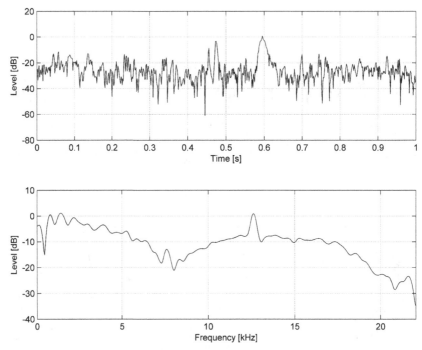

Figure 4.20   Cuts through spectrogram with the presence of a click (0.59 s, 12.57 kHz).

Confronting the information that we can extract from the two spectrogram cuts of the original spectrogram, it becomes clear that whistle detection is more than the detection of tonal sounds and also requires the tracking of the instantaneous detections of the same whistle as time progresses.

Before we enter the discussion of possible processing techniques to extract and track whistles, we should also inspect the impact of clicks on whistle detection. Figure 4.20 presents another time–frequency cut at a selected time–frequency co-ordinate, where we still see the tonal peak at 12.57 kHz but on top of a broad click-type spectrum.

Comparing the temporal time cuts of Figures 4.19 and 4.20 (top panels) one notes that in the latter figure the temporal peak (c. 0.6 s) is similar to what one would expect from a short click, whereas in Figure 4.19 the peak (c. 2.5 s) is typical of a tonal signal that needs some time to cross the frequency bin.

4.4.1   Spectral equalizing

From the spectrogram and from the spectral cuts we deduce next that the lower frequencies of the background noise also contain more energy in this dataset than the higher frequencies. As we are interested in a wide range of frequencies, a simple bandpass filter to remove unwanted noise is not an option.

The first signal-processing step is therefore to equalize the data. For the click detector we applied a low-order highpass filter to suppress the lower frequencies such that the final

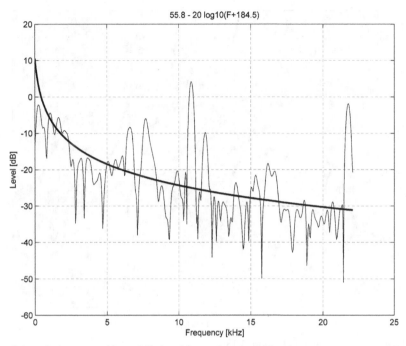

Figure 4.21   Selected spectrum with modelled ambient noise overlaid.

noise spectrum became nearly flat. Here we try another approach, that is, we try to fit a simple model to the measurements. For this we estimate the spectral noise on two different frequencies and use these two measurements to obtain a simple noise model

$$N(f) = N_0 - 20 \log(f + f_0) \tag{4.35}$$

with N_0, f_0 being two parameters to be estimated from the noise spectra. By using the term $20 \log(f)$, we assume a 6 dB/octave frequency decay law for the ambient noise.

Given two frequencies f_1, f_2 and therefore two noise estimates N_1, N_2, the parameters N_0, f_0 are estimated by

$$f_0 = \frac{q f_2 - f_1}{1 - q} \tag{4.36}$$

where

$$q = 10^{\frac{N_2 - N_1}{20}} \tag{4.37}$$

and

$$N_0 = N_2 + 20 \log 10 (f_2 + f_0) \tag{4.38}$$

The proper estimation of the noise level would be done by obtaining the mean power squared of narrowband-filtered data without any signals, be they clicks or whistles.

Figure 4.21 gives the resulting noise model overlaid on the background noise estimate. The background noise estimate was based at frequencies of 1000 and 4000 Hz, which

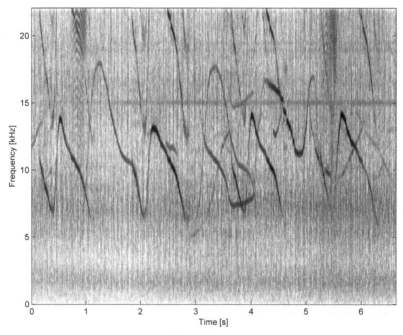

Figure 4.22   Equalized spectrogram of Figure 4.18.

may be found free of whistles. The spectral difference of 11 dB that was found between the two frequencies corresponds nearly to the theoretical decay of 6 dB/octave, giving confidence in the noise estimate.

By subtracting the estimated noise level from the spectrogram, one obtains an equalized, or better, a normalized spectrogram, as shown in Figure 4.22. The normalized spectrogram is then suited for a threshold detector.

4.4.2   Local max detector

Considering Figures 4.19, 4.20 and 4.21, one could conclude that a suitable approach to detect whistles is first to detect tonal peaks in the spectra. If the detector is at the same time based on differentiation, then slowly varying spectra of broadband clicks will contribute less than fast changes that are typical of local spectral peaks of whistles.

A natural choice to find the local maxima is based on differentiating the data and using the second derivative of the spectrum as indicator of the local maxima. In particular, we use the negative second difference to transform the equalized spectrogram into a form that is suited for threshold detection.

$$D(f) = 2P(f) - (P(f - \mathrm{d}f) + P(f + \mathrm{d}f)) \tag{4.39}$$

where $\mathrm{d}f$ is the spectral increment of the spectrogram. The sharper the peak, the greater the value of Equation 4.39 becomes.

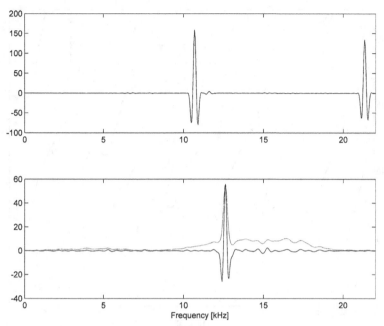

Figure 4.23   Loc-max-transformed spectra. Top panel corresponds to Figure 4.19; bottom panel corresponds to Figure 4.20. The dotted line is the original power plot, which is scaled so that its peak coincides with the loc-max-transformed spectrum.

Figure 4.23 shows the result of Equation 4.39 for the two spectral cuts of Figures 4.19 and 4.20. One clearly sees that the two functions are similar without indication of the presence of the broad click (shown as a dashed line in Figure 4.23).

Thresholding the twice-differentiated spectrograms then results in a binary spectrogram; Figure 4.24 shows the result of this operation. The different whistles are clearly visible as traces, but there are significant false detections in the form of random dots between the whistles.

The next step in the processing chain is to clean up the thresholded twice-differentiated detector display and to extract the different whistles. A simple clean-up maintains, among the thresholded values, only the spectral locations of the local maxima. Reducing the rather broad thresholded frequency band to a single value, however, increases the chances of discontinuous frequency changes in otherwise continuous whistles. Estimating the spectrogram with significant overlap (here I used an overlap of ¾ of the window length) helps to avoid small jumps in the spectrogram and to obtain smoother whistle traces.

Figure 4.25 shows the spectral local maxima of a single whistle. Even though the number of threshold crossings is now reduced, one can still see the different parts of the whistle. However, some discontinuities occur, the effect of which may or may not be a problem for whistle extraction. Frequency jumps may be genuine to the whistles and may even be characteristic of the particular species.

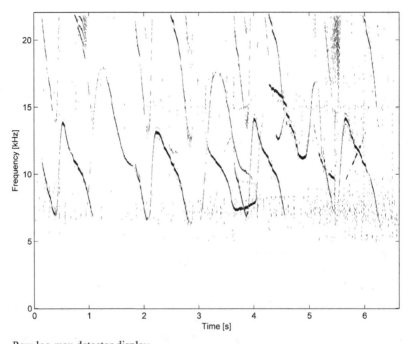

Figure 4.24   Raw loc-max detector display.

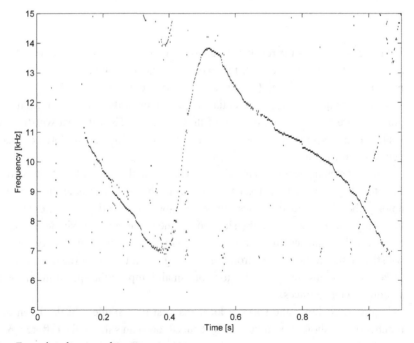

Figure 4.25   Zoom into loc-max detector output.

Using the second derivative is one method to emphasize the presence of narrow spectral whistle peaks. Another method that is worth mentioning is based on an edge detector (Gillespie, 2004) which uses the first derivative to localize both the start (positive gradient) and the stop (negative gradient) of a narrow spectral whistle peak.

Matlab code to find local maxima

```
%thresholding the 2nd derivative
th=4;
B4=(D2>th).*B2;

%find local maxima in frequency direction
M1=[0*B4(1,:);B4(2:end,:)>B4(1:end-1,:)];
M2=[B4(1:end-1,:)>B4(2:end,:);0*B4(end,:)];
M=M1.*M2; clear M1 M2 % clear intermediate matrices
```

4.4.3 Whistle tracking and extraction

Similar to the procedure developed for the click detector, it may be useful for subsequent whistle classification to extract the whistles from the dataset. For this, it is necessary to connect the different detected local spectral maxima into whistles. This task is carried out by a tracker, which takes the measurements (frequencies of local spectral maxima) for every time step and generates tracks (whistles), that is, the whistle tracker combines all the frequencies that belong to the same whistle into a unique whistle track. The key problems are to connect measurements (localized frequencies) with existing whistle tracks, to initiate new whistle tracks and to terminate whistle tracks.

A variety of tracking algorithms are available that are suited for the tracking of whistles. All methods assume that a more or less complex model describes the temporal variation of the track. They take measurements and determine the parameters of the assumed model (state of the model) by comparing the measurements (or observations) with the model predictions. Important for all tracking algorithms is that the measurements are considered to be affected by measurement errors; trackers are frequently used to improve the measurements, or to filter the observations. Here we use a simple tracking algorithm to connect local spectral measurements to one or more whistles, that is, we are interested in the implicit data association capability of trackers and not necessarily in the filtering of the measurements.

Data association

Consider a dataset where N_e whistles have so far been detected and where N_m new measurements are becoming available. The data association tries to generate unique relations between predicted (whistle) track estimates $y_e(j)$, $j = 1, \ldots, N_e$ and new (frequency) measurements $y_m(i)$, $i = 1, \ldots, N_m$, where the number of predicted estimates does not need to be equal to the number of measurements.

For example, consider the following existing track description (Table 4.1) where the top row counts the elements, each of which is a unique track, the second row gives

Table 4.1 Predicted track estimates (frequency bin)

1	2	3	4	5
10	11	12	13	14
50	**100**	**120**	**130**	**170**

Table 4.2 Measured values (frequency bin)

1	2	3	4
94	**105**	**129**	**200**

Table 4.3 Association table (5 tracks and 4 measurements)

	$n1$	1	2	3	4	5					
	L1	10	11	12	13	14					
$n2$		50	100	120	130	170	E2	I2	I1(I2)	F2	
1	94	*44*	*6*	*26*	*36*	*76*	6	2	2	1	*new*
2	105	*55*	*5*	*15*	*25*	*65*	5	2	2	0	old
3	129	*79*	*29*	*9*	*1*	*41*	1	4	3	0	old
4	200	*150*	*100*	*80*	*70*	*30*	30	5	4	1	*new*
	E1	44	5	9	1	30					
	I1	1	2	3	3	4					
	I2(I1)	2	2	4	4	5					
	F1	1	0	1	0	1					
		idle	old	*idle*	old	*idle*					

For detailed description, see text.

the unique label of the track and the bottom row denotes the predicted estimate of the track.

Next, assume the following new measurements (Table 4.2) where the top row counts again the elements and the second row contains the new measurements.

To proceed with the data association we generate a pairwise difference matrix (grey area in Table 4.3). We locate the row-wise minima in Table 4.3 and place their values and locations as E2 and I2 to the right of the difference matrix. These minima link every new measurement with the nearest existing track. We also locate the column-wise minima in Table 4.3 and place their values and locations as E1 and I1 below the difference matrix. These column-wise minima link every track to the closest measurements.

In the following we apply a simple allocation rule:

1. Tracks are continued if the row-wise minimum is also the column-wise minimum.
2. Tracks are only continued if the distance between predicted track estimates and measured value does not exceed a given threshold, otherwise the links are broken up.
3. Existing tracks that are not continued this way are considered as idle.

Table 4.4 Association result

		est		meas	label
Old	2	100	2	105	11
Old	4	130	3	129	13
Idle	1	50	*1*	*50*	10
Idle	3	120	*3*	*120*	12
Idle	5	170	*5*	*170*	14
New	*1*	*94*	1	94	**15**
New	*4*	*200*	4	200	**16**

4. Measurements that are not linked to tracks in this way are considered to start a new track.
5. Broken-up tracks result in both idle and new tracks.

From Table 4.1 we note that rule 1 requires that

$$I2(I1) = n1$$
$$I1(I2) = n2$$

(4.40)

In Table 4.3, this rule would generate the following track–measurement associations: 2–2, 4–3, 5–4 (track first and measurement second).

Rule 2 requires that the distance linking track estimates with measurement does not exceed a predefined distance. Assuming a threshold of 6 we find that in Table 4.3 the '5–4' association is linked with a distance of 30, i.e. is exceeding the threshold, and should be broken up.

Consequently, tracks 1, 3 and 5 are not connected to new measurements and are considered as idle. Likewise, measurements 1 and 4 are not linked to existing tracks and are considered as starting points of new tracks.

Applying these rules to Table 4.3 produces the results shown in Table 4.4, which shows the final set of tracks and measurements. To complete the result, we initiate new tracks by duplicating the measurements (marked with 'new' in Table 4.3) and label them with a new track number (labelled 15 and 16 in Table 4.4). In cases where tracks are potentially ending (indicated as idle in Table 4.4) the missing measurements are simulated by duplicating the last predicted estimate, or alternatively, by inserting the next predicted track estimate. Continuing this way with idle tracks, or tracks without new measurements, allows us to bridge isolated missing measurements. Keeping track of the number of missing measurements permits the final termination of idle tracks later.

Matlab code of data association function

```
function [ya ntr]=doAssoc(ye,ym,dy,yl,yb,ntr,th,mx)
% handle case that we have measurements but no tracks
if isempty(ye)
    yln=ntr+(1:length(ym))'; ntr=ntr+length(ym);
    ya=[ym ym 0*ym yln 0*ym 1+0*ym];
```

```
        return
    end
% handle case that we have tracks but no measurements
if isempty(ym)
        ya=[ye ye dy yl yb-1 2+0*ye];
        return
end
%
% standard case
% difference matrix
E=ym*ones(1,length(ye))-ones(length(ym),1)*ye';
%
% locate row-wise and column-wise minima
[E1,I1]=min(abs(E),[],1);
[E2,I2]=min(abs(E),[],2);
%
% estimate back-pointers in column form
I21=I2(I1); I21=I21(:);
I12=I1(I2); I12=I12(:);
%
% find new and idle tracks
F1=~(E1<th & I21'==(1:length(I1)));
F2=~(E2<th & I12==(1:length(I2))');
%
% allocate new track labels
if ~isempty(ym(F2))
        yln=ntr+(1:length(ym(F2)))'; ntr=ntr+length(ym(F2));
else
        yln=0*ym(F2);
end
%
%generate association matrix
% use I2(~F2), ~F2 pair for continuated track
% could use also I1(~F1), ~F1 pair
I3=I2(~F2);
%
ya=[ye(I3), ym(~F2), dy(I3), yl(I3), 0*yb(I3), 0*ym(~F2);
    ym(F2), ym(F2), 0*yln, yln, 0*ym(F2), 1+0*ym(F2);
    ye(F1), ye(F1), dy(F1), yl(F1), yb(F1)-1, 2+0*ye(F1)];
%
%prune shorties
ifl=ya(:,5)<-mx;
ya(ifl,:)=[];
%
return
```

Comment on the Matlab script

In addition to the basic allocation algorithm, the script handles cases where no measurements were taken or where no tracks are available. In addition, all idle tracks that exceed a maximal idling length are terminated and removed from the track list.

To complete the whistle detection problem we have to augment the allocation algorithm with a tracking algorithm, for which we use a simple prediction–correction scheme. After association of the new measurements to track predictions, the track estimate is corrected according to the difference between predicted track estimate y_e and measurements y_m

$$y_c = y_e + \alpha(y_m - y_e) \tag{4.41}$$

where y_c is the new best estimate of the track value and α is a gain factor. An $\alpha = 0$ ignores the new measurement and an $\alpha = 1$ ignores the predicted track estimates.

The new predicted track value is then estimated by

$$y_e = y_c + \mathrm{d}y \tag{4.42}$$

where $\mathrm{d}y$ is the result of an exponential averaged prediction–measurement difference

$$\mathrm{d}y = \beta \mathrm{d}y + (1 - \beta)(y_m - y_e) \tag{4.43}$$

and β is the exponential weighting constant.

Matlab code of tracker module

```
function [DET,ntr]=doTracking(M,iox,th,mx,gain,beta)
DET=[];
ye=[]; dy=[]; yl=[]; yb=[];
ntr=0;
%
%localize local maxima
[Mr,Mc]=find(M);
%
for io=iox
    ij=Mc==io;
    ym=Mr(ij);
    %
    [ya ntr]=doAssoc(ye,ym,dy, yl,yb,ntr,th,mx);
    %
    ye=ya(:,1);
    dy=ya(:,3);
    yl=ya(:,4); %track number
    yb=ya(:,5); %idle counter

    % correct actual measurement
    dya=gain*(ya(:,2)-ya(:,1));
    yc=ye+dya;
    %
    % store actual tracked
    DET=[DET; [io+0*yc, yc, yl, yb dy]];
    %
    %predict for next sample
    dy=beta*dy+(1-beta)*dya;
    ye=yc+dy;
end
```

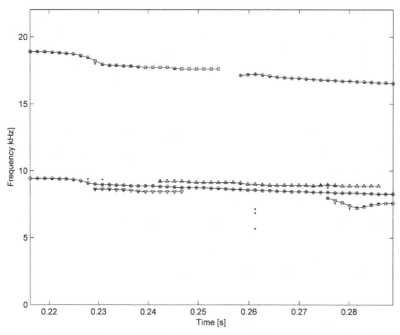

Figure 4.26   Sample of whistle fragments (isolated dots) and overlaid tracks (connected markers).

The result of the whistle detection algorithm is shown in Figure 4.26, which is based on a small sample extracted from Figure 4.25, where the measurements are shown in grey and the resulting tracks are given as punctuated lines. A complete picture will be presented in the next chapter.

From Figure 4.26, one may note that short spurious detections are successfully removed from the whistle track list, except where they are in continuation of an idle track. The association threshold th was set to 10; the maximal idle count mx was taken as 6. After tracking, all tracks that were shorter than 10 elements were eliminated.

The tracking algorithm presented above is a rather simple, but nevertheless successful, tracking algorithm. It was chosen to demonstrate the elements that are necessary for any tracker. There is a variety of possible modifications for the data association rules and the prediction and correction steps.

A useful modification addresses the case where two or more measurements are in vicinity of an existing track. Figure 4.26 shows, close to 10 kHz, two spurious tracks in parallel to the long main track. Here we resolved this problem by generating new tracks, but an interesting alternative could be to consider the additional nearby measurements as multiple measurements of the same track and to combine them into a single measurement. One could further consider the weaker of the two tracks a spurious side-lobe of the whistle, which should be removed from the measurement list.

The predictor–corrector step implemented above is very basic; algorithms that are more sophisticated are possible. For example, one could extend the predictor from a single time lag to a multiple time lag algorithm (Hamming, 1986), or one could estimate the gain factor α adaptively, as done for a Kalman filter. See Section 6.11 for more details on tracking, especially for an implementation of a Kalman filter.

5 Classification methods

The objective of this chapter is to introduce the classification of detected objects, that is, to assign the detected signals (clicks or whistles) to different cetacean species. We will concentrate on two approaches, namely classification based on 'single' and on 'multiple' signals. Single-signal-based classification exploits the features that are intrinsic to each cetacean sound emission and that may be done in both the time and the spectral domain. Multiple-signal-based classification requires the detection of more than one signal of the same species to perform the classification.

5.1 Classification basics

Classification is the process that assigns measurements to distinct classes and is of importance when there are multiple solutions to the allocation process. We have already addressed the data allocation process for the whistle tracker, where we implemented a simple and straightforward allocation algorithm. General classification problems are similar, but instead of needing to assign a single frequency to one of the whistles, we have a vector of measurements, or feature vector, which we wanted to assign to a class.

To demonstrate the problem, let us assume that our click detector generates for each detection a simple 3-element feature vector, e.g. signal length, spectral bandwidth and frequency of spectral peak. The task for the classification algorithm is then to take this feature vector and ideally to tell the user which species generated the click. That this could in principle be possible is shown by the following considerations (see Figure 2.14): a narrowband click at frequencies above 100 kHz may be coming from a harbour porpoise; a broadband click centred around 15 kHz is most likely from a sperm whale; a rather long broadband click around 40 kHz could come from a beaked whale; a very short click above 20 kHz is very probably emitted by a dolphin. The use of terms 'may be', 'most likely', 'could' indicates that classification is a probabilistic process and does not result in certainty. High-frequency monochromatic clicks are also emitted by pseudo-orcas, low-frequency clicks are also produced by pilot whales, and some dolphin species emit rather long clicks. The solution to uncertainties is either to choose different features or to add new features to improve the discrimination capabilities of the classification algorithm. For example, sperm whale clicks are multi-pulsed and the presence of this multi-pulse structure would allow the final discrimination between sperm and pilot whale clicks. Beaked whale clicks often show a weak frequency modulation, a feature that is not present in dolphin clicks.

It is fair to say that to be successful classification generally requires same *a priori* knowledge, and the classification of cetacean sounds is no exception. The paucity of this *a priori* knowledge is also the main reason why the development of classification algorithms for cetacean sounds is still in its infancy. As PAM systems have the potential to collect huge amounts of data from different species with the same hardware and software, one should expect that the development of classification methods will progress fast.

5.2 Optimal classification

Let us first address the general question of whether there is an optimal classification and what the optimal decision rules may be (Niemann, 1974). For this, we recall that the classification process takes measurements, observations or features and tries to determine the appropriate species or animal class that has generated the measured feature.

Given a set of K distinct classes

$$\Omega_k \text{ for } k = 1, \ldots, K \tag{5.1}$$

that are occupied with with *a priori* probabilities

$$p(\Omega_k) = p_k \tag{5.2}$$

Given further

$$w_m(\mathbf{c} \mid \Omega_k) = w(c_1, \ldots, c_m \mid \Omega_k) \quad k = 1, \ldots, K \tag{5.3}$$

the conditional probability density of observing the m-dimensional feature vector $\mathbf{c} = (c_1, \ldots, c_m)$ given the class Ω_k.

That is, for each class there exists a probability that this class is occupied (Equation 5.2) and may be the origin of the measurements, and a conditional probability of observing a feature vector $\mathbf{c}$ (Equation 5.3).

Let

$$\delta(\gamma_l \mid \mathbf{c}) \quad l = 1, \ldots, K \tag{5.4}$$

describe the decision γ_l for class l when $\mathbf{c}$ has been observed, whereby $\delta(\gamma_l \mid \mathbf{c})$ is a notation that is one for decision γ_l and zero otherwise. In particular, $\delta(\gamma_1 \mid \mathbf{c})$ denotes the decision that the measurement vector $\mathbf{c}$ belongs to class Ω_1 and $\delta(\gamma_2 \mid \mathbf{c})$ that it belongs to class Ω_2.

One may note that the following integral over all measurements

$$P(\gamma_l \mid \Omega_k) = \int_{R_c} \delta(\gamma_l \mid \mathbf{c}) w_m(\mathbf{c} \mid \Omega_k) d\mathbf{c} \tag{5.5}$$

is the probability of decision γ_l when class Ω_k is given.

Denote with

$$R(\gamma_j, \Omega_k) \quad k, j = 1, \ldots, K \tag{5.6}$$

the cost associated with the decision γ_j for class Ω_j, if there should have been a decision for class Ω_k, obtaining

$$V(\gamma_k \mid \delta) = \sum_{l=1}^{K} R(\gamma_k, \Omega_l) P(\gamma_k \mid \Omega_l) \tag{5.7}$$

for the cost associated with the decision rule δ for deciding for class Ω_k. It is intuitive that the costs for correct decisions are less than for wrong decisions

$$R(\gamma_k, \Omega_k) < R(\gamma_l, \Omega_k) \quad l \neq k \tag{5.8}$$

Defining an expected risk function

$$V(\delta) = \sum_{k=1}^{K} p(\Omega_k) V(\gamma_k \mid \delta) \tag{5.9}$$

we obtain by inserting 5.7 and 5.5

$$V(\delta) = \sum_{l=1}^{K} \sum_{k=1}^{K} R(\gamma_k, \Omega_l) p(\Omega_k) \int_{R_c} \delta(\gamma_l \mid \mathbf{c}) w_m(\mathbf{c} \mid \Omega_k) d\mathbf{c} \tag{5.10}$$

Assuming now that we always have a decision

$$\sum_{l=1}^{K} \delta(\gamma_l, \mathbf{c}) = 1 \tag{5.11}$$

and the measurements are always drawn from the known classes

$$\int_{R} w_m(\mathbf{c} \mid \Omega_k) d\mathbf{c} = 1 \tag{5.12}$$

then we may decompose the expected risk function

$$\begin{aligned}
V(\delta) = &\sum_{k=1}^{K} R(\gamma_k, \Omega_k) P(\Omega_k) \\
&+ \sum_{l=1}^{K} \sum_{k=1}^{K} R(\gamma_l, \Omega_k) P(\Omega_k) \int_{R} \delta(\gamma_l, \mathbf{c}) w_m(\mathbf{c} \mid \Omega_k) d\mathbf{c} \\
&- \sum_{k=1}^{K} R(\gamma_k, \Omega_k) P(\Omega_k) \int_{R} \sum_{l=1}^{K} \delta(\gamma_l, \mathbf{c}) w_m(\mathbf{c} \mid \Omega_k) d\mathbf{c}
\end{aligned} \tag{5.13}$$

or

$$V(\delta) = \sum_{k=1}^{K} R(\gamma_k, \Omega_k) P(\Omega_k)$$
$$+ \int_R \sum_{l=1}^{K} \delta(\gamma_l, \mathbf{c}) U_l(\mathbf{c}) d\mathbf{c} \tag{5.14}$$

where

$$U_l(\mathbf{c}) = \sum_{k=1}^{K} (R(\gamma_l, \Omega_k) - R(\gamma_k, \Omega_k)) w_m(\mathbf{c} \mid \Omega_k) p(\Omega_k) \tag{5.15}$$

is a test function to be minimized.

Let now $U_0(\mathbf{c}) = \min_l (U_l(\mathbf{c}))$ be the minimum value of the test function. It is clear from Equation 5.14 that we only minimize the risk function $V(\delta)$ if we decide for class Ω_o, that is, if we use

$$\delta_o(\gamma_l \mid \mathbf{c}) = \begin{cases} 1 & \text{if } l = o \\ 0 & \text{else} \end{cases} \tag{5.16}$$

as only then does the sum in the second term in Equation 5.14 reduce to the minimal value $U_0(\mathbf{c})$.

To obtain the optimal classification we have to estimate for the different classes Ω_l a test function $U_l(\mathbf{c})$ according to Equation 5.15 and to decide for that class for which the test function is minimal.

The test function given in Equation 5.15 is very general; problems that are more specific may result in slight modifications.

Assume now that the costs for the different decision functions are not very well specified, then it is in most cases justified to assume that the costs may be considered different for correct and wrong decisions but otherwise constant. We may then assume

$$R(\gamma_k, \Omega_k) = R_C \tag{5.17}$$

$$R(\gamma_l, \Omega_k) = R_F \quad k \neq j \tag{5.18}$$

and the test function becomes

$$U_l(\mathbf{c}) = \sum_{k=1}^{K} (R(\gamma_l, \Omega_k) - R(\gamma_k, \Omega_k)) w_m(\mathbf{c} \mid \Omega_k) p(\Omega_k)$$
$$= R_F \left(\sum_{k=1}^{K} w_m(\mathbf{c} \mid \Omega_k) p(\Omega_k) - w_m(\mathbf{c} \mid \Omega_l) p(\Omega_l) \right) \tag{5.19}$$

From Equation 5.19 we note that the test function becomes minimal if the product $w_m(\mathbf{c} \mid \Omega_l) p(\Omega_l)$ becomes maximal yielding to the decision rule to decide for that class Ω_o for which the product $w_m(\mathbf{c} \mid \Omega_l) p(\Omega_l)$ becomes maximal

$$w_m(\mathbf{c}\,|\,\Omega_o)p(\Omega_o) = \max_l(w_m(\mathbf{c}\,|\,\Omega_l)p(\Omega_l)) \tag{5.20}$$

or in short

$$\Omega_o = \arg\max_l\{w_m(\mathbf{c}\,|\,\Omega_l)p(\Omega_l)\} \tag{5.21}$$

Using Bayes' law, Equation 5.21 may be written differently, that is, using

$$w_m(\mathbf{c}\,|\,\Omega_k)p(\Omega_k) = w_m(\mathbf{c},\Omega_k) = w_m(\Omega_k\,|\,\mathbf{c})p(\mathbf{c}) \tag{5.22}$$

we obtain

$$w_m(\Omega_o\,|\,\mathbf{c}) = \max_l(w_m(\Omega_l\,|\,\mathbf{c})) \tag{5.23}$$

or

$$\Omega_o = \arg\max_l\{w_m(\Omega_l\,|\,\mathbf{c})\} \tag{5.24}$$

as the solution is independent of $p(\mathbf{c})$.

5.2.1   Detection as a two-class classification

For $K = 2$ and with Equation 5.20 we obtain the optimal decision for class Ω_2 if

$$w_m(\mathbf{c}\,|\,\Omega_2)p(\Omega_2) > w_m(\mathbf{c}\,|\,\Omega_1)p(\Omega_1)$$

or equivalently

$$\frac{w_m(\mathbf{c}\,|\,\Omega_2)}{w_m(\mathbf{c}\,|\,\Omega_1)} > \frac{p(\Omega_1)}{p(\Omega_2)} \tag{5.25}$$

If we consider class Ω_1 as noise and class Ω_2 as signal and assume that the noise measurements are Rayleigh distributed

$$w_m(y\,|\,\Omega_1) = \frac{y}{\sigma^2}\exp\left(-\frac{y^2}{2\sigma^2}\right) \tag{5.26}$$

and that the signal data are characterized by a constant amplitude a, which is contaminated with Rayleigh noise

$$w_m(y\,|\,\Omega_2) = \frac{y}{\sigma^2}\exp\left(-\frac{y^2 + a^2}{2\sigma^2}\right) \tag{5.27}$$

then we have the following decision rule for class Ω_2 (presence of signal)

$$\frac{w_m(y\,|\,\Omega_2)}{w_m(y\,|\,\Omega_1)} = \exp\left(-\frac{a^2}{2\sigma^2}\right) > \frac{p(\Omega_1)}{p(\Omega_2)} \tag{5.28}$$

which with $p(\Omega_1) + p(\Omega_2) = 1$ and after taking the logarithm becomes

$$\frac{a^2}{\sigma^2} > -2\ln\left(\frac{p(\Omega_1)}{1-p(\Omega_1)}\right) \tag{5.29}$$

If we choose a very small *a priori* probability of having the opportunity to classify noise, that is, $p(\Omega_1) \ll 1$, then

$$\frac{a}{\sigma} > \sqrt{-2\ln(p(\Omega_1))} \tag{5.30}$$

which becomes the Neyman–Pearson detection criterion for Rayleigh distributed noise used in Chapter 4, if we define the signal-to-noise ratio as $snr = \frac{a}{\sigma}$ and the false alarm probability $P_{FA} = p(\Omega_1)$. It should, however, be noted that Equation 5.30 assumes that the signal amplitude a is known, so using Equation 5.30 indicates that the threshold thus defined is optimal if the false alarm rate is small, that the noise is Rayleigh distributed and that we know the signal amplitude a.

5.2.2 Gaussian distributed feature vectors

Let us assume next that the feature vectors follow a multi-dimensional Gaussian distribution

$$w_m(\mathbf{c}\,|\,\Omega_l) = \frac{1}{\sqrt{(2\pi)^m |\mathbf{K}_l|}} \exp\left\{ -\frac{1}{2}(\mathbf{c}-\boldsymbol{\mu}_l)^T \mathbf{K}_l^{-1}(\mathbf{c}-\boldsymbol{\mu}_l) \right\} \tag{5.31}$$

with mean feature vector $\boldsymbol{\mu}_l$ for the classes Ω_l and covariance matrices $\mathbf{K}_l$ with determinant $|\mathbf{K}_l|$, where the superscript T denotes the vector transpose.

Using Equation 5.22 the test function becomes

$$w_m(\mathbf{c},\Omega_l) = \frac{p(\Omega_l)}{\sqrt{(2\pi)^m |\mathbf{K}_l|}} \exp\left\{ -\frac{1}{2}(\mathbf{c}-\mu_l)^T \mathbf{K}_l^{-1}(\mathbf{c}-\mu_l) \right\} \tag{5.32}$$

The optimal classification is given by Equation 5.21 and as this solution does not change by taking the logarithm

$$\Omega_o = \arg\max_l \{ w_m(\mathbf{c},\Omega_l) \} = \arg\max_l \{ \ln(w_m(\mathbf{c},\Omega_l)) \} \tag{5.33}$$

we can write

$$\ln(w_m(\mathbf{c},\Omega_l)) = -\frac{1}{2}(\mathbf{c}-\mu_l)^T \mathbf{K}_l^{-1}(\mathbf{c}-\mu_l) + \ln\left(\frac{p(\Omega_l)}{\sqrt{(2\pi)^m |\mathbf{K}_l|}}\right) \tag{5.34}$$

so that the optimal classification rule becomes:

$$\Omega_o = \arg\min_l \left\{ \frac{1}{2}(\mathbf{c}-\mu_l)^T K_l^{-1}(\mathbf{c}-\mu_l) - \ln\left(\frac{p(\Omega_l)}{\sqrt{(2\pi)^m |\mathbf{K}_l|}}\right) \right\} \tag{5.35}$$

The application of Equation 5.35 requires knowledge of the expected or mean feature vectors and the covariance matrices for each class, in addition to the *a priori* probability of the class.

Given a large set of measured feature vectors that are known to belong to the same class then it is possible to estimate both the mean feature vector and its covariance matrix. Assume that N_k observations belong to the same class Ω_k, then the class mean μ and the covariance matrix $\mathbf{K}$ are defined as

$$\mu_k = \frac{1}{N_k} \sum_{n=1}^{N_k} \mathbf{c}_n \tag{5.36}$$

and

$$\mathbf{K}_k = \frac{1}{N_k - 1} \sum_{n=1}^{N_k} (\mathbf{c}_n - \mu_k)(\mathbf{c}_n - \mu_k)^T \tag{5.37}$$

The covariance matrix is here defined as a sample covariance matrix, that is, the sum is related to the number of degrees of freedom, which is equal to the number of distances $N_k - 1$. For any finite sum, both the mean and the (sample) covariance matrix are then only estimates, which become less uncertain the larger the number of observations is.

Often it is convenient to estimate mean and covariance iteratively, e.g. to improve the estimates as new measurements are becoming available, for which the class is known.

For this we let μ_{N-1} be the estimated mean feature vector using $N-1$ measured feature vectors $\mathbf{c}_n$, $n = 1, \cdots N - 1$ of a given class, then the estimate of the mean feature vector obeys the following recursive relation

$$\mu_N = \frac{N-1}{N} \mu_{N-1} + \frac{1}{N} \mathbf{c}_N \tag{5.38}$$

and for the covariance matrix we obtain a similar recursion formula

$$\mathbf{K}_N = \frac{N-1}{N} \mathbf{K}_{N-1} + \frac{N-1}{N^2} (\mathbf{c}_N - \mu_{N-1})(\mathbf{c}_N - \mu_{N-1})^T \tag{5.39}$$

In Equation 5.35 we need the inverse of the covariance matrix and it would be convenient if one could avoid the inversion and were able to update directly the inverse of the covariance matrix. Such a method will be shown next.

Consider a matrix $\mathbf{A}$ composed of

$$\mathbf{A} = \mathbf{B} + a\mathbf{x}\mathbf{x}^T \tag{5.40}$$

where $\mathbf{x}$ is a vector; then the matrix inversion theorem yields

$$\mathbf{A}^{-1} = \mathbf{B}^{-1} - a(1 + a\mathbf{x}^T\mathbf{B}^{-1}\mathbf{x})^{-1} \mathbf{B}^{-1}\mathbf{x}\mathbf{x}^T\mathbf{B}^{-1} \tag{5.41}$$

Using Equation 5.41 the recursive estimation of the inverse covariance matrix (Equation 5.39) becomes, with $\mathbf{x} = \mathbf{c}_N - \mu_{N-1}$, $a = \frac{N-1}{N^2}$ and $\mathbf{B} = \frac{N-1}{N} \mathbf{K}_{N-1}$

$$\mathbf{K}_N^{-1} = \frac{N}{N-1}\mathbf{K}_{N-1}^{-1}$$
$$- \frac{1}{N-1}\left[1 + \frac{1}{N}(\mathbf{c}_N - \boldsymbol{\mu}_{N-1})^T\mathbf{K}_{N-1}^{-1}(\mathbf{c}_N - \boldsymbol{\mu}_{N-1})\right]^{-1} \qquad (5.42)$$
$$\left[\mathbf{K}_{N-1}^{-1}(\mathbf{c}_N - \boldsymbol{\mu}_{N-1})\right]\left[\mathbf{K}_{N-1}^{-1}(\mathbf{c}_N - \boldsymbol{\mu}_{N-1})\right]^T$$

5.2.3 Variance analysis

Classification considers in general a large dataset of multi-dimensional feature vectors belonging to multiple classes. The analysis of such multivariate datasets is closely related to the analysis of the variances and covariance of the dataset and its grouping into classes (Steinhausen and Langer, 1977).

Given a dataset $\mathbf{c}_n$, where $n = 1, \ldots, N$ runs over all samples and the individual sample is typically a m-dimensional feature vector, the total mean feature vector $\boldsymbol{\mu}_t$ is now estimated by

$$\boldsymbol{\mu}_t = \frac{1}{N}\sum_{n=1}^{N}\mathbf{c}_n \qquad (5.43)$$

Assume now that the total dataset is divided into K classes Ω_k, with $k = 1, \ldots, K$, then we obtain the mean feature vector, or class centroid, of the class Ω_k by

$$\boldsymbol{\mu}_k = \frac{1}{N_k}\sum_{n\in\Omega_k}\mathbf{c}_n \qquad (5.44)$$

where $N_k = |\Omega_k|$ is the number of samples within class Ω_k and $N = \sum_{k=1}^{K} N_k$. Obviously, we have

$$\boldsymbol{\mu}_t = \frac{1}{N}\sum_{n=1}^{N}\mathbf{c}_n = \frac{1}{N}\sum_{k=1}^{K}\sum_{n\in\Omega_k}\mathbf{c}_n = \frac{1}{N}\sum_{k=1}^{K}N_k\boldsymbol{\mu}_k \qquad (5.45)$$

The total scatter matrix of the data $\mathbf{T}$ is defined by

$$\mathbf{T} = \sum_{n=1}^{N}(\mathbf{c}_n - \boldsymbol{\mu}_t)(\mathbf{c}_n - \boldsymbol{\mu}_t)^T \qquad (5.46)$$

As

$$\mathbf{c}_n - \boldsymbol{\mu}_t = (\mathbf{c}_n - \boldsymbol{\mu}_k) + (\boldsymbol{\mu}_k - \boldsymbol{\mu}_t) \qquad (5.47)$$

and noting from Equation 5.44 that

$$\sum_{n\in\Omega_k}(\mathbf{c}_n - \boldsymbol{\mu}_k) = 0 \qquad (5.48)$$

we may decompose the total scatter matrix into two components

$$
\sum_{n=1}^{N} (\mathbf{c}_n - \boldsymbol{\mu}_t)(\mathbf{c}_n - \boldsymbol{\mu}_t)^T
$$
$$
= \sum_{k=1}^{K} \sum_{n \in \Omega_k} (\mathbf{c}_n - \boldsymbol{\mu}_k)(\mathbf{c}_n - \boldsymbol{\mu}_k)^T + \sum_{k=1}^{K} N_k (\boldsymbol{\mu}_k - \boldsymbol{\mu}_t)(\boldsymbol{\mu}_k - \boldsymbol{\mu}_t)^T
$$
(5.49)

as the cross-products disappear due to Equation 5.48.

Defining a within-class scatter matrix $\mathbf{W}_K$

$$
\mathbf{W}_K = \sum_{k=1}^{K} \sum_{n \in \Omega_k} (\mathbf{c}_n - \boldsymbol{\mu}_k)(\mathbf{c}_n - \boldsymbol{\mu}_k)^T
$$
(5.50)

and a between-class scatter matrix $\mathbf{B}_K$

$$
\mathbf{B}_K = \sum_{k=1}^{K} N_k (\boldsymbol{\mu}_k - \boldsymbol{\mu}_t)(\boldsymbol{\mu}_k - \boldsymbol{\mu}_t)^T
$$
(5.51)

then we may rewrite the fundamental result of variance analysis (Equation 5.49) as

$$
\mathbf{T} = \mathbf{W}_K + \mathbf{B}_K
$$
(5.52)

The scatter matrices are related to the covariance matrices if we divide by the number of degrees of freedom. The total covariance matrix is estimated by

$$
\mathbf{K} = \frac{1}{N-1} \sum_{n=1}^{N} (\mathbf{c}_n - \boldsymbol{\mu}_t)(\mathbf{c}_n - \boldsymbol{\mu}_t)^T
$$
(5.53)

and the within-class covariance matrix is in analogy estimated by

$$
\mathbf{K}_k = \frac{1}{N_k - 1} \sum_{n \in \Omega_k} (\mathbf{c}_n - \boldsymbol{\mu}_k)(\mathbf{c}_n - \boldsymbol{\mu}_k)^T
$$
(5.54)

The individual variances of the feature components are then the diagonals of the covariance matrices. The total variance of class Ω_k is then the sum of the diagonals, or trace of the covariance matrix

$$
v_k = \text{trace}(\mathbf{K}_k)
$$
(5.55)

and the standard deviation of class Ω_k is then given by

$$
\sigma_k = \sqrt{v_k}
$$
(5.56)

5.2.4   Principal components

Consider the general weighted distance d of a feature vector $\mathbf{c}$ from a mean class vector $\boldsymbol{\mu}$

$$
d = (\mathbf{c} - \boldsymbol{\mu})^T \mathbf{K}^{-1} (\mathbf{c} - \boldsymbol{\mu})
$$
(5.57)

The covariance matrix $\mathbf{K}$ effectively weights the influence of the different components on the distance estimate. An unweighted distance may be obtained (Steinhausen and Langer, 1977) by defining a transformed vector

$$\mathbf{z} = \mathbf{K}^{-1/2}(\mathbf{c} - \mathbf{\mu}) \tag{5.58}$$

as then the distance becomes

$$d = \mathbf{z}^T \mathbf{z} = \|\mathbf{z}\|^2 \tag{5.59}$$

To obtain the matrix $\mathbf{K}^{-1/2}$ we proceed as follows. Consider a matrix $\mathbf{U}$ that is suited to transform the matrix $\mathbf{K}$ into a diagonal form

$$\mathbf{U}^{-1}\mathbf{K}\mathbf{U} = \mathbf{\Lambda} \tag{5.60}$$

where $\mathbf{\Lambda} = \mathrm{diag}(\lambda_i)$ are the eigenvalues of the matrix $\mathbf{K}$ and $\mathbf{U}$ is composed of the eigenvectors.

For positive eigenvalues λ_i we may estimate $\mathbf{\Lambda}^{-1/2} = \mathrm{diag}(\lambda_i^{-1/2})$ and obtain

$$\mathbf{K}^{-1/2} = \mathbf{U}\mathbf{\Lambda}^{-1/2}\mathbf{U}^{-1} \tag{5.61}$$

As we only are interested in estimating the distance d_n, and if the eigenvectors are chosen to be orthonormal, i.e. $\mathbf{U}^{-1}\mathbf{U} = \mathbf{I}$, we may drop the leading matrix $\mathbf{U}$ and replace Equation 5.58 by

$$\mathbf{z} = \mathbf{\Lambda}^{-1/2}\mathbf{U}^{-1}(\mathbf{c} - \mathbf{\mu}) \tag{5.62}$$

In case that not all eigenvalues are strictly positive, or that some eigenvalues are too small compared with the largest eigenvalue, that is, there exists a lower limit of useful eigenvalues $\lambda_{\min}$, then $\mathbf{\Lambda}^{-1/2}$ is only estimated for $\lambda_i > \lambda_{\min}$ and all other diagonal values are set to zero. The result is equivalent to a pseudo-inverse with reduced rank.

The feature vectors that are transformed according to Equation 5.62 are uncorrelated and have zero mean and variance one, allowing the implementation of straightforward clustering algorithms.

5.2.5 Cluster analysis

For the optimal classification, we considered the case that we have *a priori* classified samples and that we only need to learn the mean feature vectors and covariance matrices of the different classes. In the following, we back up and assume that the individual samples of the dataset are not pre-classified and that we need to group the data first into one or more distinct classes. The task is therefore to find the right number of classes, or clusters, so that each feature vector is optimally allocated to its class.

To begin, we have to define the criterion of an optimal solution. A common approach is to exploit the scatter matrix of the dataset and to require for the optimal classification that the trace of the within-class scatter matrix becomes minimal, that is:

$$C = \text{trace}(\mathbf{W}) \rightarrow \min \tag{5.63}$$

The optimal classification can easily be achieved by implementing an algorithm that may be synthesized as follows (Steinhausen and Langer, 1977):

1. Estimate class centroids and variances for an initial class allocation.
2. Test for each element whether the optimization criterion is improved by shifting the element into another class. If so, then shift into other class and re-estimate class centroids and variances.
3. Continue with step 2, or terminate if for a given number of times no class changes have occurred.

As for every step, the sum of all within-cluster variances decreases and a new configuration is generated, and as there are only a finite number of configurations, the procedure will always terminate with a minimal variance. However, the minimum will be only local, as it is dependent on the initial conditions and the sequence of the elements in the dataset.

Step 1 estimates the class centroids (mean feature vector) for an initial partition of the ensemble of feature vectors. If no initial partition is given, any *ad hoc* partitioning is valid, e.g. uniform distribution of the samples among the classes. The only requirement is that the number of classes is defined beforehand. If no *a priori* knowledge exists on the number of classes, then the clustering should be carried out for a varying number of clusters.

Step 2 asks for shifting an element $\mathbf{c}$ from class Ω_i to class Ω_j where the variance in the destination class Ω_j reduces more than the variance in the source class Ω_i.

Let $\mathbf{W}_i$ and $\mathbf{W}_j$ be the within-class scatter matrix of Ω_i and Ω_j, respectively, before shifting the sample $\mathbf{c}$ from Ω_i to Ω_j, then the scatter matrices become after the transfer

$$\tilde{\mathbf{W}}_i = \mathbf{W}_i - \frac{N_i}{(N_i - 1)}(\mathbf{c} - \boldsymbol{\mu}_i)(\mathbf{c} - \boldsymbol{\mu}_i)^T \tag{5.64}$$

$$\tilde{\mathbf{W}}_j = \mathbf{W}_j + \frac{N_j}{(N_j + 1)}(\mathbf{c} - \boldsymbol{\mu}_j)(\mathbf{c} - \boldsymbol{\mu}_j)^T \tag{5.65}$$

From Equations 5.64 and 5.65 we deduce that the trace of the within-class scatter matrix reduces if

$$\frac{N_j}{(N_j + 1)}\|\mathbf{c} - \boldsymbol{\mu}_j\|^2 < \frac{N_i}{(N_i - 1)}\|\mathbf{c} - \boldsymbol{\mu}_i\|^2 \tag{5.66}$$

After moving the sample $\mathbf{c}$ from class Ω_i to class Ω_j we may update the class centroids by means of the following recursive relationships

$$(N_i - 1)\tilde{\boldsymbol{\mu}}_i = N_i\boldsymbol{\mu}_i - \mathbf{c} \tag{5.67}$$

$$(N_j + 1)\tilde{\boldsymbol{\mu}}_j = N_j\boldsymbol{\mu}_j + \mathbf{c} \tag{5.68}$$

and the class variances v according to

$$(N_i - 2)\tilde{v}_i = (N_i - 1)v_i + \frac{N_i}{N_i - 1}\|\mathbf{c} - \boldsymbol{\mu}_i\|^2 \tag{5.69}$$

$$N_j\tilde{v}_j = (N_j - 1)v_j + \frac{N_j}{N_j + 1}\|\mathbf{c} - \boldsymbol{\mu}_j\|^2 \tag{5.70}$$

So far we have assumed that the number of classes is known *a priori*. This is very often not the case. We therefore need a procedure to estimate the number of classes that best characterize the sampled dataset. Let us assume that the dataset is composed of samples that are drawn from K_{true} classes. Without knowledge of K_{true} we carry out the cluster analysis for a varying number of classes, say from a minimum of 2 to a maximum of K_{max}, and decide *a posteriori* on the best solution K_{opt}, which we hope is close, if not identical, to the true number of classes K_{true}.

One good candidate for such an optimality criterion is based on the F-ratio (Steinhausen and Langer, 1977). As we used the diagonal sum of the within-class scatter matrix, $\text{trace}(\mathbf{W}_K)$, to search for the optimal partitioning of our dataset among the K classes, it is intuitive to use the obtained sample variances for our search for the optimal number of classes. The sample variances of the different scatter matrices are estimated by dividing the trace (sum of diagonals) by the number of degrees of freedom.

In particular, we obtain for the (sample) variance of the complete dataset

$$v_T = \frac{\text{trace}(\mathbf{T})}{m(N - 1)} \tag{5.71}$$

for the accumulated within-class variance

$$v_{W,K} = \frac{\text{trace}(\mathbf{W}_K)}{m(N - K)} \tag{5.72}$$

for the between-class variance

$$v_{B,K} = \frac{\text{trace}(\mathbf{B}_K)}{m(K - 1)} \tag{5.73}$$

and for the F-ratio

$$F(K) = \frac{v_{B,K}}{v_{W,K}} = \frac{\text{trace}(\mathbf{B}_K)}{\text{trace}(\mathbf{W}_K)} \frac{N - K}{K - 1} \tag{5.74}$$

The F-ratio was introduced to test whether the centroids of different classes are significantly different. Here we do not want to test whether the differences of the class centroids are significant, but we want to find the optimal partitioning of the dataset resulting in class centroids that are as different as possible, that is, we search for the optimal number of classes K_{opt} that maximizes the F-ratio

$$K_{opt} = \arg\max_K\{F(K)\} \tag{5.75}$$

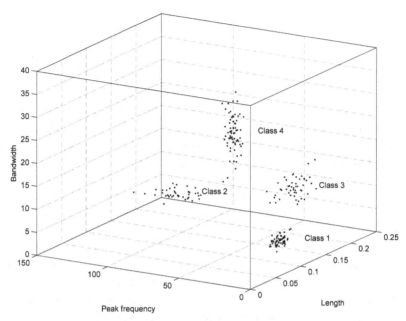

Figure 5.1   Classification by cluster analysis.

Example The following example is divided into four parts, for which I give the Matlab code.

1. Simulation of measurements
2. Transformation of measurements to obtain homogeneity of component variances
3. Cluster analysis for varying number of classes
4. Presentation of optimal results

To make the example somewhat realistic, results from a click detector were simulated. In particular, four species were assumed to be present in the dataset: sperm whale (*Pm*), Cuvier's beaked whale (*Zc*), bottlenose dolphin (*Tt*) and harbour porpoise (*Pp*). Three different types of measurement were modelled as Gaussian variables: click length, frequency of spectral peak and spectral bandwidth.

   The assumed values of mean and standard deviation may be found in the Matlab script, and the result of the simulation is shown in Figure 5.1, which shows that the four species are very well separated in the parameter space and the clicks correctly classified. The class numbers in Figure 5.1 are arbitrary; it is the final task of the user to inspect the cluster centroids and to label the different clusters with the corresponding species.

   After partitioning the dataset into different classes that are characterized by class mean μ and covariance matrix **K**, we are now in the position to apply Equation 5.35 for optimal classification of new feature vectors, which we will address in the next section of this chapter.

Matlab code for generation of Figure 5.1

```
%Scr5_1
%
%Cluster analysis
% Simulation
% simulated samples per cluster
%      Pm,    Zc,    Tt,    Pp
nc=[60        50     70     40];
% assumed centoids
mc=[0.1   0.2   0.05      0.2; %length
      15    40    30       120; %centerfrequency
       5    10    30         5];  %bandwidth
% assumed sigmas
sc=[0.01   0.02   0.005     0.01; %length
     1.5    4        3        10; %centerfrequency
       1    2        4         1]; %bandwidth
%
xx=[];
randn('state',0);
for ii=1:length(nc)

xx=[xx;ones(nc(ii),1)*mc(:,ii)'+(ones(nc(ii),1)* sc(:,ii)') ....
   *randn(nc(ii),size(mc,1))];
end

% Z-transform
zz=doTrans(xx);

%
%
Kmax=10;
F=zeros(Kmax,1);
FW=zeros(Kmax,1);
FB=zeros(Kmax,1);
for K=2: Kmax;
    % do cluster analysis for K clusters
    np=doCluster(zz,K);
    %cluster parameters
    [cm,cv,kct]=doCentroids(zz,np);
    [F(K),FW(K),FB(K)]=doFcrit(zz,np,cm,kct);
end

% find best number of clusters
K=find(F==max(F));
np=doCluster(zz,K);
%
%cluster parameters for original samples
[cm,cv,kct]=doCentroids(xx,np);
% display numbers
[kct;cv;cm]
%
```

```
% plot results
figure(1)
plot3(xx(:,1),xx(:,2),xx(:,3),'k.')
grid on
for ii=1:length(cm)
text(cm(1,ii)+0.05,cm(2,ii),cm(3,ii)-2, ...
    sprintf('Class %d',ii),'backgroundcolor','w')
end

cp=[-0.012   -4.18    0.84]*1000;
set(gca,'CameraPosition',cp)
box on
xlabel('Length')
ylabel('Peak Frequency')
zlabel('Bandwidth')
```

Matlab code for data transformation function

```
function X=doTrans(xx)
[N,M]=size(xx);
% mean data vector
S=mean(xx);
V=var(xx);
Z=(xx-repmat(S,N,1))./repmat(sqrt(V),N,1);
% data cross-correlation matrix
R=Z'*Z/N;
% eigen analysis
[U,E] = eig(R);
% transformation
X=Z*U*pinv(sqrt(E));
%diag(E)'
```

Matlab code for cluster analysis function

```
function np=doCluster(X,K)
%
[N,M]=size(X);
vv=zeros(M,1);
%
np=mod((0:N-1)',K)+1;
%
cv=zeros(1,K);
kct=zeros(1,K);
cm=zeros(M,K);
%
ind=0;
%
for kk=1:K
    nn=(np==kk);
    kct(kk)=sum(nn);
    if kct(kk)>0
```

```
                cm(:,kk)=mean(X(nn,:))';
        end
end
%
noshift=0;
loop=true;
while loop
    for ii=1:N
        noshift=noshift+1;
        if noshift>N, loop=false; break, end
        % get actual cluster
        npi=np(ii);
        if kct(npi)<=1, continue, end
        %
        %estimate variance by removing pivot sample
        vv1=sum((cm(:,npi)'-X(ii,:)).^2) ...
            *kct(npi)/(kct(npi)-1);

        % find cluster with best improvement
        vmin=inf;
        for jj=1:K
            if jj==np(ii), continue, end
            %estimate variance by adding pivot sample
            vv2=sum((cm(:,jj)'-X(ii,:)).^2) ...
                *kct(jj)/(kct(jj)+1);
            if vv2<=vmin
                vmin=vv2;
                npneu=jj;
            end
        end

        if vmin<vv1
            %change cluster
            noshift=0;
            %
            cm(:,npi) = ...
                (kct(npi)*cm(:,npi)-X(ii,:)')/(kct(npi)-1);
            cm(:,npneu) = ...
                (kct(npneu)*cm(:,npneu)+X(ii,:)')/(kct(npneu)+1);
            %
            np(ii)=npneu;
            kct(npi)=kct(npi)-1;
            kct(npneu)=kct(npneu)+1;
        end
    end
end
```

Matlab code for F-ratio criterion function

```
function [F,FW,FB]=doFcrit(X,np,cm,kct)
[N,M]=size(X);
```

```
K=max(np);

W=zeros(M);
B=zeros(M);
mcm=mean(cm,2);
vcm=cm-repmat(mcm,1,K);
for kk=1:K
    nn=(np==kk);
    vx=X(nn,:)-repmat(cm(:,kk)',kct(kk),1);
    W=W+vx'*vx;
    B=B+kct(kk)*vcm(:,kk)*vcm(:,kk)';
end
FB=sum(diag(B));
FW=sum(diag(W));
F=FB/FW*(N-K)/(K-1);
```

5.3   Cetacean classification

The implementation of an optimal classifier depends on the definition of the feature
vector. It seems obvious that the components of the feature vector should be selected in
such a way that they support the discrimination capability of the classifier. Useful feature
vectors depend on the application, which is especially the case when we are faced with
cetacean echolocation clicks or dolphin whistles.

5.3.1   Click classification

For a typical click detector we expect the feature vector to contain certain basic information: signal length, frequency of spectral peak, spectral bandwidth.

In general, the classification features should describe context-independent characteristics. For example, although we may easily measure the peak level and the total energy
for each click, both values are not suited as features as they depend on the distance of the
animal from the PAM system. However, the ratio of the two values, maximal signal
power related to integrated energy, could be seen as a measure of the shape of the signal,
which depends no more on the context than the other features. Therefore, we could add
this ratio to the feature vector.

Owing to the formal dependence of the received cetacean sound signal on the overall
geometry, i.e. owing to off-axis distortion and spectral selective transmission loss, the
previously mentioned features will vary slightly and are therefore always moderately
context-dependent. For example, the impact of off-axis distortion of echolocation clicks
is effectively a widening of the feature covariance matrix, especially when the mean
orientation of the animal is random with respect to the hydrophone location. Changes in
relative distances translate into spectrally dependent variations of received sound levels,
which may influence the ability to estimate characteristic signal features. In particular,
increasing transmission loss results in decreasing SNR, which may reach a threshold
below which some features become unreliable.

Using the individual SNR as a feature within the feature vector to control this dependency is not an option as there is no class-specific mean value. One way to ensure good feature estimates is to eliminate from the classification process all detections that have such a low SNR that the quality of the feature measurements is not sufficient for optimal classification according to Equation 5.35. This elimination, in principle, could be done during the detection process. Originally, the detection threshold was determined by the allowed probability of false alarm. Now, we could consider a new detection threshold, which is the required SNR for good classification. As this new criterion depends on the individual features, the required SNR may not be estimated as easily as for the false alarm criterion and must therefore be determined empirically.

There are, however, arguments against the use of a high SNR threshold early on in the detection process, but favour a multi-stage solution. High SNR may only be required for optimal classification according to Equation 5.35, which classifies every individual feature vector, but there may be different classification methods that still can handle parts of the feature vector that are otherwise not suited for the use of Equation 5.35. A multi-step thresholding would then allow a multi-step classification with different requirements for the quality of the feature measurements.

Another possibility to deal with SNR dependencies would be to model this dependency, and to use the model parameters as feature vectors. For example, if we note that the frequency of the spectral peak decreases significantly with decreasing SNR, e.g. due to frequency-dependent transmission loss, then it could be interesting to use the slope of this modelled tendency as feature. Using descriptive models may generate, however, uncontrollable side effects, as models are in general only approximations to reality and may introduce biases.

For the moment, we assume that the SNR is good enough to allow classification of the clicks. We assume further, that we have a sequence of clicks from the same species, e.g. Cuvier's beaked whale, as shown in Figure 4.1. As we have not yet constructed a class description, it is appropriate to apply the cluster algorithm to the detections. The data are first filtered with a high–lowpass combination to equalize the spectrum as discussed in Section 4.2.1, and then passed on to the simple click detector as discussed in Sections 4.1.1 and 4.1.3. For each detected and extracted click we estimate click time and click length as intensity-weighted mean t_C and RMS signal length τ_C according to

$$t_C = \frac{\sum\limits_{n=1}^{N} t_n |P_n|^2}{\sum\limits_{n=1}^{N} |P_n|^2} \tag{5.76}$$

and

$$\tau_C = \sqrt{\frac{\sum\limits_{n=1}^{N} (t_n - t_C)^2 |P_n|^2}{\sum\limits_{n=1}^{N} |P_n|^2}}, \tag{5.77}$$

where N is the number of samples extracted for the click, t_n is the time and P_n is the pressure of sample n.

The mean spectral peak frequency f_C and RMS bandwidth β_C are estimated by

$$f_C = \frac{\sum\limits_{m=1}^{M/2} f_m |F_m|^2}{\sum\limits_{m=1}^{M/2} |F_m|^2} \tag{5.78}$$

and

$$\beta_C = \sqrt{\frac{\sum\limits_{m=1}^{M/2} (f_m - f_C)^2 |F_m|^2}{\sum\limits_{m=1}^{M/2} |F_m|^2}}, \tag{5.79}$$

where M is the number of frequency bins of the click spectrum, and f_m and F_m are frequency and spectral value of the m-th spectral bin.

Figure 5.2 shows the results of the cluster algorithm applied to the beaked whale data. From Figure 5.2 we note that the equalization, as discussed in Section 4.2 and shown in Figure 4.11, cleaned up the picture significantly and showed fewer detections than

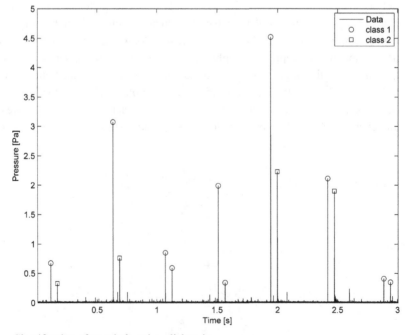

Figure 5.2   Classification of an echolocation click train.

presented in Figure 4.2. The cluster algorithm found two distinct classes, 10 in class 1 and 4 in class 2. As all clicks are Cuvier's beaked whale clicks, this division into two classes could be surprising, but as the class-2 clicks always follow shortly after a class-1 click, we may deduce that the class-2 clicks have a different quality even if they are linked to the class-1 clicks.

In fact, the class-2 clicks are the surface reflections of the class-1 clicks, which are the direct arrivals of the echolocation clicks at the hydrophone. Sound arriving at the hydrophone directly from the animal generally precedes sound that is first reflected at the surface before arriving at the hydrophone. As one might expect that surface reflections would change the feature description, it should be no surprise that the cluster algorithm divides the clicks into two classes. The remaining question is only why a correct classification is not achieved for all surface reflections.

To shed light on this problem, Table 5.1 presents all the features extracted and used for clustering (marked in bold in Table 5.1, i.e. all columns but the first three) and the clustering output (far left column, labelled C). One notes that, as expected, the mean spectral (centre) frequency f_C and the frequency of the spectral peak f_P are found around 40 kHz, and that they seem to be redundant or correlated information. Inspecting the feature table, one notes that only the shape feature P2E and the RMS signal length have bimodal values alternating between smaller and larger values. Removing the redundant peak frequency feature improves the misclassification somewhat, and only the eighth click remains in the wrong class.

We are now confronted with the following problem that surface reflections are partly classified into their own class and partly combined with direct arrivals. This is in principle

Table 5.1 Featured vector set and classes

C	t_C	E	**P2E**	τ_C	f_C	β_C	f_P
1	0.108	11.16	−14.58	37.97	35.55	7.22	36.00
2	0.163	8.73	−18.40	232.03	35.92	7.44	34.50
1	0.636	24.70	−14.94	39.67	39.97	5.61	40.50
2	0.692	16.14	−18.53	176.30	38.41	5.91	41.25
1	1.073	12.87	−14.26	32.23	35.07	7.26	34.13
1	1.129	12.08	−16.64	133.98	41.21	7.17	43.50
1	1.511	20.11	−14.13	34.19	37.74	5.38	40.13
1	1.566	5.79	−15.06	81.63	38.45	11.68	44.25
1	1.942	28.08	−14.98	41.64	41.17	6.09	41.25
2	1.998	24.69	−17.73	190.72	38.11	4.90	42.00
1	2.417	21.03	−14.52	41.40	40.12	5.94	42.38
2	2.473	23.35	−17.78	230.94	41.72	7.25	41.63
1	2.885	8.14	−15.84	61.38	40.83	7.58	42.00
1	2.941	8.67	−17.78	110.88	38.07	7.66	34.13

C is the resulting class, t_C mean click time [s], E accumulated intensity, P2E ratio peak intensity to accumulated intensity, τ_C the RMS signal length [µs], f_C the mean spectral frequency [kHz], β_C the RMS bandwidth [kHz], and f_P the frequency of the spectral peak [kHz]. The features used for clustering are marked in bold.

not a surprise, as both signals are coming from the same species and sometimes show similar features. The close relationship between direct and surface-reflected arrivals suggests, however, that one should exploit these apparent double clicks as a separate classification cue, leading to a deep diver classification scheme.

5.3.2   Deep diver classification

When dealing with echolocation clicks of deep divers, e.g. Cuvier's beaked whales, and shallow hydrophones, i.e. when the whales are echolocating well below the hydrophones, then we are confronted with significant surface reflections, as shown in Figure 5.2. The delays between direct arrivals and their surface reflections are context-dependent, that is, they depend on the relative locations of whale and hydrophones. This is, however, not a serious problem, as the relative geometry between a deep diving whale and a hydrophone does not vary very much for consecutive clicks, resulting in a very slowly varying time difference between direct arrival and surface reflection.

Although the time differences between consecutive echolocation clicks exhibit sometimes large variations, as they depend on the animal's behaviour, which may change dramatically from one second to another, the relative distance between whale and hydrophone changes only slowly. For a short period, say of the order of tens of clicks, the surface delay is nearly constant or varies linearly with time. This slowly varying time difference may now be considered as a unique classification cue describing the overall geometry.

In Figure 5.3, a further Cuvier's beaked whale click sequence was analysed by using the time delay between direct arrival and surface reflection as the only criterion to classify and label the echolocation click train. As we are for the moment only interested in detecting and classifying beaked whales by means of surface reflections, no other features are added to the feature vector. The data are again equalized as discussed in Section 4.2.1 before being passed on to the simple click detector discussed in Sections 4.1.1 and 4.1.3. The click classifier operates in two phases: first it finds the characteristic delay between direct arrival and surface reflection, and then it assigns the proper label to the individual clicks of the click sequence.

Typically, finding constant delays of a click trains is the domain of the correlation function, which peaks at the time delay, but the amount of data to be processed (here 1 s worth of data already contains 384 000 samples) makes this classical method not practicable. A simple method that uses the histogram of all possible delays within the individual detection times is, however, a reasonable alternative.

Given a time sequence of detected clicks $t_C(n)$, with $n = 1, \ldots, N$, we accumulate a histogram h_C by

$$h_C(m) = h_C(m) + 1 \tag{5.80}$$

where

$$m = \text{round}\left(a(t_C(j) - t_C(i))\right) \tag{5.81}$$

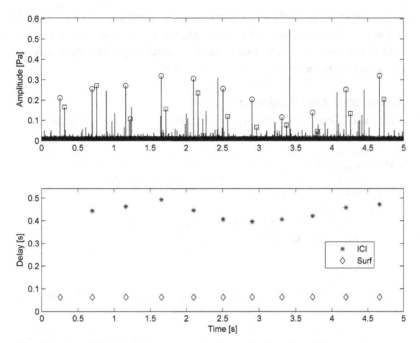

Figure 5.3 Classification of click train by means of surface reflection time delay.

with $i = 1, \ldots, N$, $j = i + 1, \ldots, N$ and a being a constant that is chosen to obtain good binning of the histogram.

If the click times are given in seconds, then a factor $a = 1000$ would result in a bin-width of 1 ms, which seems appropriate in most cases. As there are some limits to the arrival delay of the surface reflections, the processing time may be reduced by limiting the analysed delay $t_C(j) - t_C(i)$ to a maximal value. Based on geometric considerations we observe that the arrival delay of the surface reflection is maximal when the animal is exactly below the hydrophone. Knowing the hydrophone depth h, the maximum delay Δt_{max} is then estimated as the round trip time of a sound travelling from the hydrophone to the surface and returning to the hydrophone:

$$\Delta t_{max} = 2\frac{h}{c} \tag{5.82}$$

After finding possible surface reflection delays, we use a simple algorithm doAllocTrain to label the individual detections as either direct or surface reflected arrival, or keep them unallocated.

Figure 5.3 shows that the method is suited to pick up the constant arrival delay of the surface reflection. The well-behaved interval between two consecutive clicks (inter-click interval, or ICI) indicates that no click pair was missed or erroneously added. Note that, although there are 11 classified clicks with 11 surface reflection arrival delays, there are only 10 ICI estimates between these 11 clicks. The timings for the delays follow the following notation. Surface reflection delays are plotted at the time of the direct arrival; ICI is plotted at the time of the next click. For this reason there is no ICI indicated at 0.33 s. As expected, the variation in the ICI is much larger than the observed arrival delay of the surface reflection.

Matlab code for delay histogram

```
function [icm,Hc]=doDelayHisto(tcl,dt1,dt2, ac,Hcmin)
%
Hc=zeros(ceil(1.1*ac*dt2),1);
for ii=1:length(tcl)-1
    itcl = tcl>tcl(ii)+dt1 & tcl<=tcl(ii)+dt2;
    tclj = tcl(itcl);
    if ~isempty(tclj)
        ind=round(ac*(tclj(:)'-tcl(ii)));
        Hc(ind) = Hc(ind)+1;
    end
end
%
locmax=1+find(Hc(2:end-1)>Hc(1:end-2) & Hc(2:end-1) >Hc(3:end));
icm=locmax(Hc(locmax)>Hcmin)';
```

Comments on Matlab code

We declare a histogram vector Hc to be somewhat larger than required for covering all delays up to a maximal delay dt2. As Hc is defined as a row vector, the operation tclj(:)' ensures the ind vector is also a row vector. The scale factor ac defines effectively the bin-width of the histogram. For the present application a histogram bin width of 1 ms was selected by using ac = 1000. We consider potential surface delays icm as the local maxima in the histogram with more than Hcmin hits. The proper choice of Hcmin depends on the amount of data, but in general, it will exceed 2.

Matlab code for double click classification

```
function inp=doAllocDoubleClick(tcl,pcl,icm,ac, dt2,xdx_min)
%
inp=zeros(length(tcl),3);
%
pcm=0*icm;
for ii=1:length(tcl)-1
    %
    xdx=inf-0*icm;
    ij=0*icm;
  for jj=ii+1:length(tcl)
        if inp(ii,1)>0, continue,end
        tclij=(tcl(jj)-tcl(ii));
        if tclij>dt2, break,end
        % do we fit to an icm on an ici basis?
        dd=abs(ac*tclij-icm);
```

```
        [dx,kx]=min(dd);
        %
        if dx<xdx(kx)
            xdx(kx)=dx;
            ij(kx)=jj;
        end
    end
    %if best solution is within a predefined limit
    kk=find(xdx<xdx_min);
    if isempty(kk)
        continue
    elseif length(kk)==1
        kx=kk;
    else
        [dx,kx]=min(abs(pcl(ii)-pcm));
    end
    inp(ii,:)=[1+2*(kx-1),ii,ij(kx)];
    inp(ij(kx),:)=[2*kx,ii,ij(kx)];
    pcm(kx)=pcl(ii);
end
```

Comments on Matlab code

The algorithm implements a hierarchical classification scheme. Based on time delays, it first tries to classify by allocating the double clicks. In case of multiple possibilities of acceptable time delays, we compare the actual signal amplitude with the last measured values for the different delay classes and choose the one for which we observe the best fit.

5.3.3   Click train classification

So far, we have been fortunate, as the selected datasets contained the detections for only one individual. Depending on location and species, it may, however, be more likely to detect multiple animals simultaneously, resulting in an increased number of potential allocation conflicts. One way to solve allocation difficulties is to ignore the fact that the detections stem from multiple animals, and to rely on the individual click classification based on standard feature vectors. Another option is to count the whales and to partition the whale chorus into well-behaved sequences of echolocation clicks.

To differentiate between different individuals it is convenient to delay the final classification, allowing a global allocation of multiple clicks to a set of potential click trains. We are now confronted with the situation that we need a minimum amount of detections to classify multiple echolocating whales or dolphins of the same species. We have further to find features within the dataset that are unique to the individuals. If the animals click at significantly different depths then the arrival delays of the surface reflection may be consistently different, so that suface delays could be again used to differentiate individuals.

As mentioned before in Section 2.3.7, sequences of echolocation clicks, or click trains, are described by the inter-click interval (ICI). The ICI is in general slowly varying within short timescales that cover only a few clicks. As the click sequence is driven by the foraging activity of the individual animal, the instantaneous ICIs of different animals tend to be different. Another useful measurement is the received level of the click. Again, this feature is varying in time, but may be significantly different for individuals as the received sound pressure varies as a function of animal orientation and distance.

Using the ICI for classification is not trivial as the ICI can only be measured if the multiple clicks are attributed to the same animal, which, strictly speaking, is only known after classification. One solution to this problem is not to classify only to species level, but to go one step further and to keep track of individual animals of the same species (Gerard *et al.*, 2008), that is, we have to build a tracker into our classification scheme. The received sound level is then a convenient added feature to resolve classification association conflicts in case they arise.

Figure 5.4 shows the recording and detections from a scenario in which at least two sperm whales were found to be present. The data were recorded with a sampling rate of 31.25 kHz, filtered with a 1 – 14 kHz pass-band, and threshold detected with $th = 3.5$. The pruning window of the detector was set to 4 ms to eliminate multiple detections due to the multi-pulse structure of the sperm whale clicks. The data were recorded using a towed array that was deployed at about 80 m, a depth that suggests the presence of detectable surface reflections. The data of only one hydrophone was selected from the array for this analysis.

The first step is therefore an overall analysis of the data for multiple presences of delays between detections that may represent surface reflections. Applying the function doDelayHisto, we consider potential surface delays as the local maxima in the

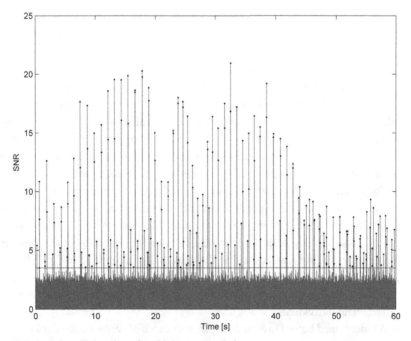

Figure 5.4   Echolocation click trains of multiple sperm whales.

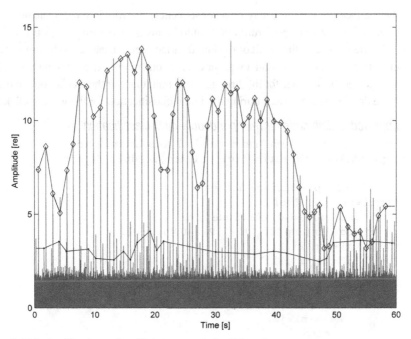

Figure 5.5 Initial classifications of multiple sperm whale click trains.

histogram with more than Hcmin hits. The proper choice of Hcmin depends on the amount of data and for the present dataset Hcmin=6 was selected, which seems adequate for a 60 s dataset. The choice of 60 s was driven by the actual file size and is slightly larger than needed. An analysis window of 20 s was found on other occasions to be adequate for analysing echolocation click trains.

Having determined the number of different surface reflection delays, we can now construct the initial set of click trains. We do this by a procedure similar to that for the Cuvier's beaked whale case (Figure 5.3) with some additions. The presence of multiple click trains may result in allocation conflicts, e.g. due to interferences resulting from the simultaneous detection of different clicks. The majority of these interferences can be resolved by considering the received sound pressure amplitude of the direct arrival in addition to the surface delay. For example, in Figure 5.4 we see one click train that is rather strong with large fluctuations in SNR and one or potentially more additional click trains that have lower SNR values.

To use the sound pressure amplitude as an additional allocation criterion we could modify the distance function in the double-click classification routine, but it is again easier to follow a hierarchical classification scheme and to use sound pressure amplitude only in cases where allocation conflicts occur. If we consider allocation conflicts as some sort of procedural error, than the selected approach is comparable with standard error handling, which detects and removes discrepancies in the results. However, once these 'errors' become regular, then instead of being 'errors' they should be considered a deficiency and should result in a modification of the allocation algorithm. Sometimes these allocation conflicts may be a consequence of errors in the data acquisition system (e.g. buffer over- or under-run) and should be investigated further for this reason also.

Figure 5.5 shows the result of the initial surface delay-based classification. We note two click trains, a strong one and a weaker one. As one might expect, the number of observed surface delays is higher for the stronger click train than for the weaker one, indicating that more detections are missed during the initial classification for the weaker click train.

Improved Matlab code for multiple double-click classification

```
function inp=doAllocTrain(tcl,pcl,icm,ac,dt2, xdx_min)
inp=doAllocDoubleClick(tcl,pcl,icm,ac,dt2,xdx_min);
%
%check multiple surface delays
for ii=1:length(tcl)
    if inp(ii,1)==0, continue, end
    if mod(abs(inp(ii,1))-1,2)==0 %have click
        j1=inp(ii,2);
        j2=inp(ii,3);
        if mod(abs(inp(j2,1))-1,2)==0
            %have multiple direct
            if pcl(j1)>pcl(j2)
                n1=j2;
            else
                n1=j1;
            end
            inp(n1,:)=[0,0,0];
        end
    end
end
%
%clean up interlaced classification
for ii=1:length(tcl)-1
    if inp(ii,1)==0, continue,end
    if mod(abs(inp(ii,1)),2)==0 %have click
        j1=inp(ii,2);
        j2=inp(ii,3);
        for jj=j1+1:j2-1
            if abs(inp(ii,1))==abs(inp(jj,1))
                %have interlaced detections
                if pcl(ii)>pcl(jj)
                    n1=jj;
                    n2=inp(jj,2);
                else
                    n1=ii;
                    n2=inp(ii,2);
                end
                inp(n1,:)=[0,0,0];
                inp(n2,:)=[0,0,0];
            end
        end
    end
end
```

Comment on the Matlab code

Two allocation conflicts are handled: multiple clicks for the same surface-reflected arrival and interlaced allocations for the same click train. In both cases, we eliminate the allocation with the weakest direct arrival.

After the initial allocation of some clicks to click trains by using surface reflection delays, we should now allocate the remaining clicks to these initiated click trains. For this we use both the received level and the inter-click interval.

To be successful we have to make some assumptions about the resulting click trains. It seems fair to assume that the ICIs before and after a click are similar or have a common trend, and the same assumption also seems reasonable for the received sound pressure amplitude. In particular, we assume that three consecutive measurements y_{n+1}, y_n, y_{n-1} are related by

$$(y_{n+1} - y_n) = (y_n - y_{n-1}) + N(0, \sigma_y) \tag{5.83}$$

or

$$y_{n+1} = 2y_n - y_{n-1} + N(0, \sigma_y) \tag{5.84}$$

where we assume a normal distributed mismatch $N(0, \sigma)$, with standard deviation σ.

We use Equation 5.84 to generate two types of prediction formulas, one where we replace variable y by the received sound pressure P of a detection and another one, where we use the ICI in lieu of the variable y

$$P_{n+1} = 2P_n - P_{n-1} + N(0, \sigma_P) \tag{5.85}$$

$$ICI_{n+1} = 2ICI_n - ICI_{n-1} + N(0, \sigma_{ICI}) \tag{5.86}$$

As the ICI is the time difference between consecutive click times t_C, Equation 5.86 may also be written as

$$t_{Cn+1} = t_{Cn-2} + 3ICI_n + N(0, \sigma_{ICI}) \tag{5.87}$$

In cases where we only deal with two measurements, e.g. when initiating a track, then Equations 5.85 and 5.87 reduce to

$$P_{n+1} = P_n + N(0, \sigma_P) \tag{5.88}$$

$$t_{Cn+1} = t_{Cn-1} + 2ICI_n + N(0, \sigma_{ICI}) \tag{5.89}$$

In absence of any ICI measurement, that is, where we have only a single click, we assume a nominal ICI_0 to predict the time of the next potential click in the click train.

$$t_{Cn+1} = t_{Cn} + ICI_0 + N(0, \sigma_{ICI}) \tag{5.90}$$

This nominal ICI_0 will be species-dependent and may vary as a function of the behavioural context. A reasonable value is given by the overall ICI expectation: say 1 s for sperm whale, 0.4 s for beaked whales, 0.2 s for pilot whales, 0.1 s for dolphins.

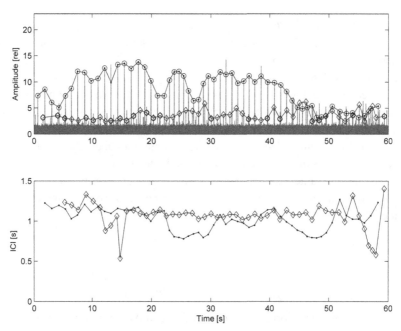

Figure 5.6   Automated click-train classifications.

To build the click trains we take each un-associated click and compare its time and sound pressure amplitude with the predictions that are made for each initiated click train. We select from all click trains the one that minimizes the distance between prediction and measurement. Again, we use a hierarchical approach and use first the time difference and then, if there are allocation conflicts, the pressure measurements.

Figure 5.6 shows the result of this automated click-train classification. The implemented algorithm is still elementary and far from complete. The main purpose is to show the concept of click-train classification and the problems that may arise when dealing with real data.

In general, Figure 5.6 shows also that ICI and amplitude are reasonable slowly varying functions for both click trains, especially for high signal-to-noise ratio, and that at low signal amplitudes, correct classification becomes increasingly difficult.

So far, we only have generated click trains, but what about species classification? This can be achieved when obtaining descriptive features of a click train (mean ICI, variation of ICI, click-train length, etc.) and classifying them with standard classification or clustering methods as discussed in Section 5.2.

Matlab code for click train classification

```
function inp=doClickTrain(inp,tcl,pcl,icm,cmin)
K=length(icm);
%
in0=0*icm;
in1=0*icm;
```

```
in2=0*icm;
% % ===================================================
for nn=1:10
    count=0;
    %
    for ii=2:length(tcl)
        % if we have a free click
        if inp(ii,1)==0
            dd1k=inf+0*icm;
            dd2k=inf+0*icm;
            ij=length(tcl);
            for kk=1:K
                % find next click
                for jj=ii+1:length(tcl)
                    if abs(inp(jj,1))==1+2*(kk-1)
                        ij=jj;
                        break
                    end
                end
                % check if still space for new click in kk
                if in0(kk)>0
                    if tcl(ij)-tcl(in0(kk))<1.5
                        continue
                    end
                end
                %
                if in2(kk)>0
                    icin=tcl(in0(kk))-tcl  (in1(kk));
                    if icin<0, icin=1/3; end
                    ddkx=abs(tcl(ii)-tcl(in2  (kk))-3*(icin));
                    if ddkx<dd1k(kk), dd1k(kk)  =ddkx; end
                end
                if dd1k(kk)>cmin && in1(kk)>0
                    icin=tcl(in0(kk))-tcl(in1  (kk));
                    if icin<0, icin=1/2; end
                    ddkx=abs(tcl(ii)-tcl(in1  (kk))-2*(icin));
                    if ddkx<dd1k(kk), dd1k(kk)  =ddkx; end
                end
                if dd1k(kk)>cmin && in0(kk)>0
                    icin=1;
                    ddkx=abs(tcl(ii)-tcl (in0(kk))-icin);
                    if ddkx<dd1k(kk), dd1k (kk)=ddkx; end
                end
                if in1(kk)>0
                    dpcl=pcl(in0(kk))-pcl (in1(kk));
                    dd2k(kk)=abs(pcl(ii)-  pcl(in1(kk))-2*dpcl);
                elseif in0(kk)>0
                    dd2k(kk)=abs(pcl(ii)-  pcl(in0(kk)));
                end
            end
            % do we have a best allocation?
```

```
            [d1x,k1x]=min(dd1k);
            [d2x,k2x]=min(dd2k);
            if k1x==k2x %have same best  allocation
                kx=k1x;
                if d1x<cmin %have a good click
                    inp(ii,:)=[1+2*(kx-1),  ii,0];
                    count=count+1;
                end
            else %have conflicting allocations
                % for the moment skip
%                    [d1x k1x,d2x,k2x pcl(ii) pcl (in0(kk))]
            end
        end
        if mod(abs(inp(ii,1))-1,2)==0, % have marked click
            kx=1+floor((abs(inp(ii,1))-1)/ 2);
            in2(kx)=in1(kx); % so shift all  back
            in1(kx)=in0(kx);
            in0(kx)=ii;
        end
    end
    if count==0, break,end
end
```

Comments on the Matlab code
The algorithm implements Equations 5.87, 5.89 and 5.90 for the predicted click time and Equations 5.85 and 5.88 for the sound pressure amplitude. The case of assignment conflicts has so far not been implemented; allocation strategies as discussed for whistle detection could be considered.

5.3.4   Whistle classification

The classification of echolocating species turned out to be complicated, as in addition to the classical feature vectors, tracking concepts also had to be considered and simple measurements had to be combined into some sort of superstructure that could be used for species-level classification. The classification of dolphin whistles is faced with another type of problem. Whistles are not only species-dependent but complex, and may be drawn from a wide and varied repertoire with and without stereotyped repetition. Furthermore, even similar whistles are not replicated exactly the same by an animal, but show differences in duration and bandwidth, which may vary from detectable but insignificant to considerable (Buck and Tyack, 1993), making classification difficult.

A quantitative method to describe whistles would take the whistles (e.g. Figure 4.18) and might try to estimate the salient features of the measured instantaneous frequency function $f(t)$. Typical parameters are (a) duration, (b) starting frequency, (c) ending frequency, (d) minimum frequency, (e) maximum frequency, (f) number of local extremes and (g) presence of harmonics.

To determine these 'global' parameters one needs to extract, or be able to extract, the complete whistle from the data. This may or may not be possible. In particular, the presence of multiple whistling animals may result in overlapping whistles that make such an extraction attempt challenging. In addition, temporary interruption of the whistle trace due to low signal-to-noise ratio or interferences could result in an incomplete description of whistles leading to potential misclassification. The presence of unresolved pulsed sounds (e.g. burst clicks) may be a further confounding factor.

If we accept that complete and clean whistles are the exception in real PAM applications, then it would be wise to envision an alternative approach that is based on the local description of whistles instead of a global parameterization. Whistles are in general described by a time-dependent instantaneous frequency function $f(t)$. A local description near time t_0 would therefore follow the Taylor's theorem of arbitrary functions

$$f(t) = f(t_0) + \frac{df}{dt}\bigg|_{t=t_0} (t - t_0) + \frac{d^2f}{dt^2}\bigg|_{t=t_0} (t - t_0)^2 + R_2(t) \tag{5.91}$$

where $R_2(t)$ is the residual error made by the approximation of stopping the series expansion at the quadratic term. If the distance between t and t_0 is small enough, then the residual error $R_2(t)$ may be small enough to be ignored, that is, we approximate the function by a quadratic functions.

Using quadratic functions to describe the local shape of whistles we effectively have to estimate three parameters, $f(t_0)$, $\frac{df}{dt}$ and $\frac{d^2f}{dt^2}$, which describe the reference frequency, slope

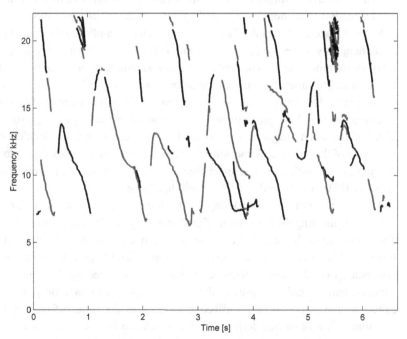

Figure 5.7   Tonal sounds of striped dolphins, using the whistle detector described in Section 4.4 (see also Figures 4.18 and 4.26). Different grey shades distinguish different snippets.

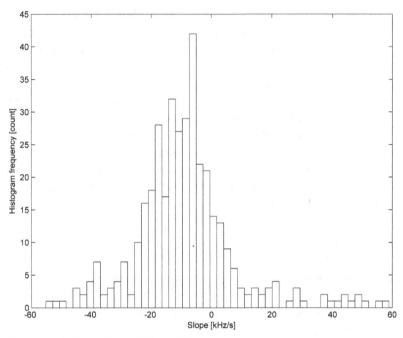

Figure 5.8   Statistics of whistle slopes of Figure 5.7.

and curvature of the whistle fragment and which may be obtained by a least mean square fit of a quadratic polynomial to small fragments of the whistle contours.

Figure 5.7 shows the result of such a whistle detector applied to the data of Figure 4.24 containing only snippets that exceed a minimal length (> 50 ms). One notes that some whistles are broken up into multiple fragments as indicated by different line styles. The fragment length varies, from the minimum imposed limit of 50 ms to a couple of hundred milliseconds. All fragments of this dataset remain above a frequency of 6 kHz and most fragments seem to have predominantly downward-oriented slopes, suggesting a skewed distribution around a negative slope value. Local extremes seem to coincide with the total excursion of the whistles. Some whistle traces at higher frequencies appear to be second harmonics of whistle traces at lower frequencies. The presence of two high-frequency bursts (20 kHz at *c*. 1 s and *c*. 5.5 s) should finally be noted.

The direct use of the (mean) slope as feature vector in a statistical classifier seems promising, allowing differentiation of predominantly up-slope and down-slope whistles. The same cannot be said for the reference frequency or the curvature of a whistle fragment. The reference frequencies of different whistle fragments tend to be uniformly distributed over the whole frequency band that is covered by the whistles. As most whistles, and not only the present dataset, are expected to have only few sharp local extremes, and as the curvature effectively measures the radius of these extremes, the curvature value, or second derivative, of most whistle fragments will be close to zero.

Figure 5.8 shows the statistics of the whistle slopes of Figure 5.7 and confirms that most whistles of that dataset are downward-oriented.

A promising set of features for a statistical whistle classification is therefore based mainly on the whistle slope and on overall minimal frequency of the whistles. However, it is not advisable to derive general feature vectors from a single dataset derived from a single species. Many more datasets from different species must be analysed with the same method to make generally valid suggestions on whistle classification methods. Scanning the literature, the reader will note that whistle classification is a field of active research with a lot of work still to be done.

6 Localization and tracking

The capability to localize and track animals in space is an important function of PAM, as it allows behavioural analysis also by acoustic means. Well-known techniques for passive range estimation are multi-hydrophone ranging, triangulation, multi-path ranging and beam-forming. Tracking finally plays an important role if one not only tries to localize acoustically active cetaceans but also tries to monitor their behaviour. For this, it is necessary to determine continuously the location of each animal, that is, to track the individuals.

The localization of sound sources requires multiple independent measurements. These measurements may stem from multiple sensors that cover large areas and are therefore suited for multi-hydrophone ranging, or are based on qualitatively different measurements, e.g. multi-path time delay and angle of arrival estimation, as used in triangulation and multi-path ranging.

In general, one can say that for any unknown parameter, e.g. range, depth, bearing, one needs at least one independent measurement. Multiple measurements, however, may or may not be independent; depending on the geometry of the hydrophones relative to the acoustically active object closely spaced hydrophones tend to provide correlated, highly dependent measurements.

In principle, there are two complementary methods for localization, one based on the estimation of time delays, and the other based on beam-forming. Time-delay-based localization uses travel-time delays that occur when different hydrophones are found at different distances from the sound source, and is commonly used with widely spaced hydrophones. Time delays may be used directly or first converted into angles to estimate the direction of arrival of the sound. Beam-forming, on the other hand, is common with closely spaced hydrophones, where multiple hydrophones are correlated to obtain the direction of arrival of the sound.

It is important to keep always in mind that reliable localization requires that the estimated location of the object is in the vicinity of real or virtual hydrophones. Large-scale extrapolation, while mathematically possible, should be avoided because localization errors may increase significantly.

6.1 Multi-hydrophone ranging

Multi-hydrophone ranging uses multiple hydrophones to locate the acoustically active animal. Based on differences in sound arrival times, it is a long-established method both in underwater and in terrestrial applications (e.g. Spiesberger and Fristrup, 1990; Spiesberger, 1999; McGregor *et al.*, 1997) and very similar to modern GPS-based navigation systems.

Given a set of hydrophones h_i, $i = 1, \cdots M$ at positions (x_i, y_i, z_i) and a whale at the position (x_w, y_w, z_w) then the distances R_{wi} between the whale and the hydrophones are given by a set of M equations

$$R_{wi}^2 = (x_i - x_w)^2 + (y_i - y_w)^2 + (z_i - z_w)^2 \tag{6.1}$$

To avoid large-scale extrapolation, we assume that the whale distances are of the same order of magnitude as the inter-hydrophone distances.

We note that to obtain the position of the whale we have to estimate three variables, that is, we need at least three equations to solve for the three unknowns if we assume that the hydrophone positions are known. We cannot measure directly the individual ranges R_i, otherwise there would be no need for range estimation, but we can measure the travel time required by the sound to travel from one hydrophone to the others. That is, we may replace the individual ranges by $R_i = R_0 + \delta R_i$, where R_0 is the now unknown range to the reference hydrophone h_0 and δR_i is the estimated acoustic distance between hydrophone h_i and the reference hydrophone h_0. The measured acoustic distance δR_i is related to a travel time difference of the sound δT_i by $\delta R_i = c\delta T_i$, where c is the effective sound speed between the two hydrophones.

Expanding Equation 6.1 for the reference hydrophone, we obtain

$$\begin{aligned} R_0^2 = {}& x_0^2 - 2x_0 x_w + x_w^2 \\ & + y_0^2 - 2y_0 y_w + y_w^2 \\ & + z_0^2 - 2z_0 z_w + z_w^2 \end{aligned} \tag{6.2}$$

and for all other hydrophones

$$\begin{aligned} R_0^2 + {}& 2R_0(\delta R_i) + (\delta R_i)^2 \\ = {}& x_i^2 - 2x_i x_w + x_w^2 \\ & + y_i^2 - 2y_i y_w + y_w^2 \\ & + z_i^2 - 2z_i z_w + z_w^2 \end{aligned} \tag{6.3}$$

Subtracting Equation 6.2 from Equation 6.3, we remove all quadratic terms of the unknowns and obtain

$$\begin{aligned} 2R_0(\delta R_i) + {}& (\delta R_i)^2 \\ = {}& x_i^2 - x_0^2 - 2(x_i - x_0)x_w \\ & + y_i^2 - y_0^2 - 2(y_i - y_0)y_w \\ & + z_i^2 - z_0^2 - 2(z_i - z_0)z_w \end{aligned} \tag{6.4}$$

We now have 4 unknowns R_0, x_w, y_w, z_w, and using 5 hydrophones we may form 4 equations to obtain a unique solution by solving the following set of linear equations

$$\left(\begin{array}{l} (x_1{}^2 - x_0{}^2) + (y_1{}^2 - y_0{}^2) + (z_1{}^2 - z_0{}^2) - (\delta R_1)^2 \\ (x_2{}^2 - x_0{}^2) + (y_2{}^2 - y_0{}^2) + (z_2{}^2 - z_0{}^2) - (\delta R_2)^2 \\ (x_3{}^2 - x_0{}^2) + (y_3{}^2 - y_0{}^2) + (z_3{}^2 - z_0{}^2) - (\delta R_3)^2 \\ (x_4{}^2 - x_0{}^2) + (y_4{}^2 - y_0{}^2) + (z_4{}^2 - z_0{}^2) - (\delta R_4)^2 \end{array} \right)$$
$$= 2 \left(\begin{array}{llll} \delta R_1 & (x_1 - x_0) & (y_1 - y_0) & (z_1 - z_0) \\ \delta R_2 & (x_2 - x_0) & (y_2 - y_0) & (z_2 - z_0) \\ \delta R_3 & (x_3 - x_0) & (y_3 - y_0) & (z_3 - z_0) \\ \delta R_4 & (x_4 - x_0) & (y_4 - y_0) & (z_4 - z_0) \end{array} \right) \left(\begin{array}{l} R_0 \\ x_w \\ y_w \\ z_w \end{array} \right) \tag{6.5}$$

This set of equations is of the form

$$\mathbf{b} = \mathbf{A}\mathbf{x} \tag{6.6}$$

which can easily be solved by standard algebra, i.e.

$$\mathbf{x} = \mathbf{A}^{-1}\mathbf{b} \tag{6.7}$$

We note that in order to solve the general localization problem with Equation 6.7 we need at least five hydrophones. If there are more hydrophones available, the problem is over-determined and suited for a least-mean-square (LMS) solution:

$$\mathbf{x} = (\mathbf{A}^T\mathbf{A})^{-1}\mathbf{A}^T\mathbf{b} \tag{6.8}$$

Using a LMS solution improves the reliability of the solution, as small errors in measuring the time delays are reduced further.

The matrix inversion indicated above requires that the matrix is not singular or ill-conditioned, that is, the determinant of the matrix should not be close to zero. In particular, it is important that not all sensors are at the same depth, otherwise the last column of matrix **a** will become identical and zero. More generally, no column of matrix **a** should be zero, that is, the five hydrophones should form a strict volumetric array.

In the case that all sensors are at the same depth, the above equation cannot be used to estimate directly the whale depth z_w and the set of equations reduces to

$$\left(\begin{array}{l} (x_1{}^2 - x_0{}^2) + (y_1{}^2 - y_0{}^2) - (\delta R_1)^2 \\ (x_2{}^2 - x_0{}^2) + (y_2{}^2 - y_0{}^2) - (\delta R_2)^2 \\ (x_3{}^2 - x_0{}^2) + (y_3{}^2 - y_0{}^2) - (\delta R_3)^2 \end{array} \right)$$
$$= 2 \left(\begin{array}{lll} \delta R_1 & (x_1 - x_0) & (y_1 - y_0) \\ \delta R_2 & (x_2 - x_0) & (y_2 - y_0) \\ \delta R_3 & (x_3 - x_0) & (y_3 - y_0) \end{array} \right) \left(\begin{array}{l} R_0 \\ x_w \\ y_w \end{array} \right) \tag{6.9}$$

where we conveniently reduced the number of necessary hydrophones to 4.

We finally estimate the whale depth z_w by

$$z_w = z_0 \pm \sqrt{R_0^2 - (x_w - x_0)^2 - (y_w - y_0)^2} \tag{6.10}$$

As is to be expected, the whale depth may be found either above or below the depths of the hydrophones and the solution is therefore not unique. In cases where the hydrophone array is close to the bottom or to the surface, one of the two solutions is not realistic and the procedure results in a *de facto* unique solution.

Although the approach described in Equation 6.5 is well suited for five or more hydrophones because it is mathematically easily implemented, the question arises, why we need four delays to estimate the position of the whale, which is completely determined by three unknown parameters, as R_0 is not an independent variable and may be estimated by using (x_w, y_w, z_w) and (x_0, y_0, z_0).

Having four hydrophones with at least one at a different depth, one may obtain another solution than given by Equation 6.5 by not solving for R_0, that is, by writing the three equations that use hydrophones $h_0 - h_3$ as

$$
\begin{pmatrix} (x_1{}^2 - x_0{}^2) + (y_1{}^2 - y_0{}^2) + (z_1{}^2 - z_0{}^2) - (\delta R_1)^2 \\ (x_2{}^2 - x_0{}^2) + (y_2{}^2 - y_0{}^2) + (z_2{}^2 - z_0{}^2) - (\delta R_2)^2 \\ (x_3{}^2 - x_0{}^2) + (y_3{}^2 - y_0{}^2) + (z_3{}^2 - z_0{}^2) - (\delta R_3)^2 \end{pmatrix} - 2 \begin{pmatrix} \delta R_1 \\ \delta R_2 \\ \delta R_3 \end{pmatrix} R_0
$$

$$
= 2 \begin{pmatrix} (x_1 - x_0) & (y_1 - y_0) & (z_1 - z_0) \\ (x_2 - x_0) & (y_2 - y_0) & (z_2 - z_0) \\ (x_3 - x_0) & (y_3 - y_0) & (z_3 - z_0) \end{pmatrix} \begin{pmatrix} x_w \\ y_w \\ z_w \end{pmatrix}
$$

$$(6.11)$$

which may be written in vector notation as

$$\mathbf{b}_0 - \mathbf{b}_1 R_0 = \mathbf{A}\mathbf{u}_w \tag{6.12}$$

where

$$\mathbf{b}_0 = \begin{pmatrix} (x_1{}^2 - x_0{}^2) + (y_1{}^2 - y_0{}^2) + (z_1{}^2 - z_0{}^2) - (\delta R_1)^2 \\ (x_2{}^2 - x_0{}^2) + (y_2{}^2 - y_0{}^2) + (z_2{}^2 - z_0{}^2) - (\delta R_2)^2 \\ (x_3{}^2 - x_0{}^2) + (y_3{}^2 - y_0{}^2) + (z_3{}^2 - z_0{}^2) - (\delta R_3)^2 \end{pmatrix} \tag{6.13}$$

$$\mathbf{b}_1 = 2 \begin{pmatrix} \delta R_1 \\ \delta R_2 \\ \delta R_3 \end{pmatrix} \tag{6.14}$$

$$\mathbf{A} = 2 \begin{pmatrix} (x_1 - x_0) & (y_1 - y_0) & (z_1 - z_0) \\ (x_2 - x_0) & (y_2 - y_0) & (z_2 - z_0) \\ (x_3 - x_0) & (y_3 - y_0) & (z_3 - z_0) \end{pmatrix} \tag{6.15}$$

and $\mathbf{u}_w$ is the vector describing the whale position.

Using $\mathbf{u}_0 = \mathbf{A}^{-1}\mathbf{b}_0$ and $\mathbf{u}_1 = \mathbf{A}^{-1}\mathbf{b}_1$, then the solution to Equation 6.12 becomes a linear function of reference range R_0, which remains an unknown parameter

$$\mathbf{u}_w = \mathbf{u}_0 - \mathbf{u}_1 R_0 \tag{6.16}$$

Recalling, however, that the range R_0 is also determined by

$$R_0{}^2 = (\mathbf{u}_w - \mathbf{h}_0)^T (\mathbf{u}_w - \mathbf{h}_0) \tag{6.17}$$

where $\mathbf{h}_0$ is the location of the reference hydrophone, we obtain with Equation 6.16

$$R_0{}^2 = (\mathbf{u}_0 - \mathbf{h}_0 - \mathbf{u}_1 R_0)^T (\mathbf{u}_0 - \mathbf{h}_0 - \mathbf{u}_1 R_0) \tag{6.18}$$

which after some rearrangements becomes a quadratic equation in R_0

$$R_0{}^2 (1 - \mathbf{u}_1{}^T \mathbf{u}_1) + 2(\mathbf{u}_0 - \mathbf{h}_0)^T \mathbf{u}_1 R_0 - (\mathbf{u}_0 - \mathbf{h}_0)^T (\mathbf{u}_0 - \mathbf{h}_0) = 0 \tag{6.19}$$

having the solution

$$R_0 = \frac{-b \pm \sqrt{b^2 + ac}}{a} \tag{6.20}$$

with

$$a = (1 - \mathbf{u}_1{}^T \mathbf{u}_1) \tag{6.21}$$

$$b = (\mathbf{u}_0 - \mathbf{h}_0)^T \mathbf{u}_1 \tag{6.22}$$

$$c = (\mathbf{u}_0 - \mathbf{h}_0)^T (\mathbf{u}_0 - \mathbf{h}_0) \tag{6.23}$$

The plus and minus signs in Equation 6.20 indicate that the solution is not always unique, but for $a > 0$ the ambiguity is resolved, as only the plus sign gives positive ranges.

Estimating the range with four hydrophones, one always obtains two solutions (Equation 6.20) and only for a limited subset of whale locations may these two solutions be reduced, resulting in the correct location of the whale. It is therefore advisable always to use five hydrophones and to apply Equation 6.8 for unique whale ranging. Alternatively, if there is external information, or if the range estimation is part of an integral tracking effort where the whale moves from an unambiguous to an ambiguous region, then the potential range ambiguity of Equation 6.20 may not be an issue any more.

For $\mathbf{u}_0 = \mathbf{h}_0$ the range R_0 is undefined, as Equation 6.19 reduces to $R_0{}^2(1 - \mathbf{u}_1{}^T \mathbf{u}_1) = 0$, indicating that either R_0 is zero or $(1 - \mathbf{u}_1{}^T \mathbf{u}_1) = 0$ for all ranges R_0. It follows then that for $\mathbf{u}_0 = \mathbf{h}_0$ the vector $\mathbf{u}_1$ becomes the unit vector pointing from the reference hydrophone to the whale, that is: $\mathbf{u}_1{}^T \mathbf{u}_1 = 1$.

For $(1 - \mathbf{u}_1{}^T \mathbf{u}_1) = 0$ and $\mathbf{u}_0 \neq \mathbf{h}_0$ the quadratic equation (Equation 6.19) becomes formally a linear equation with the solution

$$R_0 = \frac{(\mathbf{u}_0 - \mathbf{h}_0)^T (\mathbf{u}_0 - \mathbf{h}_0)}{2(\mathbf{u}_0 - \mathbf{h}_0)^T \mathbf{u}_1} \tag{6.24}$$

Matlab code to test multi-hydrophone ranging

```
%Scr6_1
% Geometry
h0=[0,0,0.5]';
h1=[2,0,0.5]';
h2=[0,2,0.5]';
```

```
h3=[0,0,-0.5]';
%
% Simulation
% whale position
w=[1,5,1]';
% true acoustic distances
R0=sqrt((w-h0)'*(w-h0));
R1=sqrt((w-h1)'*(w-h1));
R2=sqrt((w-h2)'*(w-h2));
R3=sqrt((w-h3)'*(w-h3));
%
% true acoustic differences
dR1=R1-R0;
dR2=R2-R0;
dR3=R3-R0;
%
Rs=R0; %only to compare the result
%
%Range estimation ---------------------
%
A=2*[(h1-h0)'
      (h2-h0)'
      (h3-h0)'];
%
Ainv=inv(A);
b0=[h1'*h1-dR1.^2;
    h2'*h2-dR2.^2;
    h3'*h3-dR3.^2]-h0'*h0;
%
b1=2*[dR1;dR2;dR3];
x0=Ainv*b0;
x1=Ainv*b1;
%
rr1=-(x0-h0)'*x1;
rr2=rr1*rr1+(1-x1'*x1)*(x0-h0)'*(x0-h0);
rr3=(1-x1'*x1);
R0=(rr1+sqrt(rr2))/rr3;
fprintf('%f %f\n',R0,R0-Rs)
```

6.2   Triangulation

A classical approach to estimate the range of a sound source by passive methods is triangulation, where one measures from different positions the directions to the sound source and estimates the location where these directions cross (Figure 6.1).

Given two hydrophone positions h_1, h_2 and two measured sound directions described by bearing γ_1, γ_2, and elevation angle β_1, β_2, then each sound vector may be expressed as

$$w_i = h_i + m_i R_i \tag{6.25}$$

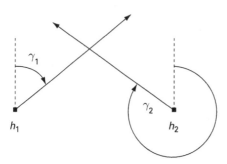

Figure 6.1   Triangulation with two vectors (presented in horizontal plane).

where

$$h_i = \begin{pmatrix} x_i \\ y_i \\ z_i \end{pmatrix} \tag{6.26}$$

and the direction m_i is given by

$$m_i = \begin{pmatrix} \sin \gamma_i \cos \beta_i \\ \cos \gamma_i \cos \beta_i \\ \sin \beta_i \end{pmatrix} \tag{6.27}$$

The intersection of the two vectors defines the whale position. In particular, we have

$$h_2 - h_1 + \begin{pmatrix} \sin \gamma_2 \cos \beta_2 \\ \cos \gamma_2 \cos \beta_2 \\ \sin \beta_2 \end{pmatrix} R_2 - \begin{pmatrix} \sin \gamma_1 \cos \beta_1 \\ \cos \gamma_1 \cos \beta_1 \\ \sin \beta_1 \end{pmatrix} R_1 = 0 \tag{6.28}$$

and by multiplying the equation appropriately for the x and y components we obtain

$$\begin{pmatrix} (x_2 - x_1) \cos \gamma_1 \\ (y_2 - y_1) \sin \gamma_1 \end{pmatrix}$$
$$+ \begin{pmatrix} \cos \gamma_1 \sin \gamma_2 \cos \beta_2 \\ \sin \gamma_1 \cos \gamma_2 \cos \beta_2 \end{pmatrix} R_2 - \begin{pmatrix} \cos \gamma_1 \sin \gamma_1 \cos \beta_1 \\ \sin \gamma_1 \cos \gamma_1 \cos \beta_1 \end{pmatrix} R_1 = 0 \tag{6.29}$$

After subtracting the two equations we eliminate the coefficient of R_1 and obtain the range estimate R_2

$$R_2 = \frac{(x_2 - x_1) \cos \gamma_1 - (y_2 - y_1) \sin \gamma_1}{\sin(\gamma_1 - \gamma_2) \cos \beta_2} \tag{6.30}$$

The whale position is then estimated by using one of the original vector equations

$$w = h_2 + m_2 R_2 \tag{6.31}$$

Here we estimated the whale position as seen from the second sensor. In theory the same solution should be found by solving for R_1. In practice, one will realize that this alternative approach may lead to slightly different whale locations. This is mainly due to slightly inconsistent estimation of the 3-D directions (Equation 6.27). Consequently, it seems better to estimate the range by using only the horizontal components of the direction vectors to ensure the existence of a solution, and also because two directional vectors that are based on real measurements might not cross in a three-dimensional space. The method is mathematically safe, as long there is a significant horizontal component in the direction from one hydrophone to the whale. If the whale is nearly above or below one hydrophone, then one could use the other hydrophone as the reference point for the triangulation.

The above approach assumes that the whale is stationary while the two directions are estimated. This is always the case when the measurements are made with an array of distributed hydrophones. However, in cases where a single moving direction finder makes its measurements at different times, this technique could still be useful if the speed of the sensor is significantly faster than the speed of the whale, that is, the whale could be considered as nearly stationary.

The triangulation requires the PAM system to be able to measure the angles describing the direction from where the sound is coming. This may be done either with directional hydrophones or with closely spaced multiple hydrophones, as we will see later in this chapter.

6.3 Multi-path ranging

Multi-hydrophone ranging requires an increased number of hydrophones covering the area of interest. Such configurations are expensive to implement and therefore rare, even though they do exist, especially in the context of military test ranges. As an alternative, multi-path ranging tries to exploit the complexity of underwater sound propagation to localize the origin of the sound emission.

Figure 6.2 sketches the principle behind multi-path ranging. It indicates on the left a hydrophone at depth h and on the right we have the sound source at depth d. The horizontal distance between hydrophone and sound source is denoted as x, and the bottom depth is b. The depth values are all negative numbers to allow a right-handed co-ordinate system, where the z-axis is upward. The dotted lines are the unfolded reflected sound paths, where R_x indicates the length of the sound path, R_0 direct arrival, R_S surface-reflected sound path, R_B bottom-reflected sound path, R_{BS} bottom–surface-reflected sound path, and R_{SB} surface–bottom-reflected sound path.

In Figure 6.2, all acoustic sound paths are presented as reflected and as unfolded paths. The unfolding is especially useful, as it not only helps to develop the different path length formulas, but also explains the basic geometric concept behind this multi-path ranging. It should be clear from Figure 6.2 that all reflected sound paths may be treated as recordings by an array of virtual hydrophones that are found exactly above or below the real hydrophone. In other words, the multiple arrivals on the real hydrophone are assumed to arrive at exactly the same times as they would arrive on the constructed vertical array. This approach holds only if the sound propagation is strictly radial, i.e. for spherical

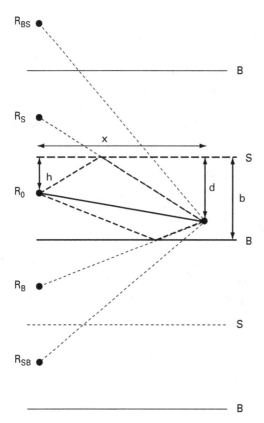

Figure 6.2   Multi-path geometry. Thick horizontal dashed line, sea surface; thick horizontal solid line, bottom; thin horizontal dashed and solid lines, reflected sea surfaces and bottoms.

waves and therefore constant sound speed and when the sea boundaries are parallel and flat.

As multi-path ranging is equivalent to the use of a vertical array, it also suffers from the drawback of vertical arrays, namely that it provides a rotationally ambiguous solution, i.e. pure multi-path ranging provides the range and depth of the sound source but not the horizontal direction or bearing.

Using straightforward geometry, we obtain a set of equations

$$R_0{}^2 = x^2 + (h - d)^2 \tag{6.32}$$

$$R_S{}^2 = x^2 + (h + d)^2 \tag{6.33}$$

$$R_B{}^2 = x^2 + ((b - d) + (b - d))^2 = x^2 + (2b - (h + d))^2 \tag{6.34}$$

$$R_{BS}{}^2 = x^2 + ((b - d) + b + h)^2 = x^2 + (2b + (h - d))^2 \tag{6.35}$$

$$R_{SB}{}^2 = x^2 + (d + b + (b - h))^2 = x^2 + (2b - (h - d))^2 \tag{6.36}$$

We could continue to add more reflection; for example, for surface–bottom–surface reflection sound paths we obtain

$$R_{SBS}{}^2 = x^2 + (2b + (h + d))^2 \tag{6.37}$$

Forming differences between different range estimates, we have

$$R_S{}^2 - R_0{}^2 = 4hd \tag{6.38}$$

$$
\begin{aligned}
R_B{}^2 - R_S{}^2 &= 4b^2 - 4b(h + d) \\
&= 4b(b - (h + d))
\end{aligned} \tag{6.39}
$$

$$
\begin{aligned}
R_B{}^2 - R_0{}^2 &= 4b^2 - 4b(h + d) + 4hd \\
&= 4(b - d)(b - h)
\end{aligned} \tag{6.40}
$$

Obviously there is a variety of other 'natural' differences

$$R_{BS}{}^2 - R_{SB}{}^2 = 8b(h - d) \tag{6.41}$$

$$R_{SBS}{}^2 - R_B{}^2 = 8b(h + d) \tag{6.42}$$

$$
\begin{aligned}
R_{BS}{}^2 - R_B{}^2 &= 4b(h - d) + 4b(h + d) - 4hd \\
&= 8bh - 4hd
\end{aligned} \tag{6.43}
$$

$$
\begin{aligned}
R_{SB}{}^2 - R_B{}^2 &= -4b(h - d) + 4b(h + d) - 4hd \\
&= 8bd - 4hd
\end{aligned} \tag{6.44}
$$

All these differences may be considered independent measurements and suited for estimating range and depth of the sound source.

6.3.1 Sound source range estimation

Combining equations for direct and surface reflected paths, that is, after having measured the arrival path difference δR_{S_0} or equivalently the time delay between the direct and the surface reflected sound path, Equation 6.38 yields

$$2(\delta R_{S\_0})R_0 + (\delta R_{S\_0})^2 = 4hd \tag{6.45}$$

and the slant range R_0 becomes

$$R_0 = \frac{4hd - (\delta R_{S\_0})^2}{2(\delta R_{S\_0})} \tag{6.46}$$

This solution requires the whale depth d to be known. Otherwise, the slant range R_0 results, for a given time delay, in a linear function of the animal's depth.

6.3.2   Sound source depth estimation

To estimate both the slant range R_0 and the depth d of the animal, we have to use another dataset. A next natural candidate is based on the bottom reflection, in particular

$$R_B{}^2 - R_0{}^2 = 4(b-d)(b-h)$$

For two unknowns R_0 and d we have now two equations

$$2(\delta R_{S\_0})R_0 + (\delta R_{S\_0})^2 = 4hd$$
$$2(\delta R_{B\_0})R_0 + (\delta R_{B\_0})^2 = 4(b-d)(b-h)$$
(6.47)

which in matrix notation becomes

$$2\begin{pmatrix} 2h & -\delta R_{S\_0} \\ 2(b-h) & -\delta R_{B\_0} \end{pmatrix}\begin{pmatrix} d \\ R_0 \end{pmatrix} = \begin{pmatrix} (\delta R_{S\_0})^2 \\ 4b(b-h) + (\delta R_{B\_0})^2 \end{pmatrix}$$
(6.48)

and which may be solved in classical way for both the slant range R_0 and the whale depth d.

This solution requires bottom depth b to be known. This information may be obtained from nautical charts, or be measured *in situ* with an echo-sounder. To estimate bottom depth also from the data we need a further equation, that is, additional reflection delay measurements. For example, one could select the following three linear equations, which exhibit no cross terms in the unknowns d and b:

$$R_S{}^2 - R_0{}^2 = 4hd$$
(6.49)

$$R_{BS}{}^2 - R_{SB}{}^2 + R_{SBS}{}^2 - R_B{}^2 = 16bh$$
(6.50)

$$R_{BS}{}^2 - R_B{}^2 = 8bh - 4hd$$
(6.51)

which after some manipulations are written in matrix notation

$$2\begin{pmatrix} 2h & 0 & -\delta R_{S\_0} \\ 0 & 8h & -\delta R_{BS\_SB} - \delta R_{SBS\_B} \\ -2h & 4h & -\delta R_{BS\_B} \end{pmatrix}\begin{pmatrix} d \\ b \\ R_0 \end{pmatrix} =$$
$$\begin{pmatrix} (\delta R_{S\_0})^2 \\ (\delta R_{BS\_SB})[\delta R_{BS\_SB} + 2\delta R_{SB\_0}] + (\delta R_{SBS\_B})[\delta R_{SBS\_B} + 2\delta R_{B\_0}] \\ (\delta R_{BS\_B})[\delta R_{BS\_B} + 2\delta R_{B\_0}] \end{pmatrix}$$
(6.52)

This set of linear equations may again be solved by standard methods. Equation 6.52 requires six different delay measurements, which may be available on only a few occasions, limiting its usefulness somewhat.

Multi-path ranging requires (a) the existence and (b) the identification of the individual sound arrivals. As multi-paths are primarily due to reflections of the sound on the boundary (sea surface and sea bottom), the existence of detectable multi-path arrivals depends heavily on the overall geometry. Significant reflections are to be expected if the sound energy of the reflected sound arrivals is comparable to the sound energy of the direct sound arrivals. For omni-directional sound emissions, this may happen if the sound source is in the vicinity of a sea boundary. For directional sounds, e.g. echolocation clicks, where the sound energy is focused in one direction, reflections occur regularly if the animal is oriented to the receiver and at the same time is pointing the sound beam towards the boundary, sea surface or bottom. Identification of multiple sea-boundary reflections are, in general, difficult but are facilitated by additional knowledge. A deep-diving whale that is known to forage close to the bottom is expected to generate early bottom reflections, especially when the distance to the bottom is less than the depth of the hydrophone. Sounds that are typically generated close to the surface result in early surface reflections. In principle, it should be possible to differentiate the first surface reflection from the first bottom reflection, because the former is characterized by a 180° phase shift of the signal due to reflection on a water–air interface, as discussed in Section 1.4.2. The correct identification of higher-order reflections, however, may become very difficult and may require additional information that, for example, could come from modelling of animal behaviour.

6.3.3 Sound source depth estimation using angle of arrival

An alternative method for estimating the depth d without using additional reflections is to measure the elevation angle of the direct arrival (Figure 6.3).

Given the elevation angle ϑ_0 of the direct path, implicitly defined by

$$(d - h) = R_0 \sin \vartheta_0 \tag{6.53}$$

and using Equation 6.45, one obtains two linear equations in d and R_0

$$d - h = R_0 \sin \vartheta \tag{6.54}$$

$$2(\delta R_{S\_0}) R_0 + (\delta R_{S\_0})^2 = 4hd \tag{6.55}$$

that is, we have the following set of linear equations

$$\begin{pmatrix} 1 & -\sin \vartheta_0 \\ 4h & -2(\delta R_{S\_0}) \end{pmatrix} \begin{pmatrix} d \\ R_0 \end{pmatrix} = \begin{pmatrix} h \\ (\delta R_{S\_0})^2 \end{pmatrix} \tag{6.56}$$

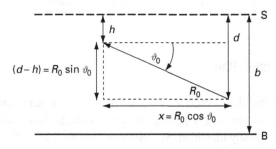

Figure 6.3 Geometry and definition of elevation angle ϑ, which in this figure is negative.

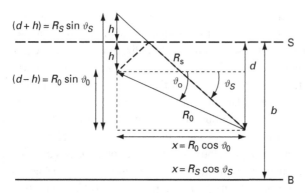

Figure 6.4   Geometry and definition of the arrival angles of direct and surface-reflected sound paths.

which may be solved by standard methods. One surface delay and one angle measurement allow complete localization of the sound source, even without knowledge of the bottom depth, making this approach a favourite for multi-path ranging.

6.3.4   Sound source depth estimation using only angles of arrival

Instead of the range difference of surface-reflected and direct sound paths, we could also use the two elevation angles.

From Figure 6.4 we conclude

$$(d - h) = R_x \tan \vartheta_0 \tag{6.57}$$

$$(d + h) = R_x \tan \vartheta_S \tag{6.58}$$

or

$$\left(\frac{1 - \tan \vartheta_0}{1 - \tan \vartheta_S} \right) \left(\frac{d}{R_x} \right) = h \left(\begin{array}{c} 1 \\ -1 \end{array} \right) \tag{6.59}$$

which may be solved by standard methods. This approach requires the capability to estimate the arrival angle of both the direct and the surface-reflected paths; this should be possible, as the surface-reflected path is trailing the direct arrival and the signal appears phase-reversed.

6.4   Ranging by acoustic modelling

Multi-path ranging as described in Section 6.3 may suffer from two assumptions: that the sound speed is constant so that sound propagates along straight lines; and that the different sound arrivals are easily identified. Model-based ranging relaxes these requirements and obtains the whale's location in a different way.

Model-based ranging techniques assume a whale's location and predict the different sound path differences one should measure for that particular geometry. The predicted measurements are then compared with the actual measurements and the prediction–measurement mismatch is used to build an ambiguity surface showing the most likely whale location by repeating the procedure for multiple hypothetical whale locations. The most likely whale location is the one where the mismatch between modelled and measured sound field is minimal.

To estimate whale location by acoustic modelling one uses standard acoustic models to predict the arrival times. A suitable acoustic model is Bellhop, the propagation loss model presented earlier in Section 3.3.1. It predicts the travel times for multiple acoustic sound paths and can account for depth-dependent sound speed profiles and range-dependent bathymetry.

The use of acoustic models to predict the temporal distribution of arriving sound energy is especially important for extended distances, where sound paths are in general refracted and only few true reflected sound paths may be identified.

Compared with the ranging methods presented so far, the model-based ranging may improve the solution by adding more information. A potential difficulty is the interpretation of a complex solution space, the ambiguity surface, to determine the true location of the animal, especially in connection with uncertain knowledge of the environmental input parameters needed for acoustic modelling. So far, whale ranging based on acoustic models is still in its infancy and requires the participation of acoustic modelling experts (e.g. Baggeroer *et al.*, 1988, 1993; Tiemann, 2008; Tiemann and Porter, 2003; Tiemann *et al.*, 2006; Thode *et al.*, 2000).

6.5 Measurement of arrival time difference

So far, we have only stated that we need a measurement of the arrival time difference between two hydrophones. One method to determine this difference is to measure precisely the times at which the same click arrives at the two hydrophones and to calculate the difference. A similar approach was used in Section 5.3.2 to determine the delay between the direct and surface-reflected arrival of echolocation clicks. This method is suitable if the signals are strong enough to be detected easily and to allow a common definition of the arrival time, whether this is the onset of the signal or the time at which the signal amplitude becomes maximal.

Another method to determine the arrival time difference cross-correlates the received sounds of the two hydrophones and determines the relative delay by finding the lag time where the cross-correlation becomes maximal. The cross-correlation function is given by the integral

$$C(\tau) = \int\limits_{-\infty}^{\infty} s_0(t)s_1(t+\tau)\mathrm{d}t \tag{6.60}$$

where s_0 and s_1 denote the time series of hydrophone h_0 and h_1, respectively, and τ is the variable lag time between the two time series. $C(\tau)$ becomes maximal only if τ corresponds to the delay between signals s_1 and s_0.

Time delay estimation by means of cross-correlation works best if the signals have distinctive temporal or spectral features. These features may be sharp onset of the signal or significant amplitude modulation, but also strong non-linear frequency modulation. The similarity of Equation 6.60 to the matched filter (Equation 4.22) emphasizes that a high time–bandwidth product may be very useful to overcome noise-induced uncertainties.

Cross-correlation for time delay estimation may be very time-consuming, as without *a priori* information on the expected delay, Equation 6.60 has to be evaluated for all possible delays, the number of which depends on the sampling frequency and the distance between the hydrophone pair. This may be a problem for very high sampling frequency and very widely spaced hydrophones. If we take, for example, a sampling rate of 500 kHz to cover all possible cetacean sounds and use a towed array with two hydrophones spaced at, say 15 m, then the maximum travel time between the two hydrophones is 10 ms (15 [m] / 1500 [m/s]). The amount of possible delays is 5000 (500 [kHz] × 10 [ms]). Considering that the delays may be positive or negative, the number of delays increases to 10 000. This may be too much processing, even for the shortest click of, say, 25 µs duration, which is equivalent to a sampled signal length of 125 samples. For typical signals of, say, 1 ms to 1 s, the computational effort may be prohibitive.

Fortunately, a brute force cross-correlation generating all possible lag times is not always necessary, especially if the cross-correlation is considered as a refinement in a multi-step direction finding process. One could first determine the approximate delay by forming the time differences of individual detections, and in a second step use the cross-correlation to refine the delay estimate. This would significantly decrease the computational effort.

When determining the time delay in the presence of multiple clicks one may experience some complication if the expected time delay between two hydrophones is greater than the ICI of an echolocation click train. Fortunately, the ICI is never really constant, allowing a rough synchronization by matching the variations in the ICI for both hydrophone measurements.

6.6   Direction finding

The simplest method of finding the direction from which the sound is coming is to use a pair of hydrophones. Depending on the hydrophone separation and the distance of the sound source, we may obtain slightly different formulas for the estimation of the direction of arrival.

Given two hydrophones, which are receiving sound from a distant sound source as shown in Figure 6.5, then standard geometry yields:

$$R_1{}^2 = R_0{}^2 + L^2 + 2R_0 L \cos\beta \tag{6.61}$$

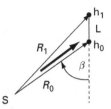

Figure 6.5 Estimation of sound arrival angle. (S is the sound source and h_1, h_2 denote the two hydrophones.

If we denote the sound path difference $\delta R_{1\_0} = R_1 - R_0$ then we obtain for the arrival angle β at hydrophone h_0

$$\cos\beta = \frac{\delta R_{1\_0}}{L} + \frac{L^2 - (\delta R_{1\_0})^2}{2R_0 L} \tag{6.62}$$

which for very large ranges, i.e. $R_0 \to \infty$ and using the relation between sound path difference and travel time difference $\delta R_{1\_0} = c\delta\tau_{1\_0}$ becomes:

$$\cos\beta = \frac{c\delta\tau_{1\_0}}{L} \tag{6.63}$$

For distant sound sources, two closely spaced hydrophones are sufficient to estimate the direction from which the sound arrives.

One may also note from Figure 6.5 that there is a rotational symmetry and one would obtain the same arrival angle β if the whole figure were flipped to the right. It should be further obvious that $R_0 \to \infty$ is equivalent to the assumption that the sound arrival angles become the same for both hydrophones, i.e. the sound may be described as a plane wave.

6.7 Three-dimensional direction finding

To eliminate all potential directional ambiguities, we need a three-dimensional direction finder. Such a device may be constructed with four hydrophones in a three-dimensional or volumetric configuration. Figure 6.6 shows a volumetric array where the four hydrophones are arranged as a tetrahedron, providing the smallest possible volumetric array.

In Section 6.1, we noted that four hydrophones are in general not sufficient to localize unambiguously a sound source, but the number is sufficient to estimate the direction to a distant sound source. Similarly to the derivation of Equation 6.63, we assume that the volumetric array is small enough to justify a plane wave assumption, that is, the direction vector to the sound source is equal for all four hydrophones.

Given four hydrophones

$$\mathbf{h}_i = \begin{pmatrix} h_{ix} \\ h_{iy} \\ h_{iz} \end{pmatrix} \quad i = 0, \cdots, 3 \tag{6.64}$$

we use three pairs of hydrophones to construct three hydrophone baseline vectors

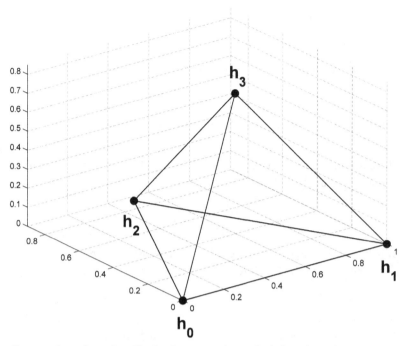

Figure 6.6   Compact three-dimensional hydrophone array in tetrahedral configuration.

$$\mathbf{d}_i = \mathbf{h}_i - \mathbf{h}_0, \quad i = 1, \cdots, 3$$

Further, we define the distance between two hydrophones $L_i = |\mathbf{d}_i|$.

To form a volumetric array we assume that the three baseline vectors are not collinear, that is, the absolute value of the dot product between two of the three vectors is not one, and the three vectors $\mathbf{d}_i$ may therefore be used to form a co-ordinate system.

The angle ϑ_i between a direction (unit) vector $\hat{\mathbf{s}}$ and the three hydrophone pairs is implicitly defined by

$$
\begin{aligned}
L_1 \cos \vartheta_1 &= \mathbf{d}_1{}^T \hat{\mathbf{s}} \\
L_2 \cos \vartheta_2 &= \mathbf{d}_2{}^T \hat{\mathbf{s}} \\
L_3 \cos \vartheta_3 &= \mathbf{d}_3{}^T \hat{\mathbf{s}}
\end{aligned}
\tag{6.65}
$$

or in matrix notation

$$
\begin{pmatrix} L_1 \cos \vartheta_1 \\ L_2 \cos \vartheta_2 \\ L_3 \cos \vartheta_3 \end{pmatrix} = \begin{pmatrix} d_{1x} & d_{1y} & d_{1z} \\ d_{2x} & d_{2y} & d_{2z} \\ d_{3x} & d_{3y} & d_{3z} \end{pmatrix} \begin{pmatrix} s_x \\ s_y \\ s_z \end{pmatrix}
\tag{6.66}
$$

Using Equation 6.63, we may replace the individual components of Equation 6.65 by

$$L_i \cos \vartheta_i = c\delta\tau_{i\_0} \tag{6.67}$$

and Equation 6.66 becomes in matrix notation

$$c(\delta\boldsymbol{\tau}) = \mathbf{D}^T\hat{\mathbf{s}} \tag{6.68}$$

where

$$\delta\boldsymbol{\tau} = \begin{pmatrix} \delta\tau_{1\_0} \\ \delta\tau_{2\_0} \\ \delta\tau_{3\_0} \end{pmatrix}, \mathbf{d}_i = \begin{pmatrix} d_{ix} \\ d_{iy} \\ d_{iz} \end{pmatrix}, \mathbf{D}^T = \begin{pmatrix} \mathbf{d}_1{}^T \\ \mathbf{d}_2{}^T \\ \mathbf{d}_3{}^T \end{pmatrix} \tag{6.69}$$

To be more general, we allow now an arbitrary rotation of hydrophone vectors, that is, the actual orientation of the hydrophone vectors $\mathbf{d}_i$ may be derived from a default orientation $\mathbf{d}_{0i}$ according to

$$\mathbf{d}_i = \mathbf{M}\mathbf{d}_{0i} \tag{6.70}$$

which in matrix notation becomes

$$\mathbf{D} = \mathbf{M}\mathbf{D}_0 \tag{6.71}$$

then we obtain by transposing properly the matrix product

$$c(\delta\boldsymbol{\tau}) = \left(\mathbf{D}_0^T\mathbf{M}^T\right)\hat{\mathbf{s}} \tag{6.72}$$

The sound direction is now estimated by standard algebra

$$\hat{\mathbf{s}} = c\left(\mathbf{D}_0^T\mathbf{M}^T\right)^{-1}(\delta\boldsymbol{\tau}) \tag{6.73}$$

and azimuth φ and elevation ϑ of the sound direction are finally estimated by solving

$$\tan\varphi = \frac{s_y}{s_x} \tag{6.74}$$

and

$$\sin\vartheta = s_z \tag{6.75}$$

or alternatively

$$\tan\vartheta = \frac{s_z}{\sqrt{s_x^2 + s_y^2}} \tag{6.76}$$

6.7.1   Rotation matrix

The inclusion of the rotation of the volumetric array into the three-dimensional direction finding technique allows the use of arbitrarily oriented volumetric arrays within a standard geographic reference system, where we define three fundamental rotations (Figure 6.7).

$\mathbf{T}_\alpha$: rotate around x-axis so that y-axis moves towards z-axis (roll)
$\mathbf{T}_\beta$: rotate around y-axis so that z-axis moves towards x-axis ($-$pitch)
$\mathbf{T}_\gamma$: rotate around z-axis so that x-axis moves towards y-axis ($90° - $heading)

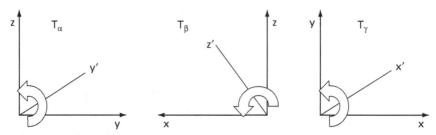

Figure 6.7   Definition of the rotation matrix.

A transformation from the (x, y, z) co-ordinate system to an (x', y', z') co-ordinate system may be considered as a sequence of the three fundamental rotations. The actual sequence of the three rotations is of importance, but one usually follows the following sequence

(1) rotate by γ so that the x-axis is in the vertical plane of x'
(2) rotate by β so that the x-axis is parallel to the x'-axis
(3) rotate by α so that the z-axis is parallel to the z'-axis

or in matrix notation

$$\mathbf{M} = \mathbf{T}_\alpha \mathbf{T}_\beta \mathbf{T}_\gamma \tag{6.77}$$

6.8   Two-dimensional constrained direction finding

Assuming next that only three hydrophones are available, so that one may not be able to form a vector $\mathbf{d}_3$, then the following procedure would still allow three-dimensional direction finding with some constraints. Three hydrophones always describe a plane, and there will be a residual ambiguity for finding a direction as one cannot decide on which side of the plane the correct sound direction should be.

Figure 6.8 may be used to define different angles between the vector towards the source S and the vector connecting the different hydrophones. Only hydrophones h_0 and h_1 are shown. Similar angles exist between source direction and hydrophones h_0 and h_2. By assumption, direction finding is done using hydrophones h_0, h_1 and h_2, and hydrophone h_3 is not available.

Following Figure 6.8, we express the directional cosines as

$$\cos \beta_i = \cos \gamma_i \cos \vartheta \quad \text{for } i = 1, 2 \tag{6.78}$$

where

$$\gamma_1 + \gamma_2 = \omega_{1,2} \tag{6.79}$$

with $\omega_{1,2}$ being the angle between the two vectors d_1, d_2.

In particular, we may write for both directional cosines

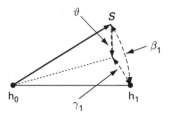

Figure 6.8 Definitions of angles for two-dimensional constrained direction finding.

$$\cos\beta_1 = \cos\gamma_1 \cos\vartheta \tag{6.80}$$

$$\cos\beta_2 = \cos\gamma_2 \cos\vartheta \tag{6.81}$$

or with Equation 6.79

$$\cos\beta_2 = \cos(\omega_{1,2} - \gamma_1) \cos\vartheta \tag{6.82}$$

and therefore

$$\cos\beta_2 = (\cos\gamma_1 \cos\omega_{1,2} + \sin\gamma_1 \sin\omega_{1,2}) \cos\vartheta \tag{6.83}$$

Using Equation 6.80 we obtain

$$\cos\beta_2 = \cos\beta_1 \cos\omega_{1,2} + \cos\beta_1 \frac{\sin\gamma_1}{\cos\gamma_1} \sin\omega_{1,2} \tag{6.84}$$

that is

$$\tan\gamma_1 = \frac{\cos\beta_2 - \cos\beta_1 \cos\omega_{1,2}}{\cos\beta_1 \sin\omega_{1,2}} \tag{6.85}$$

Having estimated γ_1, or by virtue of Equation 6.79 also γ_2, we may use either Equation 6.80 or Equation 6.81 to estimate the elevation angle ϑ. One would use Equation 6.80 if γ_1 is less than 45°, and Equation 6.81 otherwise.

There is an ambiguity as ϑ may be positive or negative, that is, above or below the hydrophone plane.

In order to use the set of linear equations (6.72) to obtain the desired sound source direction vector we would construct a third hydrophone vector by

$$\mathbf{d}_3 = \mathbf{d}_1 \times \mathbf{d}_2 \tag{6.86}$$

where × denotes the cross-product of the two vectors. We assume the missing delay to be

$$c\delta\tau_3 = \pm L_3 \sin\vartheta \tag{6.87}$$

with + sign for direction above and − sign for direction below the hydrophone plane and $L_3 = |\mathbf{d}_3|$.

Obviously, we cannot overcome the ambiguity in the location for a three-hydrophone array, but with additional information, this ambiguity may not be an issue. For example, if the three hydrophones form a nearly horizontal array close to the sea surface, then the

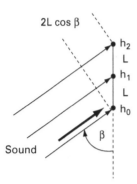

$2L\cos\beta$

h_2

L

h_1

L

h_0

β

Sound

Figure 6.9   Geometry of a 3-hydrophone line array.

direct arrival of echolocation clicks of all deep divers will come from below the array. Surface-reflected signals will always come from above the array, independent of the array depth.

6.9   Beam-forming

Beam-forming is a standard procedure when multiple hydrophones form a compact array. The primary objective of beam-forming is not so much direction finding but to exploit the observation that if the hydrophones are close enough then the received signal from a given direction should be only time-delayed versions of the same basic signal, but each hydrophone will be contaminated by different ambient noise fluctuations. Instead of estimating the time delays between the different hydrophone pairs, one uses all hydrophones and sums the time-delayed amplitudes to form what is called a beam pattern, which emphasizes coherent signals and minimizes the overall noise contribution to the array output.

To see how beam-forming works, let us assume that the sound $s(t)$ of a distant source is received with a three-element equally spaced line array as depicted in Figure 6.9. The sound shall first arrive at hydrophone h_0, which should be also our reference hydrophone, then at hydrophone h_1 and finally at hydrophone h_2. The time delays between the different hydrophones and the reference hydrophone are for a plane wave arriving with the angle β at the array estimated according to

$$\delta\tau_{n-0} = n\frac{L}{c}\cos\beta = n\delta\tau_\beta \quad n = 1,2 \tag{6.88}$$

In addition to the sound signal, the different hydrophones will receive random noise from all directions, that is, the local sound pressure measurement is composed of signal plus random noise $n(t)$

$$x(t) = s(t) + n(t) \tag{6.89}$$

which for the nth hydrophone becomes

$$x_n(t) = s(t - n\delta\tau_\beta) + n(t) \tag{6.90}$$

The minus sign in Equation 6.90 reflects the fact that the signal needs the time $n\delta\tau_\beta$ to reach the nth hydrophone.

To beam-form the array data, one compensates for the delay of the sound arrival and takes the mean value

$$
\begin{aligned}
x_{A\_3}(t, \delta\tau_\beta) &= \frac{1}{3}\sum_{n=0}^{3} x_i(t + n\delta\tau_\beta) \\
&= \frac{1}{3}\sum_{n=0}^{3} s(t) + \frac{1}{3}\sum_{n=0}^{3} n(t + n\delta\tau_\beta)
\end{aligned}
\tag{6.91}
$$

If the time delay was chosen correctly, that is, when the signal is indeed arriving from direction β, then we obtain

$$x_{A\_N}(t, \delta\tau_\beta) = s(t) + \frac{1}{N}\sum_{n=0}^{N} n(t + n\delta\tau_\beta) \tag{6.92}$$

where we generalized from 3 to N hydrophones.

It should be obvious that if we take the average of N identical signals then the result, or mean signal, should be identical to the individual signal, but averaging N different noise measurements must yield less noise than experienced by the individual hydrophones. How much noise suppression we obtain depends on how independent the different noise measurements are.

In the case of completely uncorrelated noise measurements, the expected noise amplitude decreases with the square root of the number of hydrophones. This is equivalent to the observation that, if the noise is uncorrelated, the different measurements are independent and the average noise energy that is measured by the array should remain independent of the number of hydrophones. The amount of noise suppression that may be achieved with beam-forming is related to the array gain of the hydrophone array as discussed in Section 3.5 (Equations 3.30 and 3.31).

6.10 Beam patterns

So far, we have estimated the array response when the arrival direction of the signal is known and the hydrophone data are properly delayed and averaged. What happens if we have a slight mismatch between the true and the assumed signal direction? Obviously, there is no change as far as the noise suppression is concerned, but for the signal component the results must degrade as averaging N slightly different signals generates a lower array output, or array response.

The beam pattern of a hydrophone array characterizes the response function of the array. Let us assume that the sound arrives at the line array with the angle β_0 but we form the beam in direction β.

To obtain the array response we assume now that the sound signal may be expressed as a complex valued harmonic sound wave according to

$$s(t) = \exp\{i(\omega t - kr)\} \tag{6.93}$$

with circular frequency ω, wave number k and propagation distance r.

The sound wave is now received at N different hydrophones, where ranges r_n are estimated by $r_n = r_0 + nL \cos \beta$, $n = 0, \ldots, N$

$$\begin{aligned} s_n(t) &= \exp\{i(\omega t - kr_n)\} \\ &= \exp\{i(\omega t - kr_0 - nkL \cos \beta_0)\} \end{aligned} \tag{6.94}$$

The array response x_{A_N} is then obtained by compensating for the assumed delay and averaging over all hydrophones

$$\begin{aligned} x_{A\_N}(t, \beta; \beta_0) &= \frac{1}{N} \sum_{n=0}^{N} s_n\left(t + n\frac{L}{c}\cos \beta\right) \\ &= \exp\{i(\omega t - kr_0)\} B_N(\beta, \beta_0) \end{aligned} \tag{6.95}$$

with

$$B_N(\beta, \beta_0) = \frac{1}{N} \sum_{n=0}^{N-1} \exp\{inkL(\cos \beta - \cos \beta_0)\} \tag{6.96}$$

The sum may be evaluated in closed form and with $\psi = \frac{kL}{2}(\cos \beta - \cos \beta_0)$ we obtain

$$\begin{aligned} B_N(\beta, \beta_0) &= \frac{1}{N} \sum_{n=0}^{N-1} \exp\{inkL(\cos \beta - \cos \beta_0)\} \\ &= \frac{\exp\{i2N\psi\} - 1}{N\exp\{i2\psi\} - 1} \\ &= \exp\{i(N-1)\psi\} \frac{\sin(N\psi)}{N \sin \psi} \end{aligned} \tag{6.97}$$

so that the beam-former output becomes

$$x_{A\_N}(t, \beta; \beta_0) = \exp\{i(\omega t - kr_0 + (N-1)\psi)\} \frac{\sin(N\psi)}{N \sin \psi} \tag{6.98}$$

With Equation 6.98 we can finally express the beam pattern $BP(\beta; \beta_0)$ of an N-element line array as

$$BP(\beta; \beta_0) = \left| \frac{x_{A\_N}(t, \beta, \beta_0)}{x_{A\_N}(t, \beta_0, \beta_0)} \right|^2 = \left(\frac{\sin(N\psi)}{N \sin \psi} \right)^2 \tag{6.99}$$

or

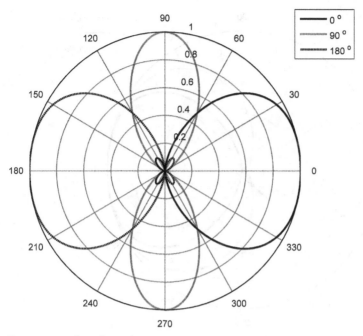

Figure 6.10 Beam pattern for a three-element array.

$$BP(\beta;\beta_0) = \left(\frac{\sin\left(N\frac{kL}{2}(\cos\beta - \cos\beta_0) \right)}{N\sin\left(\frac{kL}{2}(\cos\beta - \cos\beta_0) \right)} \right)^2 \qquad (6.100)$$

Figure 6.10 gives a visual impression of the beam pattern for a three-element array. Three different angles of arrival ($0°$, $90°$, $180°$) were assumed to generate this composite polar plot of Equation 6.100. We note immediately three important facts about beam-forming. For signals that are in line with the array, that is, they are arriving at $0°$ or $180°$, the beam pattern is much wider than the beam pattern for the signal at $90°$. We note further that all curves cross at 0.5 and while the two end-fire beams ($0°$ and $180°$) give the direction without ambiguity, the broadside beam ($90°$) shows an ambiguous solution at $270°$.

We conclude that the beam pattern does not depend directly on the angular difference $(\beta - \beta_0)$, but on the difference of the cosine of the angles $(\cos\beta - \cos\beta_0)$ as shown in Equation 6.100, with the consequence that the actual shape of the beam pattern is a function of the arrival angle of the sound. We conclude further that all directions except the two at end-fire are ambiguous, as the beam pattern does not change if one replaces β by $-\beta$.

If we define the beam-width as the angular difference for which the beam pattern reduces to 0.5, we note from Figure 6.10 that the broadside beam has a beam-width of $120°$ and the two end-fire beams have a beam-width of $240°$. Figure 6.10 indicates

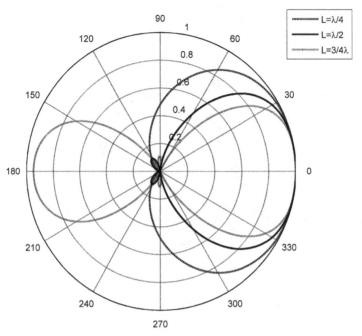

Figure 6.11   Beam pattern with different hydrophone spacing.

therefore that a three-element array may resolve three different sound sources, if one assumes as criterion for resolving two distinct sound sources that the beam patterns overlap at 0.5. This observation is easily generalized, that is, the maximum number of sound sources an array may resolve simultaneously, or detect, is equal to the number of hydrophones.

Inspecting Equation 6.100, one notes that maximum response occurs where $BP(\beta; \beta_0)$ becomes one, which is equivalent to

$$\frac{kL}{2}|\cos\beta - \cos\beta_0| = 2n\pi, \quad n = 0, 1, 2 \tag{6.101}$$

as owing to the periodicity of $\sin(x)$ with a period of 2π, $BP(\beta; \beta_0)$ will be also a periodic function. Note that $BP(\beta; \beta_0) = 1$ for $\beta = \beta_0$, as for small arguments we approximate $\sin(x) \approx x$ and therefore $\frac{\sin(Nx)}{N\sin(x)} \approx \frac{Nx}{Nx} = 1$.

A direct consequence of Equation 6.101 is that the beam pattern is unique, i.e. reaches the overall maximum only once, if

$$\max\left\{\frac{kL}{2}|\cos\beta - \cos\beta_0|\right\} < \pi \tag{6.102}$$

or with $\max|\cos\beta - \cos\beta_0| = 2$

$$L < \frac{\lambda}{2} \tag{6.103}$$

Only if the distance between the hydrophones is less than half the wavelength do we obtain a unique beam pattern. If the distance exceeds this limit, we obtain what are called grating lobes, that is, the array will have maximal response (sum up the signal coherently) for directions that do not coincide with the true direction.

Figure 6.11 shows the beam pattern for different hydrophone spacings. We note that if we reduce the spacing, the beam-width widens significantly, reducing the resolution capability of the beam-former. Increasing the hydrophone spacing over the critical limit given by Equation 6.103, a secondary peak at 180° appears, which is the grating lobe already mentioned. As we have assumed that the true angle of arrival of the sound source is 0°, the existence of a grating lobe at 180° generates a directional ambiguity.

6.10.1 Beam-forming with widely spaced arrays?

Traditionally, beam-forming will be used with a densely spaced array, where the hydrophone spacing does not exceed half the wavelength of the highest frequency of interest, that is, where Equation 6.103 holds. For such cases only, the beam pattern is unique and does not present grating lobes. This, however, is not the complete story as the above presented beam pattern has been derived for tonal plane waves without temporal limitation. Now imagine a short pulse (say, a dolphin click of 50 μs) impinging from forward end-fire ($\beta = 0$) on an array of three hydrophones that is 6 m long ($L = 3$ m). The travel time of the click from the first to the second hydrophone is 2 ms, which is 40 times the pulse length, so it is difficult to imagine that a processing scheme would erroneously result in an inverted travel direction of the dolphin click.

To see whether our argument is correct, let us assume a Gaussian pulse arriving with an angle β_0 at the array and let t_n be the time when the pulse arrives at hydrophone n

$$t_n = t + \frac{nL \cos \beta_0}{c} \tag{6.104}$$

that is, we express the waveform at the nth hydrophone (see Equation 6.94) as

$$s_n(t) = \exp\left\{-\frac{1}{2}\left(\frac{t - t_n}{\sigma}\right)^2\right\} \exp\{i(\omega t - kr_n)\}$$

$$= \exp\left\{-\frac{1}{2}\left(\frac{nkL \cos \beta_0}{\omega \sigma}\right)^2\right\} \exp\{i(\omega t - kr_0 - nkL \cos \beta_0)\} \tag{6.105}$$

then the beam pattern becomes

$$BP(\beta, \beta_0) = \left| \frac{1}{N} \sum_{n=0}^{N-1} \exp\left\{-\frac{1}{2}\left(\frac{2n\psi}{\omega \sigma}\right)^2 + i2n\psi\right\} \right|^2 \tag{6.106}$$

where $\psi = \frac{kL}{2}(\cos \beta - \cos \beta_0)$.

As Equation 6.106 does not describe a simple geometric sum, it is best to visualize the beam pattern using Matlab to see the difference between Equation 6.100 and 6.106.

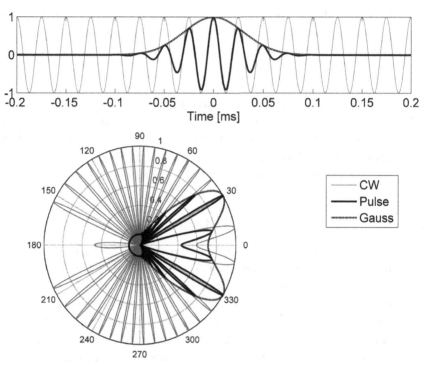

Figure 6.12   Sparse array beam pattern (bottom panel) of Gaussian pulse (top panel) arriving at an angle of 30°.

Figure 6.12 shows a short Gaussian pulse and the beam pattern of a three-hydrophone array. The signal frequency is assumed to be 40 kHz, the σ of the Gaussian amplitude function was 30 µs, and the array length was taken to be 1 m, resulting in a hydrophone spacing of 50 cm or 13.3 λ.

One notes that the beam pattern of the Gaussian pulse (solid line in Figure 6.12, bottom panel) is in principle composed of the product of the ambiguous beam pattern of the continuous wave (grey lines) with the amplitude shading function (dashed line). We can conclude that only non-oscillating pulses may lack grating lobes and that in this case the angular resolution depends on the ratio of array to effective signal length and all oscillating sound waves will have grating lobes in the vicinity of the correct direction.

Beam-forming is based on coherent summation of multiple hydrophones with the primary objective of increasing the signal-to-noise ratio and, at the same time, of determining the direction from which sound is arriving. Even if unique direction finding becomes difficult or impossible with widely spaced arrays of hydrophones, beam-forming may still be found useful for fine-scale direction finding and improvements of the signal-to-noise ratio, once rough directions have been obtained by post-detection time-delay estimation. In this sense, beam-forming with widely spaced arrays is reduced to improved direction finding and provides no advantage for detecting very weak signals embedded in noise, as the different signals have to be detected first before one determines the precise direction with a beam-forming algorithm.

Matlab code to generate Figure 6.12

```
%Scr6_5
beta=(-180:0.1:180)';
beta0=30;

N=3;
n=(0:N-1)';
%
fc=40; %kHz
sig=0.03; %ms
AL=1; % m
%
om=2*pi*fc;
lo=1.5/fc; %m
L=AL/(N-1);
dl=2*L/lo;
arg=dl*(cos(beta*pi/180)-cos(beta0*pi/180));
%
%rectangular weighting
w0=rectwin(N); w0=w0/sum(w0);
%
bp=beta*[0 0 0];
for ii=1:length(beta)
    arg1=2*n*arg(ii);
    arg2=-0.5*(n*arg(ii)/(om*sig)).^2;
  bp(ii,1)=sum(w0.*exp(arg1));
  bp(ii,2)=sum(w0.*exp(arg2+i*arg1));
  bp(ii,3)=sum(w0.*exp(arg2));
end
bp=abs(bp).^2;
%
figure(1)
set(gcf,'position',[100 100 560 600])
%subplot(211)
subplot('position',[0.1 0.75 0.8 0.2])
tt=-0.2:0.001:0.2;
hp=plot(tt,real(exp(i*om*tt)));
set(hp,'color',0.5*[1 1 1])
line(tt,real(exp(-0.5*(tt/sig).^2+i*om*tt)), ...
    'color','k','linewidth',2)
line(tt,real(exp(-0.5*(tt/sig).^2)), ...
    'color','k','linewidth',2,'linestyle','--')
set(gca,'fontsize',14)
xlabel('Time [ms]')
%
%subplot(212)
subplot('position',[0.05 0.05 0.6 0.6])
hp=polar(beta*pi/180*[1 1 1],bp);
set(gca,'fontsize',14)
set(hp(1),'color',0.5*[1 1 1],'linewidth',2)
set(hp(2),'color','k','linewidth',2)
```

```
set(hp(3),'color','k','linewidth',2,'linestyle','--')
hl=legend('CW','Pulse','Gauss');
xl=get(hl,'position');
set(hl,'position',[0.75 0.4 xl(3:4)])
```

6.10.2   Beam-forming with sampled data

Beam-forming requires very precise time delays to overlap exactly the measurements of the different hydrophones. An interpolation scheme is necessary to align the time series of the hydrophones properly for each steering direction, that is, the direction β to which the array is steered. Interpolation could be done with classical interpolation filters that up-sample the whole data stream to the desired sampling rate; and early hardware-based beam-formers implemented exactly this concept. This approach is conceptually easy, but may be computationally expensive, as more interpolated data points are generated than are in general required for beam-forming. Indeed, as interpolation is typically done by summing weighted samples of the time series, all that is necessary is to generate for each steering direction a proper set of interpolation coefficients, reducing the computational load to a minimum. However, the management of the interpolation coefficients may then be a significant computational factor of the beam-former. All solutions depend on the architecture of the processor and on whether memory access constitutes a penalty for processing speed.

6.11   Tracking

Tracking rounds up the DCLT processing suite. We have used tracking-based approaches earlier, when we discussed the detection and classification of whistles or click trains. Here we consider tracking as a subject in its own right and try to appreciate the potentials and difficulties one may encounter in tracking one or more objects that may or may not interact with each other.

6.11.1   Basics of tracking

All tracking algorithms use more or less precise measurements obtained from one or more objects (or animals, to remain within the scope of this book) to maintain an estimate of the current behavioural state of the animal. The behavioural states include kinematic elements (location, velocity, etc.), parameters that describe the acoustic activity (click rate, whistle frequency, etc.) and also any species-typical features.

The measurements are in general, noise-corrupted direct or indirect observations of the behavioural state, such as direct estimate of the location, range and bearing, time difference of arrival of a signal between two hydrophones, click rate, or frequency of tonal signals. The measurements are not necessarily raw data points but are in general the outputs from a detection, classification and localization system.

In Section 4.4, we tracked different whistles in the frequency domain. The measurements where spectral estimates of the whistles and the tracks described the frequency-dependence of the whistles. At any time, there could be multiple whistles, but there was, at each time step, at most only one estimate for each whistle frequency. More classical tracking applications address the location estimation of moving vehicles. Here the vehicle state is described, for example, by its location and velocity and it is not uncommon to have multiple measurements for the same track.

Tracking implies that measurements are associated with individual objects, allowing the inclusion of measurements typically used for classification. This suggests that, for tracking individual objects, not only geographic measurements may be useful, but also classification feature vectors.

Generally speaking, a tracker is a sequential processor that takes the measurements as they become available and tries to update the states of the objects that are to be monitored. The processing may occur in real time, that is, at the time of the measurement, or retrospectively for measurements that were taken in the past.

In principle, all trackers will follow the following processing scheme, the implementation of which may vary from trivial to very sophisticated algorithms:

- take measurements that are potentially pre-processed
- associate these measurements with available tracks
  - continue existing tracks
  - initiate new tracks
  - terminate old tracks
  - merge and split tracks
- filter and predict new track states

The available data may directly qualify as measurements, but in most situations the measured raw data require a significant amount of pre-processing to be useful for tracking. The objective of this processing is then to simplify the subsequent tracking steps, be it the data association or track maintenance steps.

Data association is the next important step where multiple measurements may compete with one or more tracks. In general, we cannot assume that there is an easy one-to-one relationship between measurements and available tracks. There may be more measurements than tracks, or more tracks than measurements, but even if the numbers coincide, it may be possible that not all measurements associate with the available tracks, and that some tracks are not continued and new tracks could be initiated. We have seen such a case when discussing whistle detection (Section 4.4.3).

Track maintenance ensures that existing tracks are updated using the new measurements. Where necessary, the measurements may lead to the initiation of new tracks, or tracks for which no measurements are available may be terminated. The termination of tracks may be immediately or deferred, that is, a track is only terminated if no new measurements are associated with it for a given number of time steps. The track maintenance must further deal with special situations where tracks merge, and single measurements could be used for two existing tracks, or where tracks split and two new tracks could relate to a single existing track.

The final step of the tracker uses the measurement to correct optimally the actual states of the tracks. This is typically done with a standard filter algorithm, which may be recursive or be implemented as a batch filter. Once the filtered track state is obtained, the tracker predicts the new state, which then becomes a valid prediction for the new measurement cycle.

6.11.2   Data association

Correct data association is essential for a multi-data multi-object tracker. As mentioned, we cannot assume that there is a unique one-to-one relationship between measurements and available track states, or that the numbers of measurements coincide with the number of tracks. This problem has already been encountered during the tracking of multiple whistles in the time–frequency domain of a spectrogram. In that example, multiple spectral peaks were associated with existing whistles by measuring the linear distance between the actual measurements and predicted whistle frequency state. In general, one would replace the simple estimation of the distance between measurements and state variables with a vector distance measure.

There are two ways of associating measurements with tracks, which have been used for generating the association table (Table 4.3) of the whistle detector; one method takes each measurement and finds a best track, whereas the other finds the best measurement for each track. Consequently, we are confronted with the same decisions as discussed in Section 4.4.3.

Data association can also be viewed as a classification problem, where one tries to classify the new measurements with respect to existing tracks. In that sense, one would consider the behavioural state vector of the track as the feature vector of a classifier and could apply Equation 5.35 for Gaussian distributed state vectors.

The data association is traditionally performed for every new set of measurements, or in tracking theory terms, at every epoch time. Whenever a complete dataset becomes available the data association is carried out. Sometimes, however, it seems better to defer the data association and to wait for a finite amount of new data before associating the measurements with the tracks *en bloc*. The objective of such deferral is to improve the association performance, in particular to reduce the number of erroneous track initializations and track terminations, and to handle track merging and splitting situations better.

After the measurement-to-track association operation, some track maintenance is required, whereby tracks are confirmed, initiated or terminated. Tracks are confirmed if there is a one-to-one relationship between existing tracks and new measurements. A simple approach to track initiation is to start a new track whenever a measurement cannot be assigned to an existing track. Tracks for which no new measurements are available and which cannot be confirmed are subject to termination. A common strategy is to delete a track only if it is not updated within some reasonable time interval.

The problem of track initiation and track termination is related to the sequential probability ratio tests (SPRT) used in Section 4.1.5 in the form of the Page test detector, where one defers the decision until the situation becomes clear enough to justify a decision. This leads to a sequential approach where a track score function is estimated and compared to two thresholds, an upper threshold where track confirmation is declared and a lower threshold where tracks are deleted. In all other cases decisions on the tracks are deferred, while continuing to update the score function.

6.11.3   Data filter and state prediction

The purpose of the data filter and the state prediction is to correct or update the old track state using the new measurements and to predict the track state one step into the future to allow a new round of data association.

In Section 4.4.3, a very simple filter was implemented (Equation 4.41) to combine a predicted track estimate with a new measurement to obtain the best estimate of the track state. Here, we learn about the Kalman filter, which can be considered as the workhorse of tracking applications (see e.g. Grewal and Andrews, 2008; Minkler and Minkler, 1993).

The Kalman filter was developed to estimate optimally the state of a dynamic system with noisy measurements and therefore residual uncertainties about the state of the system.

In the following we assume a discrete or sampled dynamic system for which the state vector $\mathbf{x}$ changes with time according to

$$\mathbf{x}(k+1) = \mathbf{A}(k)\mathbf{x}(k) + \mathbf{B}(k)\mathbf{u}(k) + \mathbf{w}(k) \tag{6.107}$$

where

$$\mathbf{A}(k) = \mathbf{\Phi}(t_{k+1}, t_k) \tag{6.108}$$

is a linear transition matrix of the state vector from time t_k to time t_{k+1}. In addition to the state update, Equation 6.107 contains the term $\mathbf{B}(k)\mathbf{u}(k)$, which describes external control variables, and $\mathbf{w}(k)$ describing the system noise, or residual uncertainty about the state update model.

The measurements are assumed to depend linearly on the state vector

$$\mathbf{y}(k+1) = \mathbf{C}(k+1)\mathbf{x}(k+1) + \mathbf{v}(k) \tag{6.109}$$

with $\mathbf{C}(k+1)$ denoting the transfer matrix and $\mathbf{v}(k)$ being the measurement noise.

The expected error covariance matrices for state model and measurements are assumed to be

$$E\{\mathbf{w}(k)\mathbf{w}^T(k)\} = \mathbf{Q}(k) \tag{6.110}$$

$$E\{\mathbf{v}(k)\mathbf{v}^T(k)\} = \mathbf{R}(k) \tag{6.111}$$

and state model and measurement noise are assumed to be uncorrelated.

The Kalman filter implements the following algorithm. Starting with an initial estimate

$$\hat{\mathbf{x}}(0) = \hat{\mathbf{x}}_0 \tag{6.112}$$

and with an initial state uncertainty covariance matrix

$$\mathbf{P}(0) = \mathbf{P}_0 \tag{6.113}$$

the new state vector is predicted according to

$$\hat{\mathbf{x}}^*(k+1) = \mathbf{A}(k)\hat{\mathbf{x}}(k) + \mathbf{B}(k)\mathbf{u}(k) \tag{6.114}$$

and the prediction error covariance matrix

$$\mathbf{P}^*(k+1) = \mathbf{A}(k)\mathbf{P}(k)\mathbf{A}^T(k) + \mathbf{Q}(k) \tag{6.115}$$

The predicted system state corresponds to a predicted measurement

$$\mathbf{y}^*(k+1) = \mathbf{C}(k+1)\hat{\mathbf{x}}^*(k+1) \tag{6.116}$$

The Kalman filter now uses the difference between predicted measurement $\mathbf{y}^*(k+1)$ and actual measurements $\mathbf{y}(k+1)$ to correct the predicted state obtaining the optimal filtered system state

$$\hat{\mathbf{x}}(k+1) = \hat{\mathbf{x}}^*(k+1) + \mathbf{K}(k+1)[\mathbf{y}(k+1) - \mathbf{y}^*(k+1)] \tag{6.117}$$

where $\mathbf{K}(k+1)$ is the so-called Kalman gain, which is estimated by

$$\mathbf{K}(k+1) = \mathbf{P}(k+1)\mathbf{C}^T(k+1)\mathbf{E}^{-1}(k+1) \tag{6.118}$$

where

$$\mathbf{E}(k+1) = \mathbf{C}(k+1)\mathbf{P}^*(k+1)\mathbf{C}^T(k+1) + \mathbf{R}(k+1) \tag{6.119}$$

and

$$\mathbf{P}(k+1) = [\mathbf{I} - \mathbf{K}(k+1)\mathbf{C}(k+1)]\mathbf{P}^*(k+1) \tag{6.120}$$

If the Kalman filter is used to filter a larger time interval then the following backward smoothing algorithm developed by Rauch, Tung and Striebel (Rauch *et al.*, 1965) may be applied to improve the filter performance. The algorithm runs backwards through the data and uses the predicted and corrected state uncertainty covariance matrix $\mathbf{P}^*(k)$ and $\mathbf{P}(k)$.

Initial values:

$$\hat{\mathbf{x}}(N, N) = \hat{\mathbf{x}}(N) \tag{6.121}$$

$$\mathbf{P}(N, N) = \mathbf{P}(N) \tag{6.122}$$

Smoothing algorithm:

$$\hat{\mathbf{x}}(k, N) = \hat{\mathbf{x}}(k) + \mathbf{V}(k)[\hat{\mathbf{x}}(k+1, N) - \mathbf{A}(k)\hat{\mathbf{x}}(k)] \tag{6.123}$$

$$\mathbf{V}(k) = \mathbf{P}(k)\mathbf{A}^T(k)\mathbf{P}^{*-1}(k+1) \tag{6.124}$$

$$\mathbf{P}(k, N) = \mathbf{P}(k) + \mathbf{V}(k)[\mathbf{P}(k+1, N) - \mathbf{P}^*(k+1)]\mathbf{V}^T(k) \tag{6.125}$$

Comparing the filter equation (Equation 6.117) with Equation 4.41, which was used for whistle tracking, one realizes that the latter may be seen as a special case where the Kalman gain is kept constant. Further, instead of a one-step linear state update model of

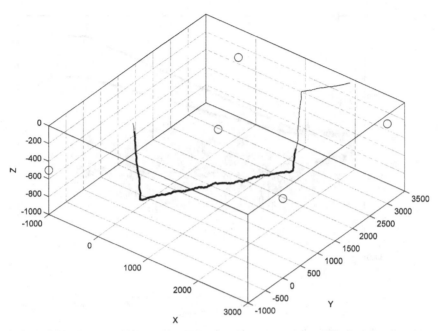

Figure 6.13 Simulated sperm whale foraging dive (solid line), with echolocation activity marked as dotted line. The circles indicate the location of the five hydrophones.

the Kalman filter, the whistle detector implements an exponential weighted update (Equation 4.43) of the filter gain. It could therefore be interesting to implement a Kalman filter for the whistle tracking also.

Example To test the Kalman filter we simulate a deep-diving, echolocating sperm whale. The whale is assumed to dive in 10 min to a depth of 900 m, where it stays for 30 min before ascending again to the surface. The whale is acoustically active for about 42 min and to keep the problem simple the ICI is assumed to be a constant 1 s. Five hydrophones are modelled to allow direct localization by using Equation 6.7. Figure 6.13 shows a 3-D view of the dive and hydrophones. This example is of particular interest as the five-hydrophone localization problem is easily solved by standard algebra, as discussed in Section 6.1.

As the Matlab script below shows, the stepwise linear whale motion was disturbed by Gaussian system noise; the delay measurements were also assumed to be erroneous. The consequence of this is that the direct solution is rather noisy, especially for the depth estimate. To construct the Kalman filter we consider as the state vector the solution vector of Equation 6.7 augmented with the velocity vector

$$\mathbf{x} = (R_0, x_W, y_W, z_W, v_R, v_x, v_y, v_z)^T \tag{6.126}$$

The state transition matrix $\mathbf{A}$ (Equation 6.108) becomes

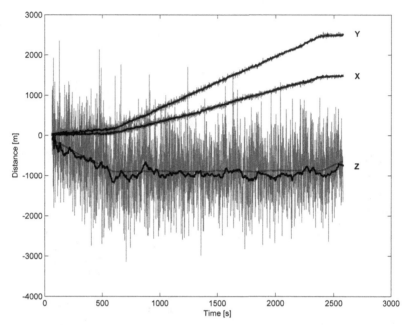

Figure 6.14 Result of Kalman filtering of the time delay measurements. The x, y, z components are given as thick solid lines. Grey lines, equivalent solution using the direct inversion (Equation 6.7). The dashed lines are the original simulated x, y, z components of the whale track (not visible for the x and y components).

$$
\mathbf{A} = 2 \begin{pmatrix}
1 & 0 & 0 & 0 & \mathrm{d}t & 0 & 0 & 0 \\
0 & 1 & 0 & 0 & 0 & \mathrm{d}t & 0 & 0 \\
0 & 0 & 1 & 0 & 0 & 0 & \mathrm{d}t & 0 \\
0 & 0 & 0 & 1 & 0 & 0 & 0 & \mathrm{d}t \\
0 & 0 & 0 & 0 & 1 & 0 & 0 & 0 \\
0 & 0 & 0 & 0 & 0 & 1 & 0 & 0 \\
0 & 0 & 0 & 0 & 0 & 0 & 1 & 0 \\
0 & 0 & 0 & 0 & 0 & 0 & 0 & 1
\end{pmatrix}
\tag{6.127}
$$

indicating that state variables v_R, v_x, v_y, v_z are modelled as 'constant', that is they are assumed to vary only in the course of the filter.

The measurement prediction matrix is given by

$$
\mathbf{C} = \begin{pmatrix}
\delta R_1 & x_1 - x_0 & y_1 - y_0 & z_1 - z_0 & 0 & 0 & 0 & 0 \\
\delta R_2 & x_2 - x_0 & y_2 - y_0 & z_2 - z_0 & 0 & 0 & 0 & 0 \\
\delta R_3 & x_3 - x_0 & y_3 - y_0 & z_3 - z_0 & 0 & 0 & 0 & 0 \\
\delta R_4 & x_4 - x_0 & y_4 - y_0 & z_4 - z_0 & 0 & 0 & 0 & 0
\end{pmatrix}
\tag{6.128}
$$

and the measurement vector $\mathbf{y}$ of Equation 6.109 is given by vector $\mathbf{b}$ of Equation 6.6.

Figure 6.14 shows the result of the direct inversion (in grey) and the Kalman filter (different solid lines labelled with x, y, and z). One may note that whereas the x and y components of the whale track are also reasonably estimated by the direct inversion, the depth estimate (z component) is rather noisy for the direct solution, but very close to the correct whale track (dashed line) for the Kalman filter estimate (solid line).

The example shows clearly that using a Kalman filter may also be recommended where direct solutions are available. Using five hydrophones to localize a deep-diving whale should be sufficient, as the solution is based on a simple matrix inversion. Measurement errors are, however, always a concern; in this example, the consequence of measurement errors was a very noisy depth estimation. The Kalman filter was nevertheless capable of providing a state estimate very close to the simulated animal track.

Matlab code for example

```
%Scr6_6
% simulation
tt=0:3600;
vv=1.5+0*tt;
beta=0*tt;
gamma=30+0*tt;

beta(tt<10*60)=-80;
beta(tt>=10*60 & tt<40*60)=0;
beta(tt>=40*60 & tt<50*60)=80;
beta(tt>=50*60)=0;

randn('state',0); %to reproduce random numbers

sigs=1;
dx=vv.*sin(gamma*pi/180.*cos(beta*pi/180) ...
    +sigs*randn(size(tt)));
dy=vv.*cos(gamma*pi/180).*cos(beta*pi/180) ...
    +sigs*randn(size(tt)));
dz=vv.*sin(beta*pi/180) ...
    +sigs*randn(size(tt)));

xx=cumsum(dx);
yy=cumsum(dy);
zz=cumsum(dz); zz(zz>0)=0;

tcl=tt(tt>60 & tt<43*60);

xcl=interp1(tt,xx,tcl);
ycl=interp1(tt,yy,tcl);
zcl=interp1(tt,zz,tcl);

xo=[xcl;ycl;zcl];

h0=[-1000,-1000,-500];
h1=[3000,0,-100];
h2=[0,3000,-100];
```

```
h3=[3000,3000,-100];
h4=[1000,1000,-100];

%plot track and hydrophones
figure(1)
hold off
hp=plot3(xx,yy,zz,'k',xcl,ycl,zcl,'k.');
set(hp(1),'color',0.5*[1 1 1])
grid on
set(gca,'CameraPosition',[11000 -13000 6000])
%
hold on
plot3([h0(1) h1(1) h2(1) h3(1) h4(1)],...
      [h0(2) h1(2) h2(2) h3(2) h4(2)],...
      [h0(3) h1(3) h2(3) h3(3) h4(3)],...
      'ko','markersize',10)
hold off
%
xlabel('X')
ylabel('Y')
zlabel('Z')
box on

% simulate measurements
%obtain acoustic ranges
R0=sqrt((xcl-h0(1)).^2+(ycl-h0(2)).^2+(zcl-h0(3)).^2);
R1=sqrt((xcl-h1(1)).^2+(ycl-h1(2)).^2+(zcl-h1(3)).^2);
R2=sqrt((xcl-h2(1)).^2+(ycl-h2(2)).^2+(zcl-h2(3)).^2);
R3=sqrt((xcl-h3(1)).^2+(ycl-h3(2)).^2+(zcl-h3(3)).^2);
R4=sqrt((xcl-h4(1)).^2+(ycl-h4(2)).^2+(zcl-h4(3)).^2);
%
%obtain measured delays
sigm=2*15;
dR1=R1-R0+sigm*randn(size(R0));
dR2=R2-R0+sigm*randn(size(R0));
dR3=R3-R0+sigm*randn(size(R0));
dR4=R4-R0+sigm*randn(size(R0));
%
if 0
figure(2)
plot(tcl,dR1,'-',tcl,dR2,'-',tcl,dR3,'-',tcl, dR4,'-')
xlabel('Time [s]')
ylabel('Delay [m]')
end
%
% initiate Kalman filter
dt=diff(tcl); dt=[dt(1) dt];
% state transfer matrix
A=[1  0  0  0  dt(1)  0      0      0;
   0  1  0  0  0      dt(1)  0      0;
   0  0  1  0  0      0      dt(1) 0;
```

```
        0  0  0  1  0      0      0    dt(1);
        0  0  0  0  1      0      0    0;
        0  0  0  0  0      1      0    0;
        0  0  0  0  0      0      1    0;
        0  0  0  0  0      0      0    1];
% measurement vector
C=2*[  0  h1-h0 0 0  0  0;
       0  h2-h0 0 0  0  0;
       0  h3-h0 0 0  0  0;
       0  h4-h0 0 0  0  0];
%
% need this for direct inversion
b0=[h1*h1';
    h2*h2';
    h3*h3';
    h4*h4']-h0*h0';

C0=2*[0 h1-h0;
      0 h2-h0;
      0 h3-h0;
      0 h4-h0];
%store here determinant
D=0*tcl;

%Kalman filter
%initialize state vector
xh=[0 0 0 0 0 0 0 0]'*tcl;
xh(:,1) = [1500 6 15 -100 1 1 1 -1]';
xh0=xh(1:4,:);
%
% measurement error covariance matrix
R=(1e+4)^2*eye(4);
%
Q=diag([1 1 1 .01 .1 .1 .1 .01].^2);
%
I=eye(8);
P=1*I;
for ii=2:length(tcl)
    %adjust state model
    A(1,5)=dt(ii);
    A(2,6)=dt(ii);
    A(3,7)=dt(ii);
    A(4,8)=dt(ii);
    %
    %adjust measurement model
    dR=[dR1(ii) dR2(ii) dR3(ii) dR4(ii)]';
    C(:,1)=2*dR;
    y=b0-dR.^2;
    %
    % prediction
    xhs=A*xh(:,ii-1); % state
```

```
        ys=C*xh(:,ii-1); %measurement
        Ps=A*P*A'+Q; % covariance matrix
        %
        % correction
        K=Ps*C'*inv(C*Ps*C'+R);
        xh(:,ii)=xhs+K*(y-ys);
        P=(I-K*C)*Ps;
        %
        % direct solution for comparison only
        C0(:,1)=C(:,1);
        D(ii)=det(C0);
        xh0(:,ii)=pinv(C0)*y;
        %
end

if 0
figure(3)
hp=plot(tcl,xh0(1,:),tcl,xh(1,:),'r',tcl,R0,'b');
set(hp(1),'color',0.5*[1 1 1]);
set(hp(2:3),'linewidth',2)
xlabel('Time [s]')
ylabel('Range [m]')
end

figure(4)
hp=plot(tcl,xh0(2:4,:),'k',tcl,xo','k:',tcl,xh (2:4,:),'k');
set(hp(1:3),'color',0.5*[1 1 1]);
set(hp(4:end),'linewidth',2)
text(tcl(end)+100,xh(2,end),'X','fontweight', 'bold')
text(tcl(end)+100,xh(3,end),'Y','fontweight',' bold')
text(tcl(end)+100,xh(4,end),'Z','fontweight', 'bold')
xlabel('Time [s]')
ylabel('Distance [m]')
```

Part III

Passive acoustic monitoring (putting it all together)

After presenting the tools required to carry out PAM, the third part of this book discusses the operational aspect of how PAM may be used. PAM is a rather new direction in cetacean research, which so far has mostly been visually oriented. However, with the availability of affordable instrumentation and the increasing cost of human-operated ship-based surveys, PAM is becoming increasingly interesting for both cetacean stock management and risk mitigation, especially when carried out with small autonomous platforms.

The implementation of PAM systems depends very much on the application and operational circumstances. Chapter 7 addresses some PAM applications and indicates a few difficulties one may encounter during the analysis of acoustic data, at least when applying classical techniques known from visually oriented applications. Chapter 8 addresses the detection function, which plays an important role in assessing the population density using survey methods. Chapter 9 introduces computer simulation as a technique to emulate real-world PAM with the intention of demonstrating, not only the usefulness of simulations to generate data, but also how to analyse PAM system performance. The last chapter of this book addresses some systems issues that are fundamental in any PAM implementation, and discusses relevant aspects of PAM hardware and software.

7 Applications of passive acoustic monitoring

The implementation of any PAM system depends very much on the details of its operation. So far, the monitoring of wildlife has been done largely by visual observation; consequently, there is still work to be done to develop a proper theory or methodology that is specific for passive acoustics. Most PAM activities try to adapt their results to standard analysis techniques that have proven successful in visually based monitoring. It may be that this is all that is required, but on the other hand, only a few passive acoustic activities have been successfully analysed with such standard tools. In this chapter, I therefore try to address some standard methods suggested for PAM applications and the difficulties one may encounter during the data analysis.

7.1 Abundance estimation

Abundance estimation is a standard application in animal ecology, where one tries to estimate the total number of animals in a given area, or equivalently to obtain an estimate of the density of the species of interest.

In general, we define the density of a species by dividing the number of individuals by the area they occupy

$$D = \frac{N}{A} \tag{7.1}$$

where N is number of animals present in an area A.

Let us now assume that the area A is known, then estimating the density is equivalent to estimating the population N. To estimate N, we typically carry out a survey, which may result in the detection of n individuals of the species of interest. As the number of detected animals n is proportional to number N of existing animals

$$n = P_N N \tag{7.2}$$

where the proportionality constant P_N may be interpreted as the probability of detecting any of the animals, we obtain the basic formula of abundance estimation

$$N = \frac{n}{P_N} \tag{7.3}$$

and for the density estimate

$$D = \frac{n}{A P_N} \tag{7.4}$$

Equation 7.4 applies to cases where one detects at least one animal ($n > 0$) and where one knows the probability P_N of detecting each one of these N animals. Note, that in order to define correctly an ecologically meaningful animal density, the size of area A should be related to the habitat of the N animals. Otherwise, one could bring the density to very small values by increasing the area A to include areas that are known to be unsuited for the species of interest.

Probability of detecting an animal

This density estimator (Equation 7.4) is very general and may necessitate some adaptation with respect to type of survey or data collection. However, let us first consider the probability of detecting an animal in a more abstract way to facilitate the survey-specific modifications to Equation 7.4.

Assume a total number of N animals of the species of interest that occupy the area of interest A. Assume further that not all animals may be detected during the survey. The reasons for not being detected may be multiple; they may be due to the behaviour of the animal (e.g. animal is out of sight), the performance of the observer (e.g. difficulties of observing in bad weather), or the implementation of the survey (does not cover all areas occupied by the animals). In other words, the number of detected animals n is related to the number of existing animals N by

$$n = P_N N = P_{\text{area}} P_{\text{avail}} P_{\text{det}} N \tag{7.5}$$

where P_{area} is the probability that the animal is within the sampled area (proportion of sampled area to area of interest), P_{avail} is the probability that the animal is available for being detected (e.g. acoustically active, in the case of PAM), given its presence in the sampling area, and P_{det} is the conditional probability that an animal is detected given its availability.

Independent of the reasons why an animal is detected or not, we may attribute to each of the existing animals a detection index $\delta_i, i = 1, \cdots, N$ which becomes 1 if the animal is detected and zero if not.

Summing up all detection indices, we obtain the number n of detected animals.

$$n = \sum_{i=1}^{N} \delta_i \tag{7.6}$$

Equation 7.6 does not change, if we write

$$n = P_N N \tag{7.7}$$

where

$$P_N = \frac{1}{N} \sum_{i=1}^{N} \delta_i \tag{7.8}$$

is simply an average of the detection indices δ_i.

By virtue of Equation 7.7, the detection probability P_N denotes the fraction of existing animals that are detected, and Equation 7.8 indicates that the detection probability is an average or mean value.

Equation 7.7 indicates further that, given an animal count n and given an estimate of the detection probability $\hat{P}_N$, then the number of existing animals is estimated by

$$\hat{N} = \frac{n}{\hat{P}_N} \qquad (7.9)$$

where the symbol "^" denotes now an estimator (or estimate) of the corresponding quantity.

As the animal count n is the result of the survey, all that needs to be done is to obtain an estimate of the average detection probability $\hat{P}_N$. Obtaining this quantity is in quintessence the objective of a statistical survey analysis. At first glance, the definition of the average detection probability (Equation 7.8) seems trivial and of little use as it depends on the number of animals N, the number one wanted to estimate.

Let us expand more on the detection variable and ask ourselves what this variable means for real surveys. Obviously, to be more specific we must make some assumptions. A weak assumption is that animals are detected if they are close to the observer and that they are not detected if they are very distant and that the critical distance up to which an animal may be detected is species-specific and constant. With these assumptions, we can express the detection probability as

$$P_N = \frac{1}{N} \sum_{i=1}^{N} \delta(r_i < r_0) \qquad (7.10)$$

where r_i is the distance of the ith animal, r_0 is the maximum distance at which the animals may be detected, and $\delta(r_i < r_0)$ is the binary detection function, which becomes one for $r_i < r_0$ and is zero otherwise.

Equation 7.10 simply says that to obtain the detection probability we have to count animals that are closer then the critical distance r_0. We therefore need to know the distances r_i of the individual animals. We should now realize that, by definition, we cannot know the distances of every animal, as otherwise there would be no need for a survey to estimate the number of animals. For the same reason we also cannot know the total number of animals. The way out of this dilemma is to treat the problem as a statistical one and assume that, although we do not know the absolute numbers, we have an idea of the distribution of the animals. The animals may be uniformly distributed in the area of interest; they may be clustered according to some negative binomial distribution; or they may follow any other distribution for which no closed-form mathematical description exists. Therefore, we next assume that, although we do not know the absolute number, we may estimate the relative density of the animals. That is, we replace Equation 7.10 by

$$P_N = \int_0^\infty w_{ri}(r)\delta(r < r_0)\,\mathrm{d}r \qquad (7.11)$$

where $w_{ri}(r)$ describes the probability density function (PDF), that is, the product $w_{ri}(r)\mathrm{d}r$ is the probability of encountering an animal in the distance interval $[r, r+\mathrm{d}r]$ obeying the condition

$$\int_0^\infty w_{ri}(r)\mathrm{d}r = 1 \tag{7.12}$$

Equation 7.11 assumes that the critical distance r_0 is constant during the whole survey for all whales of a given species. This is, however, very unlikely as all observations suffer from uncertainties, mostly due to environment variability and observer imperfections. It is therefore convenient to replace the deterministic detection function $\delta(r<r_0)$ by a statistical average

$$\delta(r<r_0) =: g(r) = \int_0^\infty w_{r0}(y)\delta(r<y)\mathrm{d}y \tag{7.13}$$

where $w_{r0}(y)$ is the PDF of the maximum detection range by the observer for the animal species. We denote this function as the detection function $g(r)$.

Inserting Equation 7.13 into Equation 7.11 we finally obtain for the detection probability

$$P_N = \int_0^\infty w_{ri}(r)\left(\int_0^\infty w_{r0}(y)\delta(r<y)\mathrm{d}y\right)\mathrm{d}r \tag{7.14}$$

In summary, we may express the number of detected animals

$$n = P_N N \tag{7.15}$$

and the total or unconditional detection probability P_N of each animal as

$$P_N = \int_0^\infty h(x)\mathrm{d}x \tag{7.16}$$

with

$$h(x) = w_{ri}(x)g(x) \tag{7.17}$$

where

$$g(x) = \int_0^\infty w_{r0}(y)\delta(x<y)\mathrm{d}y \tag{7.18}$$

We note that $w_{ri}(x)$ is a global description of the animal's behaviour (where is it and how does it behave) and $w_{r0}(y)$ characterizes mainly the performance of the sensor in detecting the animal. The function $g(x)$ is also known as the detection function; apart from a multiplicative constant, it could be estimated from the detection data.

Line transects

During a line transect the observer moves through an area of interest, covering a strip of total length L and width $2w$, that is, surveying a total area $A = 2wL$. In cases where P_N may be expressed as function of the distance from the transect line we may write the unconditional probability of detecting an object within the strip width as

$$P_N = \frac{1}{w} \int_0^w g(x)\mathrm{d}x \qquad (7.19)$$

where $g(x)$ is the detection function and x is the perpendicular distance of the detected object from the transect line.

Equation 7.19 relates to Equations 7.16 and 7.17 by setting the animal density constant

$$w_{ri}(x) = \frac{1}{w} \qquad (7.20)$$

The integral in Equation 7.19 is a number and may be used to define an effective detection area A_{eff}, within which the animals are effectively detected with probability $g(0)$

$$2L \int_0^w g(x)\mathrm{d}x = g(0)A_{\mathrm{eff}} \qquad (7.21)$$

The density estimator of a line transect (Equation 7.4) becomes, with Equation 7.21

$$D = \frac{n}{g(0)A_{\mathrm{eff}}} \qquad (7.22)$$

The term $g(0)$ is the probability of detecting an object at distance zero. It cannot be estimated from the distance data alone and is therefore, in most distance sampling applications, assumed to be one, that is, it is assumed that all animals at distance zero, or on the track line, are detected with certainty. However, Equation 7.21 also holds if $g(0)$ is known to be smaller than one, as long as it is known. Sometimes, this $g(0)$ may be assessed by reasoning independent of distance data. For example, Barlow (1999) used dive patterns of long-diving whales and determined, for Cuvier's beaked whale, $g(0) = 0.23$ when surveyed with a specific ship-based visual line transect setup and $g(0) = 0.07$ for aerial surveys (Barlow *et al.*, 2006).

From Equation 7.5 we can conclude that $g(0) < 1$ may occur if some animals are not available for detection, $P_{\mathrm{avail}} < 1$, or if the survey system is not perfect for zero distances and $P_{\mathrm{det}} < 1$. The third term P_{area} is, by virtue of Equation 7.17, not part of the detection function $g(x)$ but must be considered when estimating the total abundance. Considering PAM, it is fair to assume that most systems are implemented in such a way that $P_{\mathrm{det}} = 1$ for zero distances, that is, one should be able to detect the sound that is emitted by whales or dolphins when very close to the animal. Consequently, $g(0)$ reflects the probability P_{avail} that the animal is acoustically active and therefore available for detection.

Equation 7.22 holds also when the line transect is carried out by using multiple disconnected track lines, where the effective detection area becomes the sum of all individual detection areas. Complications could arise only if the effective detection areas overlap in such a way that detection opportunities are no longer independent. In such cases some care is necessary when estimating the effective detection area A_{eff}, as within this total effective area all detections are assumed independent.

A further complications is due to moving animals as one could envision that, say, the same pod of sperm whales that moves slowly through an ocean basin is detected multiple times when the line transect crosses multiple times the path of motion of the whales. Whereas visual methods using photo identification would most likely recognize the recurring detection, this cannot be said for acoustic methods, resulting in an overestimation of the abundance for this surveyed species.

Point transects

During so-called point transects, the observer does not move but tries to detect objects that surround the observer's fixed location, that is, the area A is expressed as a disk with radius w around the observer and $A = \pi w^2$.

The unconditional probability of detecting an object up to a radius w around the observer is given by

$$P_N = \frac{2}{w^2} \int_0^w rg(r)\mathrm{d}r \tag{7.23}$$

where $g(r)$ is the radial detection function describing the probability of detecting an object at radius r. One might expect $g(r)$ to be identical to the $g(x)$ of Equation 7.18 if we replace the radius r by the distance x, as the detection function should depend only on the relative distance between observer and animal independently of the motion of the observer.

Equation 7.23 relates to Equations 7.16 and 7.17 by letting the animal density increase with range from the observer

$$w_{ri}(r) = \frac{2r}{w^2} \tag{7.24}$$

This radial increase of detection opportunities is equivalent to a uniform distribution of animals in space.

By analogy to the line transect, we can estimate an effective detection area A_{eff}, in which animals are found with probability $g(0)$, by

$$2\pi \int_0^w rg_n(r)\mathrm{d}r = g(0)A_{\text{eff}} \tag{7.25}$$

so that the density estimator becomes

$$D = \frac{n}{g(0)A_{\text{eff}}} \tag{7.26}$$

where A_{eff} is again the effective detection area, in which animals are found with probability $g(0)$.

Similar to the line transect, Equation 7.26 may be extended to multiple point transects as long as the total effective detection area is estimated carefully.

7.1.2 Acoustic cue counting

To obtain the density of animals, Equations 7.1 and 7.4 need as input the number n of detected individuals. Although it is reasonable for visual surveys to count individuals, this is not so trivial for passive acoustics. In general, PAM systems do initially detect only acoustic signals from one or more animals and individuals of the same species may only be counted after more or less sophisticated processing or with expensive hardware. For example, multi-path ranging or multiple hydrophones may be used to resolve spatially different acoustically active animals. However, there is no guarantee that this method succeeds and allows us to count individual animals.

Acoustic cue counting seems a straightforward method, as cue counting also plays an important role in monitoring elusive terrestrial fauna. Typical acoustic cues are individual echolocation clicks of deep-diving whales. Let the number of emitted clicks n_{cl} of N animals be

$$n_{\text{cl}} = N(T\rho_{\text{cl}}) \tag{7.27}$$

where ρ_{cl} is the click rate of a single animal and T is the observation time.

The total number of detected clicks of the PAM system is then the number of correct classified detections $n_{\text{det\_cl}}$ plus the number of noise-driven false alarms n_{FA} plus the number of wrongly classified detections n_{FC}

$$n_{\text{det}} = n_{\text{det\_cl}} + n_{\text{FA}} + n_{\text{FC}} \tag{7.28}$$

The false detections in Equation 7.28 are divided into two contributions, n_{FA} and n_{FC}, reflecting the two relevant steps in a PAM system: detection of transients and classification of cetacean sound. False alarms are typical for detectors and driven by noise; classification errors may occur in the presence of multiple sound sources. It is useful to consider the two cases separately, as noise-driven false alarms may also occur in the absence of any sound source, whereas false classification requires the presence of sound sources.

To expand Equation 7.28, we note that the number of correct detections $n_{\text{det\_cl}}$ is proportional to the number of emitted clicks

$$n_{\text{det\_cl}} = P_{\text{det\_cl}} n_{\text{cl}} \tag{7.29}$$

where $P_{\text{det\_cl}}$ is the probability of detecting an emitted click.

The number of false alarms n_{FA} is proportional to the observation time

$$n_{\text{FA}} = (T\rho_{\text{FA}}) \tag{7.30}$$

where ρ_{FA} is the false alarm rate of the detector.

The number of false classified clicks n_{FC} depends in the presence of confounding species and may be summarized by

$$n_{\text{FC}} = (T\rho_{\text{FC}}) \qquad (7.31)$$

where ρ_{FC} is the effective false classification rate related to the observation time T.

The false alarm rate of the detector ρ_{FA} depends on the system setup and not on the number of animals that are present in the area. It is more or less constant and varies only when the noise statistics change. The effective false classification rate ρ_{FC}, on the other hand, depends on the presence of confounding animals and is therefore highly context-dependent.

Inserting all parts into the density equation, one obtains

$$
\begin{aligned}
D &= \frac{N}{AP_N} \\
&= \frac{n_{\text{cl}}}{AP_N(T\rho_{\text{cl}})} = \frac{n_{\text{det\_cl}}}{AP_N(T\rho_{\text{cl}})P_{\text{det\_cl}}} \\
&= \frac{n_{\text{det}} - (T\rho_{\text{FA}}) - (T\rho_{\text{FC}})}{A(P_N P_{\text{det\_cl}})(T\rho_{\text{cl}})}
\end{aligned}
\qquad (7.32)
$$

Dividing numerator and denominator by the observation time T, equation 7.32 becomes

$$D = \frac{\rho_{\text{det}} - \rho_{\text{FA}} - \rho_{\text{FC}}}{AP_{\text{ac}}\rho_{\text{cl}}} \qquad (7.33)$$

where ρ_{det} is the rate of detections $\rho_{\text{det}} = \frac{n_{\text{det}}}{T}$ and P_{ac} is the probability of detecting the acoustic cues $P_{\text{ac}} = P_N P_{\text{det\_cl}}$.

The false alarm rate ρ_{FA} is in general constant and independent of the rate of click detection, but the detector may be tuned so that ρ_{FA} is small compared with the detection rate ρ_{det}. The rate of false classifications ρ_{FC} depends on the presence of confounding species, which may or may not be related to the species of interest.

In cases where the false alarm rate may be ignored and the false classification rate is related to the cue detection rate (e.g. Marques *et al.*, 2009), e.g. where one may assume that a constant fraction of the detected cues are always misclassified, $\rho_{\text{FC}} = c\rho_{\text{det}}$, then Equation 7.33 simplifies to

$$D = \frac{\rho_{\text{det}}(1 - c)}{AP_{\text{ac}}\rho_{\text{cl}}} \qquad (7.34)$$

Using individual clicks as cues to infer the number of whales assumes that the number of clicks is proportional to the number of whales (Equation 7.27) and that the clicks from all whales are detected with the same probability (Equation 7.29). This necessitates that the distances of individual whales to a receiver have the same distribution, or equivalently, that all whales have similar detection functions.

7.1.3   Acoustic dive counting

As an alternative to counting individual acoustic emissions and relating the resulting count to the number of animals, one could try to count acoustic activities. That is, we define as the cue an acoustically compact period. For example, knowing that deep-diving odontocetes echolocate mainly during deep foraging dives, we could consider the acoustic detection of a foraging dive as a cue for the presence of a whale. This would be equivalent to counting only the characteristic fluke-ups of sperm whales for abundance estimation, and not the individual blows.

Using foraging dives as cues requires more signal processing, as the PAM system detects primarily the clicks of an echolocating whale and not the dive. On the other hand, false alarms and false classifications do not play such a large role in dive counting as they do in click counting. In addition, the rate of foraging dives is less variable for most deep divers than the rate or number of clicks emitted during a foraging dive.

As dive counting is also a form of cue counting, the density may be estimated by analogy to Equation 7.34, by reinterpretation of the different variables

$$D = \frac{\rho_{\text{det}}}{A P_{\text{ac}} \rho_{\text{dive}}} \tag{7.35}$$

where ρ_{det} is now the rate of dive detections, P_{ac} is the probability of detecting a dive acoustically and ρ_{dive} is the rate of foraging dives of the species of interest.

The variable c in Equation 7.34 represents the fraction of misclassified events (clicks), which in most cases may be set to zero for dive counting, as the frequency of false decisions for dives is typically very low.

7.1.4 Density estimation of clustered or grouped objects

The density estimations presented so far assume that the individual detections are statistically independent. However, if animals move in groups, then the detection/recording unit should be the cluster (group, pod) and not the individual animal, as their individual detections are no longer independent. Furthermore, as detecting large pods of animals is sometimes easier than detecting an isolated individual, one should expect that the detection function best suited for grouped animals is different from the function used for individuals. In particular, one should expect that large pods might be detected at greater distances than smaller pods or individuals.

To infer animal abundance from cluster densities one needs as additional information an estimate of the size of the cluster. During visual surveys, this information may be estimated during the survey, but in the case of PAM, estimating the cluster size may be difficult. This is mainly because we observe the animals only indirectly by detecting the acoustic emissions of the individuals and estimating the cluster size requires the ability to count individuals. It seems, however, appropriate to assume that acoustically active group sizes are equal to visually determined group sizes and to use in PAM applications the visually estimated group sizes. Of course, as group sizes may vary significantly, visually determined group size may also be very uncertain and even not relevant for PAM, especially when determined in different temporal and spatial situations.

7.1.5 Moving whales during line and point transects

Both examples discussed above, line and point transects, assume that surveyed objects (whales and dolphins) can be considered as immobile. Whales, however, are known to

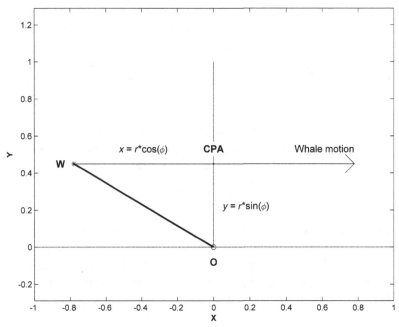

Figure 7.1   Relative whale motion, where O denotes the observer and W is the whale moving from left to right along the arrow. CPA stands for closest point of approach and is the perpendicular distance between whale track and observer. The angle φ is defined between the positive x-axis and the direction from O to W.

move more or less continuously through the oceans. The interesting aspect of detecting moving whales is that in both survey cases (line and point transects) the relative distance between whale and observer changes. It should therefore be possible to treat both cases with a single model, as the only difference between these two cases is the relative speed of movement between whale and observer. In the case of line transects the relative speed is the vector addition of both the whale and the vessel speed; in the case of point transects the relative speed is only due to the whale's motion. Of course, we must assume that the motion of the whale is not influenced by the motion or the presence of the observer. Otherwise, the results would be biased to higher estimates if the whales were attracted by the observer or to lower values if the animals were avoiding the observer.

To simplify the discussion let us consider that the observer is fixed and that the whale moves at a distance y past the observer, as shown in Figure 7.1, and that the whale depth is constant. For notational convenience, the whale is assumed to move in a positive direction from negative to positive x values, and it is further assumed that the whale is at the same depth as the hydrophone. Let us now assume that the speeds of whale and observer are constant and that the detection opportunities are uniformly distributed in time. This is equivalent to the notion that the spatial distribution of the detection opportunities is homogeneous along the line of motion. The question to be answered now is what the detection probability becomes as the whale passes by the observer (see also Buckland *et al.*, 2004).

Inspecting Figure 7.1, we note that for an approaching whale the range between whale and observer decreases according to Pythagoras:

$$r = \sqrt{x^2 + y^2} \tag{7.36}$$

This reaches its minimum distance at the CPA (closest point of approach) $r = y$, after which the distance to the whale increases again.

Let us assume that the whale, which is moving parallel to the x-axis from minus infinity to x, was not detected at position x,y with probability $Q(x,y)$. Let further

$$h(x, y)\mathrm{d}x = \text{Probability of detection in interval}[x + \mathrm{d}x, x] \tag{7.37}$$

then the probability of not having detected the whale at position $(x + \mathrm{d}x, y)$ becomes

$$Q(x + \mathrm{d}x, y) = Q(x, y)(1 - h(x, y)\mathrm{d}x) \tag{7.38}$$

Equation 7.38 may then be written as a differential equation

$$\frac{\mathrm{d}Q(x, y)}{\mathrm{d}x} = -Q(x, y)h(x, y) \tag{7.39}$$

the solution of which is straightforward

$$Q(x, y) = \exp\left\{ -\int_{-\infty}^{x} h(u, y)\mathrm{d}u \right\} \tag{7.40}$$

whereby we assume that the whale was not detected at very large ranges, that is, we have set $Q(-\infty, y) = 1$.

At the CPA the probability of having missed the whale is then

$$Q(0, y) = \exp\left\{ -\int_{-\infty}^{0} h(u, y)\mathrm{d}u \right\} \tag{7.41}$$

As the whale continues to move beyond the observer, the probability of detection increases further so that the total probability of not having detected the whale after its passage becomes

$$Q(\infty, y) = Q(0, y) \exp\left\{ -\int_{0}^{\infty} h(u, y)\mathrm{d}u \right\}$$
$$= \exp\left\{ -\int_{-\infty}^{\infty} h(u, y)\mathrm{d}u \right\} \tag{7.42}$$

Consequently, the probability of detecting a moving whale as it passes the observer at a distance y becomes

$$g(y) = 1 - Q(\infty, y)$$

$$= 1 - \exp\left\{ -\int_{-\infty}^{\infty} h(u, y)du \right\}$$ (7.43)

Equation 7.43 constitutes the cumulative detection function for whales that pass the observer at a distance y.

For omni-directional sensors, as with most PAM systems, and for constant whale depth, the conditional probability of detection of a whale (detection performance) will depend only on the relative range between whale and observer, that is

$$h(x, y) = P_{\text{det}}(r)$$ (7.44)

where $P_{\text{det}}(r)$ is the range-dependent detection probability with r estimated by Equation 7.36.

To better reflect the range dependency of P_{det}, we change the integration variable in Equation 7.43 from x to r and, using the fact that Equation 7.36 implies $r dr = x dx = \sqrt{r^2 - y^2} dx$, that the range integration is defined only for ranges $r > y$, and that the range is only positive, we obtain for the probability of detecting a passing whale

$$g(y) = 1 - \exp\left\{ -2\int_{y}^{\infty} \frac{r}{\sqrt{r^2 - y^2}} P_{\text{det}}(r)dr \right\}$$ (7.45)

7.2   Mitigation: absence estimation

The estimation of cetacean absence is a typical question for risk mitigation, where one wants to minimize the impact of anthropogenic activity on whales and dolphins. During surveys that are aimed at abundance estimates, one covers some area in a given time, counts the detected animals and estimates the population N or density D. The better the survey is designed, the more accurate will be the result. Absence estimation, however, is faced with the following problem. If we survey an area for given amount of time and find an animal, then we know there is at least one animal in the area. If, however, we do not detect an animal, we cannot conclude that there is no animal in the area. We simply may not have observed for long enough to see the animal. This is mainly because animals may not be detected all the time. For example, a visual observer will miss all submerged animals, and passive acoustics will not detect animals that are not acoustically active. Obviously, if one observes long enough in an area that is big enough to cover the full range of animal's behaviour, an animal that is present in the area should be detected with probability close to one and if no animals are detected, then one may conclude with high confidence that animals are absent. The question becomes one of how much observation is good enough, and this is a statistical problem (see, e.g., Peterson and Bayley, 2004).

Not detecting an animal may be explained by two mutually exclusive situations

- animals are present but not detected
- animals are absent and no decision for presence is made

We may write the total probability of not detecting an animal as

$$P(C_0) = P(C_0|A)P(A) + P(C_0|\sim A)P(\sim A) \tag{7.46}$$

where C_0 denotes the event that a PAM decision results in the absence of an animal, and A indicates that one or more animals are present, with $\sim A$ meaning that no animals are present.

What we now want to know is $P(A|C_0)$, the *a posteriori* probability of having one or more animals present (event A), under the condition that no animal was detected (conditional event C_0).

Using Bayes' theorem

$$P(C_0|A)P(A) = P(A|C_0)P(C_0) \tag{7.47}$$

we obtain for the *a posteriori* probability

$$P(A|C_0) = \frac{P(C_0|A)P(A)}{P(C_0|A)P(A) + P(C_0|\sim A)P(\sim A)} \tag{7.48}$$

Realizing that $P(C_0|A)$ is the probability of missing an animal and therefore

$$P(C_0|A) = 1 - P_{\text{det}} \tag{7.49}$$

and that $P(C_0|\sim A)$ is the correct decision on animal absence, which is related to the probability of false alarm by

$$P(C_0|\sim A) = 1 - P_{\text{FA}} \tag{7.50}$$

we may rewrite Equation 7.48 as

$$P(A|C_0) = \frac{(1 - P_{\text{det}})P(A)}{1 - P_{\text{det}}P(A) - P_{\text{FA}}(1 - P(A))} \tag{7.51}$$

where we used the fact that animals are either present or absent, that is $P(A) + P(\sim A) = 1$.

If we assume that the probability of false alarm P_{FA} is very small, as it should be for a promising PAM system, then we can approximate Equation 7.51 by

$$P(A|C_0) = \frac{(1 - P_{\text{det}})P(A)}{1 - P_{\text{det}}P(A)} \tag{7.52}$$

or

$$P(A|C_0) = 1 - \frac{1 - P(A)}{1 - P_{\text{det}}P(A)} \tag{7.53}$$

For a given *a priori* probability $P(A)$ that one or more animals are present in the area of interest, Equation 7.51 varies with the detection probability P_{det} and the probability of

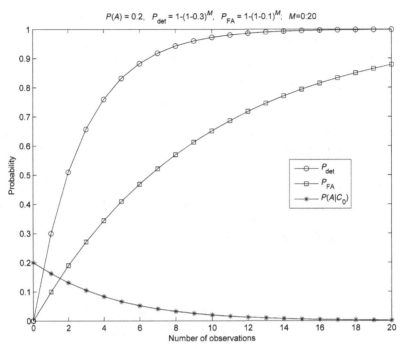

Figure 7.2   Probability of presence when no animal is detected.

false alarm P_{FA}. Let both probabilities, P_{det} and P_{FA}, be the result of multiple independent observations M, then we may write

$$P_{det} = 1 - (1 - p_{det})^M \qquad (7.54)$$

and

$$P_{FA} = 1 - (1 - p_{fa})^M \qquad (7.55)$$

where p_{det} and p_{fa} are the detection and false alarm probabilities of the individual observations.

Figure 7.2 shows the dependency of the *a posteriori* probability (Equation 7.51) as a function of increasing number of independent observations. The detection and false alarm probabilities of the individual observations are in this example assumed to be 0.3 and 0.1, respectively. For M = zero, i.e. no observation, the result corresponds to the assumed *a priori* probability $P(A) = 0.2$, that is, the probability of finding, without any observation, a whale in the area is 0.2. As the number of observations increase, the probability of both detection and false alarm approaches one, but the probability of animal presence, if none has been detected, decreases from the *a priori* value towards zero.

So far, we have considered the number of independent observations as repeated measurements. The same formula holds, however, if instead of multiple measurements, we have multiple animal locations, as long as the detections of these animals are independent events. Therefore, M is really the product of independent measurements and independent animal locations.

To answer the question about how many observations one should make to be confident that the area is cetacean-free, one first needs to establish the *a priori* probability that there could be some animals in the area. Further, we need to know both the detection and false alarm probability of the individual observations, and finally we should have input on the threshold for the *a posteriori* probability that must be crossed for considering the area cetacean-free. Obviously, this last threshold must be determined by policy makers that are interested in risk mitigation.

7.3   Habitat and behaviour analysis

While abundance and absence estimation aim for a single number (abundance, density, or probability of absence), recent cetacean research aims to understand why whales and dolphins are detected at certain locations but not in other parts of the world. In this sense, habitat analysis is driven by the empirical observation that cetaceans are not uniformly distributed throughout the world, but their distribution shows spatial and temporal heterogeneity.

Abundance and absence estimation, discussed before, assumed that the *a priori* probability of animal presence is homogeneous throughout the area of interest. Habitat analysis, on the other hand, assumes that presence or abundance is a function of space and time. To be able to characterize these spatial and temporal dependences, it is necessary to carry out multiple measurements both in space and in time.

Habitat analysis can be seen as a modelling effort that has as output abundance or probability of presence as a function of some describing, or explaining, variables, which are also called covariates. The temporal and spatial variation of presence, or occupancy, in a habitat is strongly related to the animal's behaviour. Temporal variations include diurnal variation, e.g. different behaviour by day and night, seasonal variation, e.g. winter versus summer habitat, and also include habitat preferences as function of life history. Spatial variations are related to food presence, water temperature and geographic particularities, either directly or via the food chain, but they are also related to different stages in the animal's life history.

Typically, habitat analysis tries to correlate survey results, e.g. presence or absence of animals, with a variety of covariates to establish positive or negative correlation. Developing proper and efficient sampling strategies may become very difficult if the number of covariates increases. It seems obvious that the sampling interval has to be fine enough so that changes in the model variables may uniquely be identified. This requirement is equivalent to the Nyquist theorem in data acquisition systems, which requires that sampling must be fine enough to cover significant features at least twice (one wavelength covers a significant feature of a sound wave).

Although it is convenient to present habitats finally as maps, that is, as a function of spatial variables, geographic features are not necessarily a key variable for explaining habitat occupancies. Habitat usage is mostly driven by animal behaviour and consequently it would make sense to select as habitat descriptors variables that may influence this animal behaviour. To give an example: foraging cetaceans require the presence of food, and deep-diving sperm and beaked whales forage on deep-sea prey, e.g. squid. The amount of squid would therefore be a good covariate for assessing the habitat occupancy of deep divers. However, this example is also a good one to demonstrate that biologically relevant covariates may sometimes be completely impracticable. It is already hard to detect deep divers; it is still harder to measure independently the abundance of deep-sea squid. In fact, some of the deep-sea squid species are mostly known from the stomach contents of stranded or harvested sperm and beaked whales or other deep divers.

Even if we cannot know such deep-sea biological covariates, it is clear that deep-diving foraging habitats are less variable in time and more influenced by slowly varying ocean climatology, contrary to habitats of whales and dolphins that forage most of the time in shallow waters and are confronted with diurnal and seasonal changes in the ocean. Deep-sea currents and consistent upwelling, which bring benthic nutrients into the photic zone close to the ocean surface, are therefore good candidates for covariates that describe the foraging habitats of deep divers.

Studying these remote and potentially hostile habitats could become the realm of autonomous PAM systems that are designed to descend to the foraging layers, carrying out not only acoustic but also oceanographic measurements. Such autonomous systems should ideally measure not only the animal's acoustic activity but also covariates that are related to the animal's presence in the sampled habitat, allowing a cost-effective complex sampling of cetacean habitats both in space and in time.

Modelling covariates

Let $\mathbf{x}$ be a vector of K covariates of the habitat description with $\mathbf{x} = \{x_0, \cdots, x_K\}$ and let y be some measure of the presence of animals in the habitat, which in general is a function of the covariates

$$y = f(\mathbf{x}) \tag{7.56}$$

The simplest functionality is a linear additive relation (linear additive model), where we may write

$$y_i = \sum_{k=0}^{K} \beta_k x_{ki} \tag{7.57}$$

or in matrix notation

$$\mathbf{y} = \mathbf{X}\boldsymbol{\beta} \tag{7.58}$$

Assuming that the inverse $(\mathbf{X}^T\mathbf{X})^{-1}$ exists, that is, if the covariates are linearly independent, then we obtain the solution by forming the LMS estimate

$$\boldsymbol{\beta} = (\mathbf{X}^T\mathbf{X})^{-1}\mathbf{X}^T\mathbf{y} \tag{7.59}$$

In cases where the covariates are linearly dependent, the inverse $(\mathbf{X}^T\mathbf{X})^{-1}$ does not exist. If the dependent variables cannot be identified by inspections, than one could use the pseudo- or Moore–Penrose inverse to obtain a solution for the covariates, using for example the `pinv` function in Matlab, but this would hide potential problems with the selection of our covariates. As we wanted to describe the habitat with the best choice of covariates, a direct analysis and treatment of the problems is preferable.

Equation 7.59 describes the ordinary least mean squares (LMS) regression, which could be considered as the workhorse of data analysis, but which may be too general for our purpose of habitat analysis, as there are no limits or constraints on the solution. As the covariates are only approximations to the true dependencies of the whale presence, the solution of Equation 7.59 may result in unrealistic presence probabilities when reinserted in Equation 7.58. For example, if the presence measure, or response variable (y in Equation 7.56), corresponds to the probability p_i of detecting a whale, then this probability is constrained to values between zero and one. A solution based on noisy measurements of the covariates may result in y values that are outside these boundaries and therefore meaningless. It is therefore desirable to modify the LMS procedure to guarantee meaningful solutions.

To continue with the example, let p_i, be the probability of detecting a whale in a given habitat, then the odds of success, i.e. of finding the whale, are given by the ratio of probability of success to probability of failure

$$\text{odds} = \frac{p_i}{1 - p_i} \tag{7.60}$$

If we now assume that the odds of success are simply the product of exponential functions of the covariates

$$\frac{p_i}{1 - p_i} = \prod_{k=1}^{K} \exp\{\beta_k x_{ki}\} \tag{7.61}$$

then we obtain a linear form similar to Equation 7.57 by taking the natural logarithm of the odds

$$y = \ln\left(\frac{p_i}{1 - p_i}\right) = \sum_{k=0}^{K} \beta_k x_{ki} \tag{7.62}$$

The logarithm of the odds of success is also known as the logistic function, or logit of p_i. It is further also called a link function in the terminology of generalized linear models (GLM) as it provides the relation, or link, between the linear predictor y and the response variable p_i.

Using the logit function transforms a response variable that is bound between zero and one into a linear predictor that may vary between minus and plus infinity, allowing a straightforward and safe application of the LMS regression as presented in Equation 7.59.

To generalize the mathematical framework further, software tools have been developed to replace the simple linear additive model with a generalized additive model (GAM) (Hastie and Tibshirani, 1990) where Equation 7.57 is replaced by a more general notation

$$y_i = \sum_{k=0}^{K} f_k(x_{ki}) \tag{7.63}$$

where the different additive components may now be unspecified functions of the covariates.

In simple terms, GAM does not require the functional dependence of covariates to be specified, but uses the inter-covariate scatter matrices to generate a generic functional dependency.

7.3.2   Acoustics and habitat analysis

Different habitats may not only be characterized by discriminating features, but may also be associated with specific animal behaviour. As there is a relation between acoustic activity and animal behaviour, acoustics can play an important role in establishing the behavioural state of the animals in relation to the habitat selection. Using acoustics to differentiate between foraging and socializing is straightforward for most species, especially for odontocetes, as the acoustic activity during foraging (echolocation clicks) is easily differentiated from social sounds (whistles, stereotyped clicks, etc.). Autonomous passive acoustic systems may be especially suited to monitoring the acoustic behaviour of the animals without observer interference over long timescales and in large areas.

7.4   Monitoring rare and elusive species

In most survey methods, statistical analysis plays an important role in inferring the presence of missed objects. This may be the case when the area inhabited by the species of interest is too large to be surveyed completely, or when the species is not available for detection (a silent whale cannot be detected by listening), or when the detection performance of the monitoring system is not perfect. In other words, the number of detected animals n is related to the number of existing animals N by Equation 7.2. Animals may be considered as rare and elusive, at least for the purpose of surveys, if one of the three probabilities in Equation 7.5, i.e. P_{area}, P_{avail} or P_{det}, is very small, resulting in a very small overall detection probability P_N.

A small P_N is not desirable for at least two reasons. First, owing to lack of successful detections, it is difficult to estimate the true value of P_N. Second, the remaining uncertainties in the estimate of P_N result in such a variation of the abundance estimate that the survey lacks confidence.

It seems obvious that, depending on which of the three probabilities drives the overall detection probability down, there should be a different approach to increase this overall detection probability to reasonable values. As the conditional detection probability P_{det} describes essentially the system performance, improving it requires a change in the setup of the detection procedure. The other two probabilities, P_{area} and P_{avail}, may be improved by operational changes, that is, by adapting the survey to the particularities of the monitored species.

The probability of covering an animal may easily be improved by increasing the number of sample units, i.e. by covering better the habitat occupied by the animal. This is easier said than done and the reasons why the whole area is not surveyed are manifold.

Surveys can be very expensive and this is especially true for ship-based surveys with continuous coverage of the suspected habitat. The survey cost may be reduced if *a priori* information on the species habitat exists and when the survey is adapted to the expected habitat by proper stratification of the survey. Stratification of the survey then allows a better use of resources without introducing biases in the monitoring task. If the species is highly mobile, and most cetaceans are mobile, then one could increase the probability of encountering an animal while reducing the survey cost, by placing PAM systems at selected locations and counting the animals that are passing by these listening stations.

The probability that animals are available for detection depends on the animals' behaviour, that is, when they are acoustically active and when silent. In the case of PAM the probability may be improved by listening for long enough to bridge the time span when animals are silent. The need to listen for extended periods implies, however, either slowly moving surveys, or stationary systems. As a rule of thumb, one can state that the maximal encountered speed (animal + sensor) should be less than the quotient that results from dividing the diameter of the detection area by the time for which the animal is expected to be silent.

Having stationary or slowly moving PAM systems may, however, be subject to a biased estimate of animal presence/abundance, as the risk of multiple encounters of the same animal exists. Again, the expected behaviour of the species of interest is key to determining the sampling strategy.

Overall, the total detection probability may be increased by adapting the survey strategy (duration and locations of sampling) to the animal's behaviour. If this behaviour is known to depend on covariates then statistical analysis is facilitated if the sampling densities correspond to the expected statistics of the covariates.

7.5 Logging acoustic information

PAM applications are notorious for producing a vast quantity of data; the collection of terabytes of acoustic data is not uncommon during a month-long acoustic survey. Good acoustic information logging facilitates not only data processing post-survey, but may provide the proper detection history that is required by the PAM application.

Acoustic logging collects the occurrence of specific acoustic events and should contain three different types of information: meta data describing the context of the event (e.g. time of event, location of the observer, depth of receiver), description and status of PAM system (e.g. individual carrying out the PAM operation, on or off effort, relevant PAM characteristics, environmental conditions), and acoustic description of event (e.g. type, loudness, duration, quantity of sound, direction or location of sound source).

Associating locations with acoustic events allows the use of GIS (geographic information systems) to visualize acoustic detections in an intuitive way. Even if the location of the animal is the datum of interest, it is also important to log continuously the location of the observer. This allows us not only to assess the effort of the PAM operations but also to have an approximate estimate of the animal's location in case the PAM system fails to provide a precise animal location.

Logs are usually products that are generated by human observers, but they may also be produced by automated data processing schemes. In fact, a log could be considered as useful if the elaboration of the logged information does not require human intervention but can be carried out by computer software. One of the most important aspects in designing a logging tool is to standardize the way in which events are recorded. This is of particular relevance if different operators are involved in the PAM application, as different logging styles may impede efficient analysis of the acoustic information log.

As expected, logging of acoustic information can be done in different ways. A straightforward method is to timestamp sequentially every single event. This is easily implemented and very flexible, and much of the work may be delegated to the computer. Locations of the PAM system, for example, could be acquired continuously from GPS receivers and it is therefore easy to log the PAM position at regular intervals. Timestamp logging of acoustic events is practicable if the events are either rare or well defined (e.g. start of a click train) and may become difficult when the event is not well defined, or only known in hindsight. For example, it is sometimes not easy to establish when an echolocation click train of a sperm whale has ended, as sperm whales regularly interrupt the echolocation click trains by more or less extended pauses. In addition, if multiple events occur in a short time the logging effort for human operators may be too high to be practicable.

An alternative logging method is to aggregate the information during short periods, and to report at regular intervals, say every minute, only a summary of the events of interest. This procedure is easier for the human operator, especially if the quality and quantity of the report is limited and restricted to the information of interest. The method is furthermore suited for autonomous PAM systems where the communication bandwidth is limited. To generate aggregate logs one needs, however, the complete DCLT signal processing chain, the implementation of which may or may not be easily available.

8 Detection functions

We saw in the previous chapter that the detection function $g(r)$ plays an important role in assessing the population density by using survey methods. This function reflects the observation that the surveyed objects, here acoustically active whales and dolphins, are harder to detect if they are at greater distances or, equivalently, that the detection capability of any observer is usually limited in range.

8.1 Empirical detection function

There is a variety of methods for obtaining the detection function, but in most survey applications, one starts with empirical histograms of detected whales as a function of distance from the observer and obtains the detection function $g(r)$ by fitting an appropriate model to the data.

To model the detection function, the data are frequently approximated by a hazard function

$$g(r) = 1 - \exp\left\{-\left(\frac{r}{\sigma}\right)^{-b}\right\}$$ (8.1)

where the two parameters b and σ are to be estimated by fitting the hazard function to the empirical detection function, or histogram.

Alternatively, a half-normal function may be used

$$g(r) = \exp\left\{-\frac{1}{2}\left(\frac{r}{\sigma}\right)^2\right\}$$ (8.2)

where both functions (Equations 8.1 and 8.2) are defined only for positive ranges r.

Whereas the hazard function has two parameters for fitting the model to the empirical detection histograms, the half-normal model has only one parameter, reducing its flexibility to fit the data. If more parameters are needed to better fit the data, the selected basic detection function could be extended by multiplying it with a series of corrective terms (Buckland *et al.*, 2001).

The detection functions as modelled by Equations 8.1 and 8.2 are really conditional probabilities as they imply that the objects are detectable, if the whales are within the survey area and acoustically active in the case of PAM. This can be seen from the fact that both models become one at zero ranges, $g(0) = 1$. This makes sense, as the purpose of the detection function is to characterize the performance of the observer and not the behaviour of the surveyed objects. A $g(0) < 1$ may be modelled by multiplying Equations 8.1 and 8.2 by a constant $g_0 < 1$.

8.2   LMS parameter estimation

Given an empirical detection function in the form of a histogram $h(r)$ and a modelled detection function $g(r; \mathbf{q})$ where $\mathbf{q}$ is a vector of parameters that should be estimated, then we may define the mean square error $E(\mathbf{q})$ by

$$E(\mathbf{q}) = \int_0^\infty (h(r) - g(r; \mathbf{q}))^2 \mathrm{d}r \tag{8.3}$$

The LMS parameter estimation obtains a best estimate of the parameters $\mathbf{q}$, say $\hat{\mathbf{q}}$, by minimizing the quadratic error

$$E(\hat{\mathbf{q}}) = \min_q \{E(\mathbf{q})\} \tag{8.4}$$

The solution vector $\hat{\mathbf{q}}$ may be found via an exhaustive search, with standard gradient methods, or by other LMS techniques.

If the model is a non-linear function of the parameters, then the parameter estimation becomes somewhat complex, but most data analysis tools provide a means to carry out a non-linear least squares fit.

8.3   Sonar equation-based modelling of the detection function

While mathematically useful for theoretical analysis, the hazard function, or other functions that approximate the detection function, do not provide satisfactory insight into, or allow assessment of the detection process. This may not be necessary for the sole estimation of abundance, but may be useful in guiding the improvement of survey instrumentations. For example, PAM uses more or less sophisticated technology where different technical parameters may have different impacts on the detection function, and it would be desirable to have a feedback from the application for improvements in the system design. Modelling the detection function may also be necessary when attempting to estimate habitat-dependent covariates, because it is then necessary to differentiate between habitat covariates and PAM system parameters.

From the sonar equation we know that an acoustically active whale may be detected if the signal-to-noise ratio (*SNR*) of the received sound exceeds a given threshold *TH*, that is, the detection function is nothing else than the probability that the *SNR* exceeds *TH*

$$g(r) = P_{\text{det}}(r) = \Pr\{SNR(r) > TH\} = \int\limits_{x > TH} w_{SNR}(x; r)\mathrm{d}x \qquad (8.5)$$

where $w_{SNR}(x; r)$ is the PDF of SNR at the range r and where the integral runs over all SNR values (variable x) that exceed TH.

Using the sonar equation for passive sonar (Equation 3.2), we may expand the received SNR

$$SNR(\vartheta, r) = SL(\vartheta) - TL(r) - NL_0 - 10 \log B \qquad (8.6)$$

where array and processing gain (AG, PG, respectively) have been set to zero for simplicity.

Before presenting the formula for the probability of detection, it is convenient to introduce a term that in sonar applications is called signal excess, SE, which is the difference between received SNR and the detection threshold TH:

$$\begin{aligned} SE(\vartheta, r) &= SNR(\vartheta, r) - TH \\ &= SL_0 - DL(\vartheta) - TL(r) - NL_0 - 10 \log B - TH \qquad (8.7) \\ &= FOM_0 - TL(r) - DL(\vartheta) \end{aligned}$$

where SL_0 is the on-axis, or nominal, source level. FOM_0 is what is called the figure of merit of the PAM system, combining source and system quantities.

We may realize that in terms of Equation 8.7 detections only occur if FOM_0 exceeds the sum of propagation loss $TL(r)$ and off-axis attenuation $DL(\vartheta)$. As mentioned in Chapter 3, this off-axis attenuation is especially relevant for echolocation clicks. There are good reasons why this may be not the case for communication signals, the calls of baleen whales and the whistles of dolphins, but this has not been well quantified. For the time being, we consider the most general case of possible off-axis attenuation.

As acoustically active whales may have a variable (on-axis) source level we should also consider the nominal source level SL as a random variable centred around SL_0 with PDF $w_{SL}(x)$, where x varies over all possible source levels. By varying the source level, the FOM is no longer constant but becomes a function of the source level. As we may assume that the source level variation is independent of the off-axis angle, we estimate the effective detection probability by integrating over all possible source levels

$$g(r) = \int\limits_{-\infty}^{\infty} w_{SL}(x) \Pr\{FOM(x) - TL(r) > DL(\vartheta)\}\mathrm{d}x \qquad (8.8)$$

To estimate the probability $\Pr\{FOM(x) - TL(r) > DL(\vartheta)\}$ that the on-axis signal excess exceeds the off-axis attenuation we need to quantify the off-axis distribution $w_{OA}(\vartheta, r)$, so that

$$\Pr\{FOM(x) - TL(r) > DL(\vartheta)\} = \int\limits_{FOM(x) - TL(r) > DL(\vartheta)} w_{OA}(\vartheta, r)d\vartheta \qquad (8.9)$$

Inserting Equation 8.9 into Equation 8.8, we obtain for the detection function

$$g(r) = \int\limits_{-\infty}^{\infty} w_{SL}(x) \left(\int\limits_0^{DL^{-1}(FOM(x)-TL(r))} w_{OA}(\vartheta, r) d\vartheta \right) dx \tag{8.10}$$

where the variable x runs over all possible source levels and the off-axis angle ϑ varies from zero (on-axis) to the maximal value given by the DL^{-1} function.

In order to model the detection function in terms of the sonar equation we have to specify two distribution functions: $w_{SL}(x)$, which is the PDF of the on-axis source level, and $w_{OA}(\vartheta, r)$, the PDF describing the off-axis distribution. It should be noted that the detection function depends on range r via the range-dependent $TL(r)$ and also the range-dependent off-axis PDF $w_{OA}(\vartheta, r)$.

For the on-axis source level, it is convenient to assume that the PDF is normally distributed around a nominal average value SL_0 and standard deviation σ_{SL}

$$w_{SL}(x) = \frac{1}{\sqrt{2\pi}\sigma_{SL}} \exp\left\{ -\frac{1}{2}\left(\frac{x - SL_0}{\sigma_{SL}} \right)^2 \right\} \tag{8.11}$$

This Equation 8.11 is a suitable model if no better information becomes available on the distribution of the source level. In cases where more insight into the sound generation process indicates deterministic or statistical features, modifications to Equation 8.11 could be necessary. A source level variation of $\pm$ 10 dB is frequently observed for echolocation clicks, resulting in a σ_{SL} of about 5 dB (10 dB = 2 σ_{SL}).

8.3.1   Modelling the off-axis attenuation

Obviously, the off-axis PDF is only of interest if the sound sources are directional and not omni-directional. This is especially appropriate for echolocation clicks used for foraging, as communication signals are expected to be broadcast without (or with minimal) directional preference.

As shown in Section 3.2, the off-axis attenuation of a typical broadband echolocation click may be approximated by

$$DL(\vartheta) = C_1 \frac{(C_2 \sin \vartheta)^2}{1 + C_2 \sin \vartheta + (C_2 \sin \vartheta_1)^2} \quad 0 \le \vartheta \le \frac{\pi}{2} \tag{8.12}$$

where DL is measured in dB and C_1 and C_2 are two parameters to be estimated either from the detection data or independently. Typical values are $C_1 = 47$ dB and $C_2 = 0.218ka$, where ka is related to the expected directivity index (Equation 2.24). As a directivity index of 25 dB is a reasonable assumption for most echolocating whales and dolphins, an off-axis attenuation that is modelled with $ka = 17.8$, as shown in Figure 3.1, may even be considered as a universal first estimate.

Obviously, we have

$$0 \le DL < DL_{\max} \tag{8.13}$$

where

$$DL_{\max} = C_1 \frac{C_2{}^2}{1 + C_2 + C_2{}^2} \tag{8.14}$$

as the off-axis attenuation varies for angles between zero and 90°. For angles that exceed 90°, that is, for off-axis angles behind the animal, we assume the off-axis attenuation to be constant and equal to the value at 90°. This last assumption is only justified in absence of better knowledge. It may very well be that signals received behind the animal are attenuated by the animal's body. In addition, the assumption does not hold for the sperm whale, where sound is originally transmitted backwards, resulting in different directivity patterns and therefore different off-axis attenuation for each pulse within the clicks.

To integrate Equation 8.10, we need the inverse function of the off-axis attenuation, that is, we need a function that results, for any off-axis attenuation, in the off-axis angle. For this, we solve Equation 8.12 for the angle ϑ and obtain

$$\vartheta(DL) = DL^{-1}(DL(\vartheta))$$

$$= \sin^{-1}\left(\frac{1}{2C_2}\left(\frac{DL}{DL - C_1}\right)\left\{-\sqrt{1 - 4\left(\frac{DL - C_1}{DL}\right)} - 1\right\}\right) \tag{8.15}$$

which is valid for $DL < DL_{\max}$.

8.3.2 Off-axis distributions of echolocation clicks

Limiting the following discussion to directional echolocation clicks, we should differentiate two cases. First, we estimate the detection probability of an arbitrary click; second, we consider the detection probability of the clicks when the animal points generally towards the receiver, that is, when we expect to receive the loudest clicks.

Detecting an arbitrary echolocation click

If we assume that there is no preference in the off-axis angle and that the angles ϑ between click and hydrophone directions are uncorrelated then the off-axis angle ϑ should be uniformly distributed on the sphere, suggesting that the off-axis distribution of arbitrary clicks may be modelled as

$$w_{OA\_A}(\vartheta) = \frac{1}{2}\sin\vartheta \tag{8.16}$$

Equation 8.16 is the exact distribution function for off-axis angles that are uniformly distributed on the sphere and real distributions may differ somewhat, but Equation 8.16 is a reasonable first approximation.

Detection of near on-axis echolocation clicks

The loudest clicks are detected if the animal moves or points towards the receiver, that is, if the receiver is on-axis with respect to the whale orientation. As echolocating whales or dolphins are searching for food, it is intuitive to assume that the animal will illuminate a forward sector that is somewhat larger than the acoustic beam-width. This will result in small variations in off-axis values around the main acoustic beam direction. This variation may be due to small oscillating changes in swim direction or explicitly generated by head motion.

To model these small deviations of acoustic beam directions, we assume that the head is moving randomly relative to a main acoustic axis (or nominal on-axis). We assume further that the searching animal moves the head without horizontal or vertical preference; then it is convenient to describe the head motion by a two-dimensional Gaussian distribution around the nominal on-axis. The resulting off-axis values will therefore be Rayleigh distributed:

$$w_{OA\_H}(\vartheta) = \frac{\vartheta}{\sigma_H{}^2}\exp\left\{-\frac{1}{2}\left(\frac{\vartheta}{\sigma_H}\right)^2\right\} \tag{8.17}$$

Equations 8.16 and 8.17 present two different ways of modelling the off-axis distribution, indicating the importance of the definition of the objective of the detection process. As these off-axis statistics depend on the way in which we treat the echolocation clicks, it seems obvious that the detection function will depend on the implemented signal processing method. In particular, we expect that the general form of the detection function will be application- and receiver-specific. This lack of generality indicates that it will be difficult to specify a single detection function for all PAM applications. On the other hand, using the particularities of the PAM system would allow an improved or even optimal design of the detection function.

8.3.3   Sonar equation based detection function

Considering the case where we are interested in the detection of any echolocation click, that is, we use Equation 8.16 for the off-axis PDF, we obtain for the integral of Equation 8.9

$$\int_0^{DL^{-1}(FOM(x)-TL(r))} \sin(\vartheta)\mathrm{d}\vartheta = \frac{1}{2}\left(1 - \cos\left(DL^{-1}(FOM(x) - TL(r))\right)\right) \tag{8.18}$$

where $FOM(x) - TL(r)$ is limited to values less than DL_{max}. For values greater than DL_{max}, that is, the whale will be detected from all angles, and Equation 8.18 should integrate to one.

The on-axis figure of merit $FOM(x)$ as defined implicitly in Equation 8.8, may be written as

$$\begin{aligned} FOM(x) &= x - (NL_0 + 10\log B + TH) \\ &= x - F_0 \end{aligned} \tag{8.19}$$

that is, we separate the variable source level from constants that describe environmental and PAM system parameters. The nominal figure of merit is then given by letting the variable x become the nominal source level SL_0

$$FOM_0 = SL_0 - F_0 \tag{8.20}$$

Before proceeding with the integration of Equations 8.9 or 8.10, it is convenient to differentiate three cases:

$$x < F_0 + TL(r) \tag{8.21}$$

where no detection is possible;

$$F_0 + TL(r) < x < F_0 + TL(r) + DL_{\max} \tag{8.22}$$

where Equation 8.18 applies; and

$$x > F_0 + TL(r) + DL_{\max} \tag{8.23}$$

where detection always occurs, so that we obtain for the detection function

$$g(r) = \int_{F_0+TL(r)+DL_{\max}}^{\infty} w_{SL}(x)\mathrm{d}x$$

$$+ \int_{F_0+TL(r)}^{F_0+TL(r)+DL_{\max}} w_{SL}(x)\left(\int_0^{DL^{-1}(x-F_0-TL(r))} w_{OA}(\vartheta, r)\mathrm{d}\vartheta\right)\mathrm{d}x \tag{8.24}$$

where the integral that corresponds to Equation 8.21 has been omitted from the equation as it does not contribute to the detection function.

Now using Equation 8.18, we obtain

$$g(TL) = \int_{F_0+TL+DL_{\max}}^{\infty} w_{SL}(x)\mathrm{d}x$$

$$+ \frac{1}{2}\int_{F_0+TL}^{F_0+TL+DL_{\max}} w_{SL}(x)\left(1 - \cos\left(DL^{-1}(x - F_0 - TL)\right)\right)\mathrm{d}x \tag{8.25}$$

where we use the fact that the transmission loss TL is now the only function of range and that it is convenient to consider the detection function as a function of permissible TL and not as a function of range r to emphasize that the integral may be carried out without knowledge of the functional dependences of the transmission loss.

With the following notation

$$\int_{x_0}^{\infty} w_{SL}(x)\mathrm{d}x = \Phi(x_0) \tag{8.26}$$

we may write the detection function as

$$g(TL) = \Phi(F_0 + TL + DL_{\max})$$

$$+ \frac{1}{2} \int_{F_0+TL}^{F_0+TL+DL_{\max}} w_{SL}(x)\left(1 - \cos\left(DL^{-1}(x - F_0 - TL)\right)\right)\mathrm{d}x \qquad (8.27)$$

or, by letting $u = x - F_0 - TL$

$$g(TL) = \Phi(F_0 + TL + DL_{\max})$$

$$+ \frac{1}{2} \int_0^{DL_{\max}} w_{SL}(u + F_0 + TL)\left(1 - \cos\left(DL^{-1}(u)\right)\right)\mathrm{d}u \qquad (8.28)$$

The detection function is composed of two components, one integrating over all source levels for which detections occur, independent of off-axis angles, and a second term that handles the cases where the off-axis angles matter.

Using the Gaussian source level distribution (Equation 8.11) and with Equation 8.20, we obtain

$$w_{SL}(u + F_0 + TL) = \frac{1}{\sqrt{2\pi}\sigma_{SL}} \exp\left\{ -\frac{1}{2}\left(\frac{u + TL - FOM_0}{\sigma_{SL}}\right)^2 \right\} \qquad (8.29)$$

With a nominal $FOM_0 = 116\,\mathrm{dB}$ and a standard deviation of $\sigma_{SL} = 5\,\mathrm{dB}$ we obtain a probability of detection that is shown in Figure 8.1.

Up to a transmission loss of about 70 dB, that is, for $TL < FOM_0 - DL_{\max}$, all clicks are detected and for transmission loss greater some 116 dB, or $TL > FOM_0$, none of the clicks is detected. Increasing FOM_0 will shift the whole curve to the right; decreasing FOM_0 will shift the curve to smaller TL values. Figure 8.1 is strictly related to the PAM system parameters and holds for all variations in the system description, as long the FOM_0 remains the same. Ignoring the off-axis attenuation (the solid line in Figure 8.1) results in a reasonable estimate for the situation where all clicks are detected, but underestimates the TL for which no clicks are detected. This is to be expected, as higher permitted TL will always result in greater detection ranges.

Using, for the transmission loss estimation, a spherical spreading law with an absorption coefficient of 9.5 dB/km, which is typical for 40 kHz signals, Figure 8.1 translates into Figure 8.2, where the detection function is plotted as a function of range.

The advantage of using the sonar equation to model the detection function is that it immediately gives a feeling for what the PAM system can achieve and what will be hard to implement.

The detection function presented in Figure 8.2 is similar to a hazard function $h(r) = 1 - \exp\left\{-\left(\frac{r}{1647}\right)^{-4.531}\right\}$, which, however, provides less insight into the detection performance of the PAM system. Only the 1647 m indicates the range where the hazard function becomes 0.63.

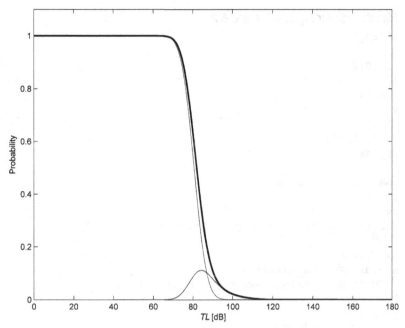

Figure 8.1   Detection probabilities as function of allowed transmission loss. The dashed curve indicates the first component of Equation 8.27, the thin curve the second component of the equation and the thick curve the sum of both components.

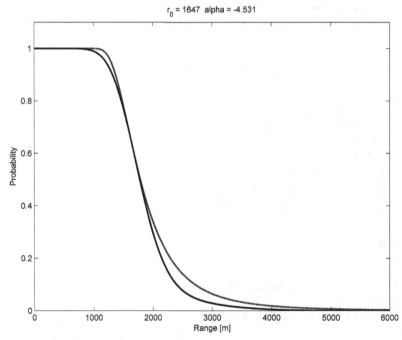

Figure 8.2   Detection function $g(r)$ for a nominal FOM_0 of 116 dB. The dashed line is an approximation of a hazard function to the modelled detection function.

Matlab code for Figures 8.1 and 8.2

```matlab
%Scr8_1
%
TL=(0:0.1:180)';
%
FOM=116;
%
C1=47;
C2=0.218*17.8;
DL_max=C1*C2^2/(1+C2+C2^2);

%Gaussian weighting
sig=5;
Wx=@(x,xo) ...
   1/(sqrt(2*pi)*sig)*exp(-1/2*((x-xo)/sig).^2);

dx=0.1;
x=0:dx:(FOM+DL_max+4*sig);
%FOM+DL_max+4*sig is assumed to be infinity
%
g1=0*TL;
for ii=1:length(TL);
    W=Wx(x+TL(ii),FOM-DL_max);
    g1(ii)=sum(W)*dx;
end
%
th=asin(1/(2*C2)*(x./(x-C1))...
    .*real(-sqrt(1-4*((x-C1)./x))-1));
%
th(1)=0; % for x=0;
th(x>=DL_max)=pi; % for x>=DL_max
%
x2=x(x<DL_max);
x3=x(x>=DL_max);
th2=th(x<DL_max);
g2=0*TL;
g3=0*TL;
for ii=1:length(TL)
    W2=Wx(x2+TL(ii),FOM);
    g2(ii)=0.5*sum(W2.*(1-cos(th2)))*dx;
    W3=Wx(x3+TL(ii),FOM);
    g3(ii)=sum(W3)*dx;
end
gr=g3+g2;

%inverting the spherical spreading law
R=1:0.1:6000;
TLR=20*log10(R)+9.5*R/1000;

r=interp1(TLR,R,TL);
y=real(log(-log(1-gr)));
```

```
x=log(r);
%yo=log(-log(1-0.6321));
yo=0;
i1=find(y<=yo,1,'first');
alpha=(y(i1)-y(i1-1))/(x(i1)-x(i1-1));
x0=x(i1-1)+(yo-y(i1-1))/alpha;
r0=exp(x0);
%
figure(1)
plot(TL,g3,'k--',TL,g2,'k-')
line(TL,gr,'color','k','linewidth',2)
ylabel('Probability')
xlabel('TL [dB]')
ylim([0 1.1])

figure(2)
plot(r,gr,'k','linewidth',2)
line(r,1-exp(-(r/r0).^alpha),...
    'color','k','linestyle','--','linewidth',2)
ylabel('Probability')
xlabel('Range [m]')
ylim([0 1.1])
title(sprintf('r_0 = %.0f alpha = %.3f',r0,alpha));
return
```

8.4   Modelling animal behaviour

The animal's behaviour enters the detection function in two instances: the acoustic activity determines the detection opportunities, as only acoustically active animals may be detected by PAM systems; and the animal's motion determines the off-axis distribution, which is part of the model for the detection function.

Whales and dolphins move in the oceans for a specific purpose (e.g. migration, foraging, socializing, resting), that is, an animal may be in one of its multiple behavioural states. It is reasonable to assume that these states are mutually exclusive, and that at any moment in time an animal will be in one of these behavioural states. It is therefore possible to define a state vector describing the probability that an animal may be found in one of the behavioural states. Table 8.1 gives a possible state vector of a hypothetical animal.

Table 8.1 Behavioural state vector of a hypothetical animal

Resting	Socializing	Migrating	Foraging
10%	10%	10%	70%

Table 8.2 Transition matrix between behavioural states

Previous\Actual	Resting	Socializing	Migrating	Foraging
Resting	0%	50%	10%	40%
Socializing	50%	0%	10%	40%
Migrating	5%	35%	0	60%
Foraging	5%	5%	90%	0

Table 8.3 Simplified acoustic state vector

Loquacious	Silent
70%	30%

It seems further obvious that the animal may change its behavioural state. If we compile the probabilities of possible transitions into a table then we obtain a transition table. Table 8.2 gives the transition matrix for a hypothetical animal. The different rows describe the transition probability from the indicated state to all other states. Obviously, the sum of all probabilities in each row must add up to 100%. Animals will always transition from the actual state to another behavioural state. For example, the table gives 40% probability that the hypothetical animal will start foraging after having rested.

It is clear that the behavioural state vector is always somewhat arbitrarily defined and that the differences between the different states are not always clearly recognizable. Although the animal's behaviour, as described in Tables 8.1 and 8.2, seems plausible, it may not be the best characterization for the purpose of PAM performance analysis. As we are concerned with the acoustic detection of cetaceans, we may reduce the behavioural states of the animals to acoustically active and inactive, or silent, as shown in Table 8.3.

A transition table is not necessary for such a binary behavioural state vector, as the transitions occur with 100% probability. Acoustic state tables for cetaceans that exhibit different types of sound emission must differentiate these sound classes and also provide a transition matrix, or restrict themselves to the sound category of interest, that is, the sound class used for detecting the species. For example, if a characteristic up-call is used to detect right whales, say by using a matched filter, then the acoustic state vector should compare the times when we may detect right whale up-calls to the times when the detector should remain silent. Obviously, the animal may emit other types of sound, e.g. tonal moans, but for the PAM system that is tuned to detect up-calls, the whale is silent. The same observation holds for dolphin whistles versus echolocation clicks, etc.

To estimate the impact of silence on the detection function we consider the detection probability of a whale passing a PAM system at a distance y (Equation 7.37), which is given by

$$h(x, y)\mathrm{d}x = \Pr\{\text{detection in interval } [x + \mathrm{d}x, x]\} \tag{8.30}$$

In order to be detected the whale must be acoustically active, but we cannot know when the passing animal initiates its acoustic active phase. However, if the whale is moving slowly enough it should become acoustically active with high probability during its passage.

The probability of not having detected the whale at position $(x + dx, y)$ depends now on the acoustic state of the animal, whether the animal is acoustically active or not

$$Q(x + dx, y) = Q(x, y)[(1 - h(x, y)dx)P_{ac} + (1 - P_{ac})] \qquad (8.31)$$

or

$$Q(x + dx, y) = Q(x, y)[(1 - P_{ac}h(x, y)dx)] \qquad (8.32)$$

where P_{ac} is the probability that the animal is acoustically active, which is assumed to be independent of the location (x, y), and $h(x, y)$ is the PDF of detecting an acoustically active animal at location (x, y).

Equation 8.31 states that a whale remains undetected, either if it is not acoustically active $(1 - P_{ac})$, or it is not detected while acoustically active $(1 - h(x, y)dx)P_{ac}$. By integrating Equation 8.32 we obtain the detection function

$$g(y) = 1 - \exp\left\{ -P_{ac} \int_{-\infty}^{\infty} h(x, y)dx \right\} \qquad (8.33)$$

8.4.1   Acoustic activity of cetaceans

Describing the acoustic activity of cetaceans with a binary state variable seems rather simplistic. In fact, the early chapters of this book reveal an enormous complexity in cetacean sound. However, little information exists that relates specific acoustic activity to animal behaviour. This is mainly due to the difficulty of documenting and studying acoustics in synchrony with other behavioural aspects. The classical audiovisual methods are only suited for animals that are kept in captivity or that spend most of their time at or close to the surface. For underwater acoustic activity, especially during deep dives, new technology is needed. The appearance of complex electronic tags that record both audio and motion behaviour helped to give insight into the behavioural context of acoustic activities. Among popular audio tags are the digital acoustic recording tags (DTAG) developed at WHOI (Johnson and Tyack, 2003) and the bioacoustic probes (Bprobe) developed by Burgess *et al.* (1998). By recording the sound emitted, or received, by the tagged animal in synchrony with its motion, conclusions may be reached regarding the acoustic activity in a realistic, nearly undisturbed behavioural context. How far tagged animals behave normally depends on the dimension of the tag in relation to the size of the animal, and on the design of the tag. Technological advances in electronic miniaturization in recent years have allowed the tag developers to overcome size constraints and to develop tags that are also suited for the smallest cetacean species.

As acoustic tags have been developed only recently, and as their deployment on free-ranging cetaceans is still more an achievement than day-to-day practice, analysing the acoustic behaviour of wild cetaceans is still in its infancy and statistical results on acoustic behaviour are scarce.

Table 8.4 Selected acoustic state vectors

Species	Loquacious	Silent	Reference
Pm	58%	42%	Watwood *et al.*, 2006
Zc	26%	74%	Tyack *et al.*, 2006
Md	19%	81%	Tyack *et al.*, 2006

Based on preliminary DTAG analysis, Table 8.4 presents indicative acoustic state vectors for three non-whistling whales: sperm whale (*Pm*), Cuvier's beaked whale (*Zc*) and Blainville's beaked whale (*Md*). The restriction to non-whistling toothed whales allows us to take echolocation clicks as sole acoustic events.

8.5   Modelling the influence of animal motion

We have seen that animal motion influences the detection function via range and off-axis dependences. Ideally, one wants to use a closed expression for off-axis distributions $w_\vartheta(\vartheta, r)$, similar to the ones given in Section 8.3, but such formulas may not always be obtainable, because the off-axis angle is generally a complicated function of different parameters, as shown next.

The animal's motional behaviour, that is, the three-dimensional motion during acoustic activity relative to the location of the PAM system, determines the off-axis angle ϑ as follows

$$\vartheta(\gamma, \beta, \eta) = \cos^{-1}(\sin\gamma\cos\beta\cos\eta + \sin\beta\sin\eta) \tag{8.34}$$

where γ is the whale's aspect or heading (measured clockwise relative to the observer), β represents the whale's pitch (positive pitch is measured upwards), η is the elevation angle of the hydrophone as seen from the whale, which is given by

$$\eta = \tan^{-1}\left(\frac{d-h}{R_H}\right) \tag{8.35}$$

where d is the depth of the whale, h is the depth of the hydrophone and R_H is the horizontal range from whale to hydrophone.

As, by convention, whale and hydrophone depth are expressed with negative values, the elevation angle is positive if the whale is at a shallower depth than the hydrophone.

8.5.1   Distribution models for depth, pitch and heading

In the following, we consider only echolocating whales and dolphins that forage at a certain depth and assume that the three quantities, animal depth, pitch and heading, may be described by random variables (Zimmer *et al.*, 2008).

First, we model the whale depth during a dive as normally distributed around a mean foraging depth d_0.

$$w(d) = \frac{1}{\sqrt{2\pi}\sigma_d} \exp\left(-\frac{1}{2}\left(\frac{d-d_0}{\sigma_d}\right)^2\right) \tag{8.36}$$

The distribution of whale pitch is conveniently modelled as a circular normal, or von Mises, distribution (Fisher, 1993) above and below the horizontal:

$$w(\beta) = \frac{1}{2\pi I_0(\kappa)} \exp(\kappa\cos(\beta)) \tag{8.37}$$

where κ is a parameter related to the variance of the distribution by $\text{var}(\beta) = 1 - I_1(\kappa)^2/I_0(\kappa)^2$, and $I_j(x)$ is the modified Bessel function of order j.

The heading distribution of a general dive can be modelled as a superposition of two circular normal distributions relative to a mean whale heading γ_0 and $\gamma_0 + 180$.

Such dual-modal distributions allow for the inversion of the whales during their foraging activity; that is, the animals are allowed to exploit the prey patch by moving back and forth.

If we assume that the vertical and horizontal search patterns of the animal are equivalent, then one could use the same parameter for both the pitch and the heading distribution.

$$w(\gamma) = \frac{1}{2\pi I_0(\kappa)}\{u\exp(\kappa\cos(\gamma-\gamma_0)) + (1-u)\exp(-\kappa\cos(\gamma-\gamma_0))\} \tag{8.38}$$

The parameter u controls the proportion of track inversion by the whale and is between zero and one. As the general orientation of the hydrophone with respect to the whale motion is unknown, we assume γ_0 to be uniformly distributed in the range $-180° < \gamma_0 < 180°$.

The actual foraging behaviour is very poorly understood for most cetaceans, and only in recent years have behavioural tags started to reveal how whales and dolphins move while foraging. Equation 8.38 was found to describe reasonably well the foraging behaviour of Cuvier's beaked whales (Zimmer et al., 2008). One should expect that the continuing tag-based analysis of animal behaviour will augment our detailed knowledge of animal motion and will improve its statistical description.

9 Simulating sampling strategies

This chapter introduces computer simulation as a technique to emulate real-world PAM and should demonstrate that simulations are useful not only to generate data, but also to analyse the system's performance.

PAM is really a field effort where one takes a single hydrophone or an array of hydrophones, embarks on a boat or ship, moves over the oceans in a hopeful, well-planned way and collects data to detect the presence or assess the absence of acoustically active whales and dolphins.

Simulation implies experimentation also, but instead of using a real-world PAM system, we use a computer and experiment with a model of the PAM application. The term model in this context means that we deal with a simplified description of the real-world system, where the simplifications are in general made to allow easy discussion of the result. The distribution models for whale depth, pitch and heading, presented in Section 8.5.1, are such simplifications. Similar to real-world applications, computer simulations will generate different results each time they are executed.

Using models as a substitute for real-world systems has a consequence that the results are only indicative of real-world results. If our model parameters are not a reasonable characterization of the real-world system, then we should not expect the simulation results to be realistic.

The simulations we will be using in this chapter involve the sampling of values of stochastic variables from their distributions. To become more specific, recall the detection probability (Equations 7.14–7.16), which when combined read

$$n = \int_0^\infty n(r)\mathrm{d}r = NP_N = N \int_0^\infty w_{ri}(r)g(r)\mathrm{d}r \tag{9.1}$$

with N being the number of animals, $w_{ri}(r)$ being the PDF of encountering an animal at range r and $g(r)$ being the detection function, or the probability of detecting this animal at range r. The detection function is here assumed to be only a function of the range/distance to the animal.

As we have seen in the previous chapters, the detection function may be formulated in different ways. For example, Equation 7.13 gives

$$g(r) = \int_0^{\infty} w_{r0}(y)\delta(r<y)dy \tag{9.2}$$

The purpose of the simulation is then twofold: to generate samples that emulate the animal distribution PDF $w_{ri}(r)$ and to implement the detection process $\delta(r<y)$ for each simulated sample.

In the following, we will first simulate a point survey, which is relatively easy to implement without generating trivial results. Being a straightforward example, the point survey of a uniformly distributed object in combination with a half-normal detection function could be assessed without stochastic simulation. However, other and more specific detection functions may be implemented. In fact, complex models for detection functions are one of the main reasons for simulations.

9.1   Modelling a point survey detection probability

Before we simulate a point survey, we should obtain some idea of what type of data we should expect from the simulation for a point survey. This can be done in a deterministic way without the use of random variables.

Assume some animals to be uniformly distributed in a circular area of interest having a radius of R_{max} = 7000 m. The probability density function for having an animal at a distance r from the observer is then proportional to the distance

$$w_{ri}(r) \propto r \tag{9.3}$$

To obtain the proportionality constant, we note that by definition the animals of interest are not outside a circle with radius R_{max}, and we obtain for the presence PDF of each individual

$$w_{ri}(r) = \frac{2r}{R_{max}^2}, \text{ for } r<R_{max} \tag{9.4}$$

and

$$w_{ri}(r) = 0, \text{ for } r>R_{max} \tag{9.5}$$

To describe the sensor performance we assume for simplicity that the detection range r_0 is normally distributed with a standard deviation σ = 800 m around a mean value r_0 = 4000 m

$$w_{r0}(r) = \frac{1}{\sqrt{2\pi}\sigma}\exp\left\{-\frac{1}{2}\left(\frac{r-r_0}{\sigma}\right)^2\right\} \tag{9.6}$$

resulting in a detection function that is shown in Figure 9.1. Using the 3σ criterion, we may already note from the figure that at short ranges (<1600 m), all animals are detected, at a range of 4000 m 50% of the animals should be detected, while for ranges greater than 6400 m hardly any animals will be detected.

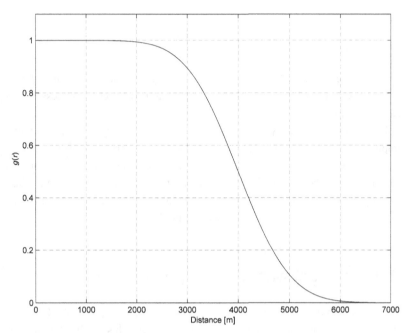

Figure 9.1   Detection function with Gaussian distribution of critical detection ranges.

The number of animals detected in a range interval dx now becomes

$$n(r) = Nw_{ri}(r)g(r)\mathrm{d}x \qquad (9.7)$$

which is depicted in Figure 9.2 for $N = 100$ and d$x = 500$ m.

We note that initially the expected number of detected animal increases linearly, but at greater distances the expected number of detected animals should drop back to zero.

Integrating over all ranges, we obtain the expected total number of detected animals, which becomes $n = 33.9$ animals, resulting in a overall detection probability $P_N = 0.34$, that is, for every animal detected in the search area there will be about two animals that are missed.

A total detection probability of 34% results in an effective detection area (see also Equation 7.25) of 52.33 km^2, approximately one third of the total search area of about 154 km^2, or alternatively, the effective detection range is about 4082 m. The effective detection range is not equal to the modelled 50% detection range of the detection function owing to the increasing animal distribution as a function of range (Equation 9.4).

Matlab code

```
%Scr9_1
NA=100;
dr=1;
r=0:dr:7000;
```

```
wri=r; wri=wri/(sum(wri)*dr);
%
r0=4000; cv=0.2; sig=cv*r0;
wr0=1/(sqrt(2*pi)*sig)*exp(-0.5*((r-r0)/sig).^2);
%
gr=1-cumsum(wr0)*dr;
P_n=cumsum(wri.*gr)*dr;
%
figure(1)
plot(r,gr,'k'),ylim([0 1.1])
grid on
xlabel('Distance [m]')
ylabel('g(r)')
%
dx=500; %range interval
idx=dx/dr;
ihx=1:idx:7000;
hx=diff(P_n(ihx));
rx=r(ihx)-dx/2;
%
figure(2)
plot(rx(2:end),NA*hx,'ko-')
grid on
xlabel('Distance [m]')
ylabel('n(r)')
```

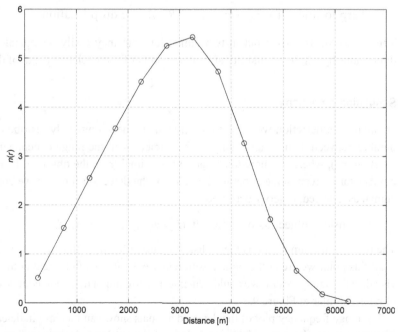

Figure 9.2   Number of detected animals as a function of range. Modelled range interval $dx = 500$ m. Assumed total number of animals present in the search area $N = 100$.

9.2   Simulating a point survey

The simulation of the point survey is now divided into two steps. First, we distribute N objects uniformly in a circular area of interest surrounding the observer. For the point survey, we assume rotational invariance and consequently characterize the objects only by their range from the observer. Then, we use the detection function to decide for each object whether it is detected or not. What we obtain is then a list of ranges from the observer to the detected objects.

9.2.1   Simulating object locations

The simplest and most intuitive way to distribute N objects uniformly in a circular area of interest is to generate uniform distributed x- and y-co-ordinates according to

$$x_i = 2R_{\max}(U - 0.5) \tag{9.8}$$

$$y_i = 2R_{\max}(U - 0.5) \tag{9.9}$$

where U is a random number that is uniformly distributed between 0 and 1. The co-ordinates (x_i, y_i) are then uniformly distributed between $-R_{\max}$ and $+R_{\max}$. If the distance from the observer, which is assumed to be at the origin (0,0), is less than $R_{\max}$, the location is retained; otherwise it is discarded. The procedure is carried out until the number of objects within the design circle reaches the desired number of objects

$$\text{keep location, if } r_i = \sqrt{x_i^2 + y_i^2} \le R_{\max}, \text{ else drop location} \tag{9.10}$$

The simulation is carried out in two dimensions but may easily be extended to three dimensions by adding an additional random number for the relative depth of the animal.

9.2.2   Simulating detections

To simulate a detection we draw for each animal a new, uniformly distributed random number between 0 and 1 and compare the number with the range dependent detection function $g(r)$, where r is now the range of the animal from the observer. If the random number for a given range is below the value of the detection function we consider this animal as detected, otherwise as missed

$$\text{consider object as detected, if } U \le g(r_i), \text{ else as missed} \tag{9.11}$$

The result of the simulation is then a list of ranges of animals that are detected.

At this point we are at the same position as we would be after a real point survey; to visualize the survey result, we could generate a histogram of detection counts as function of range, similar to Figure 9.3.

The range frequency plot of the detected animals shows an increase in detections to a maximum range between 3000 and 3500 m followed by a decline in detections. No animals were detected beyond 5500 m.

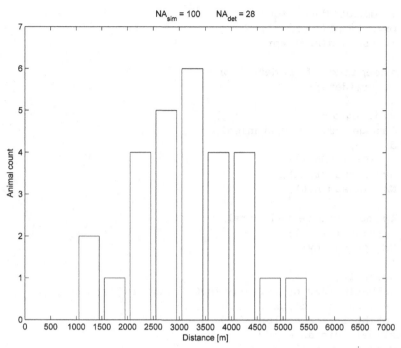

Figure 9.3   Range frequency plot of detected animals.

Matlab code for Figure 9.3

```
%Scr9_2
NA=100; Rmax=7000;
dr=1; r=0:dr:Rmax;
%
r0=4000; cv=0.2; sig=cv*r0;
rs=r0;
wr0=1/(sqrt(2*pi)*sig)*exp(-0.5*((r-rs)/sig).^2);

gr=1-cumsum(wr0)*dr;
%
%simulate animal ranges
na=1.5*NA;
%
rand('state', 2);

% animal ranges
xx=2*Rmax*(rand(na,1)-0.5);
yy=2*Rmax*(rand(na,1)-0.5);

rr=sqrt(xx.^2+yy.^2);
%
rr(rr>Rmax)=[]; rr(NA+1:end)=[];
nr=length(rr); % nr should be NA
```

```
% simulate detections
pr=interp1(r,gr,rr);
idets=rand(nr,1)<pr;
%
% keep range of all detections
ri=rr(idets);

% histograms
% ranges of simulated animals
dx=500;
hx=(0:dx:Rmax)';
hy0=histc(rr,hx);
NA_sim=sum(hy0);

% ranges of detected animals
hy=histc(ri,hx);
NA_det=sum(hy);
%
figure(1)
hb=bar(hx+250,hy);xlim([0 7000]),ylim([0 7])
set(gca,'xtick',hx);
set(hb(1),'facecolor','w')
xlabel('Distance [m]')
ylabel('Animal count')
title(...
    sprintf('NA_{sim} = %d        NA_{det} = %d',...
            NA_sim,NA_det))
```

Comment on Matlab code

The random number sequence generated by this script is always the same sequence. This is achieved by the call "`rand('state', 2);`", which determines the first number of the random number sequence. To avoid this repetitive behaviour, this Matlab statement should be commented or removed.

9.3   Point survey abundance estimation

Assume we have a survey result as shown in Figure 9.3. Assume we know further that the survey was a point survey where the observer noted the distance to all detected animals without directional preference. Therefore, the question is now how many animals are within the area of interest.

By summing up all animal counts, we obtain a total number of 28 animals. Having the data coming from a point survey, we could assume that the PDF of detecting an animal, and therefore the animal count, increases linearly with distance (Equation 9.3). From Figure 9.3 we may deduce that between 3500 and 5500 m the number of detections decreases, most likely due to the range limitation of the detection function. Taking 4500 m (the mean between 3500 and 5500) as the effective detection radius, we would obtain an animal density of 28 animals/63.6 km², or 44 animals/100 km². Considering a search area with 7 km radius, the detection probability becomes $P_N = (4.5/7)^2 = 0.41$, resulting in a total abundance estimate $\hat{N} = 28/0.41 = 68$ animals. Obviously, this estimate is very crude and we should be able to do much better than that.

A better estimate is based on a model for the observed animal counts

$$n(r) = Nf(r) = Nw_{ri}(r)g(r)\mathrm{d}x \tag{9.12}$$

where N is the number of animals within the search area $(r < R_{max})$ with PDF

$$w_{ri}(r) = \frac{2r}{R_{max}^2}, \quad \text{for } r < R_{max} \text{ and zero otherwise} \tag{9.13}$$

and the assumed detection function

$$g(r) = \frac{1}{\sqrt{2\pi}\sigma} \int_0^r \exp\left\{ -\frac{1}{2}\left(\frac{x - r_0}{\sigma}\right)^2 \right\} \mathrm{d}x \tag{9.14}$$

The unknown variables that are to be estimated are now animal count N, mean detection range r_0 and standard deviation σ of the detection range.

Figure 9.4 shows the result of the non-linear parameter estimation. The estimated model parameters are $\hat{N} = 89$ animals, the mean detection range r_0 becomes 4196 m and

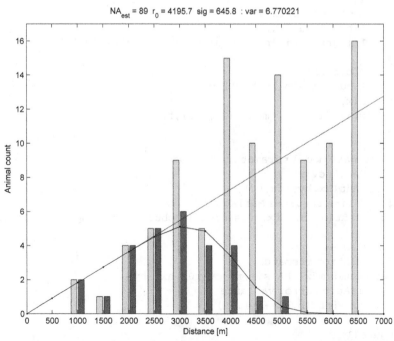

Figure 9.4   Abundance estimation. The light grey bars count the simulated animal ranges, the dark grey bars count the detected animals as a function of range. The dashed line is the estimated animal density, and the connected dots are the fitted estimate of the detected animals.

the standard deviation σ is 646 m. As we simulated the animal locations, we know the true animal distribution. Figure 9.4 also shows, for comparison, the range distributions of all simulated locations.

Matlab code for Figure 9.4

```
%Scr9_3
% abundance estimation
% requires script scr9_2
Scr9_2
% or detection vector ri;
%
r0=4000; cv=0.2; sig=cv*r0;
%
dx=500; Rmax=7000;
hx=(0:dx:Rmax)';
% ranges of simulated animals
hy0=histc(rr,hx);
NA_sim=sum(hy0);
% ranges of detected animals
hy=histc(ri,hx);
NA_det=sum(hy);

% PDFs
wri=@(a,hx) ...
    (a*dx)*2*hx/(Rmax^2);
wro=@(hx,r0,sig) ...
    1/(sqrt(2*pi)*sig)*exp(-0.5*((hx-r0)/(sig)).^2);
%
mgx=@(b,c,hx) ...
    1-cumsum(wro(hx,b,c))*dx;
mfx=@(x,hx) ...
    wri(x(1),hx).*mgx(x(2),x(3),hx);
%
%
% estimate model parameters from data
b0=[NA,r0,cv*r0];
b=nlinfit(hx,hy,mfx,b0);
% next line requires Matlab 7.5
%b=nlinfit(hx,hy,mfx,b0,statset('robust','on','WgtFun','logistic'));

% Variances
% simulation mismatch
varo=sum((mfx([NA_sim,r0,cv*r0],hx)-hy).^2);
% estimation error variance
var=sum((mfx(b,hx)-hy).^2);

figure(1)
hold off
hb=bar(hx,[hy0 hy]);xlim([0 7000]),ylim([0 17])
hold on
```

```
plot(hx,wri(b(1),hx),'k--',hx,mfx(b,hx),'k.-')
hold off
set(hb(1),'facecolor',0.9*[1 1 1])
set(hb(2),'facecolor',0.5*[1 1 1])
xlabel('Distance [m]')
ylabel('Animal count')
title(...
sprintf('NA_{est} = %.0f r_0 = %.1f sig = %.1f : var = %f',...
        b,var))
return
```

The next question is about the quality of this estimate. How good is this estimate? How would the result change if we were to repeat this simulation or experiment? To obtain an answer we repeat the same experiment 1000 times (simulation followed by non-linear parameter estimation) and obtain a random animal count estimate, the distribution of which is shown in Figure 9.5. We note that the animal count peaks at a value close to 100 but is widely distributed, varying from about 50 to over 250, and the distribution is asymmetric with a significant tail of high animal counts.

As the maximum PDF value of Figure 9.5 is close to the true animal count, the estimation process may be considered as asymptotically unbiased. However, this may not be a surprise, because the abundance estimation used the shape of the detection function that correctly describes the detection process.

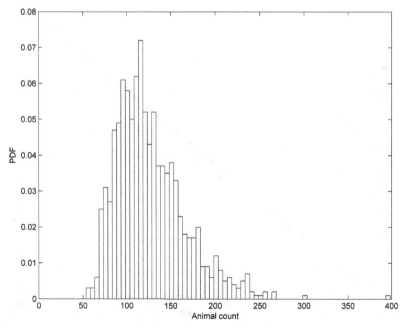

Figure 9.5   Animal count distribution based on multiple simulation of Figure 9.4.

9.4   Distance distribution of echolocation clicks of randomly moving animals

Consider next the following scenario. An echolocating whale initiates an acoustically active foraging phase at a random position within the search area. During foraging the animal moves with a constant speed of 1.5 m/s randomly around an arbitrary direction, emitting 3600 clicks at a rate of 2 clicks/s (any similarity with Cuvier's beaked whales is intentional). What we want to know is the range distribution of the individual clicks.

Figure 9.6 shows the simulation result, indicating that the PDF of animal positions while emitting clicks is linearly increasing with the distance from the observer, if one ignores the boundary effects at ranges above 5000 m.

Only one animal at a time is simulated and the experiment is repeated 10 000 times with random start location and direction of motion. The click numbers presented in Figure 9.6 therefore give the expected number of clicks as a function of range for an animal with unknown start location and direction of motion.

From the point survey we know that the number of possible whale locations varies according to

$$w_{ri}(r) = \frac{2r}{R_{\max}^2}, \quad \text{for } r < R_{\max} \text{ and zero otherwise} \tag{9.15}$$

Consequently, the number of clicks varies according to

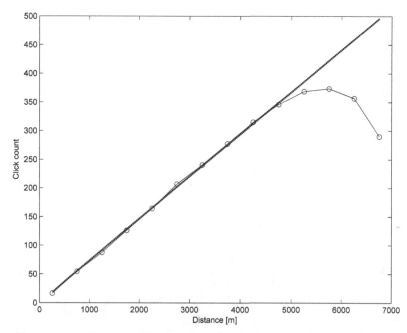

Figure 9.6   Click count as a function of distance of a randomly moving whale (range bin 500 m).

$$n_{cl}(r) = N_{cl}w_{ri}(r) = \frac{2r}{R_{max}^2}N_{cl}, \quad \text{for } r < R_{max} \tag{9.16}$$

as shown in Figure 9.6, where the dashed line describes the functionality given in Equation 9.16. For this particular simulation, the total number of clicks N_{cl} was selected to be 3600 clicks.

This example shows that if the start position of an echolocation track is uniformly distributed in the search area, so also will be the locations of each click, as long the animal's motion is not influenced by the observer. This result is, with hindsight, intuitive, but it is reassuring that simulation can be used to demonstrate this relationship.

Matlab code for Figure 9.6

```
%Scr9_5
%
Rmax=7000;
%
vo=1.5; %speed [m/s]
nsim=3600; %number of clicks / dive
%
clear Ho %comment it skip simulation
%
%heading distribution
dg=0.01;
gam=-pi:dg:pi;
k=5;
wg=1/(2*pi*besseli(0,k))*exp(k*cos(gam));
cwg=cumsum(wg)*dg;
%
nrep=10000;
dx=500;
hx=0:dx:Rmax;
%
if ~exist('Ho','var')
    Ho=zeros(length(hx),nrep);
    Hi=zeros(length(hx),nrep);
    tic
    rand('state',1)
    for ii=1:nrep
        % simulate initial track position
        % randomly located within circle
        ro=inf;
        while ro>Rmax
            rdo=rand(1,2);
            xo=2*Rmax*(rdo(1,1)-0.5);
            yo=2*Rmax*(rdo(1,2)-0.5);
            ro=sqrt(xo^2+yo^2);
        end
        %simulate initial direction
        gamo=2*pi*rand(1,1);
        %
```

```
        %simulate animal track
        ncl=rand(nsim,2);

        %simulate animal track
        arg=ncl(:,1);
        icwg=cwg>0.0001 & cwg<0.9999;
        args=gamo+...
            interp1(cwg(icwg),gam(icwg),arg,'lin','extrap');
        %
        %assume constant speed between clicks and 2 clicks/s
        dro=vo*0.5;
        dxa=dro*cos(args);
        dya=dro*sin(args);
        xx=xo+[0;cumsum(dxa)];
        yy=yo+[0;cumsum(dya)];
        rr=sqrt(xx.^2+yy.^2);
        %
        % keep range distribution of all animals
        hy=histc(rr,hx);
        Ho(:,ii)=hy(:);
    end
    toc
end
%
hyo=mean(Ho,2);
hxx=hx(1:end-1)+dx/2;
%
Nnorm=2*dx*nsim/Rmax^2;
%
figure(1)
plot(hxx,hyo(1:end-1),'ko-')
line(hxx, hxx*Nnorm, ...
    'color','k','linestyle','-','linewidth',2)
xlabel('Distance [m]')
ylabel('Click count')
```

To simulate the detection of echolocation clicks of randomly moving animals it is therefore sufficient to draw random numbers according to the PDF given in Equation 9.15 and to decide for every resulting range whether a detection occurred or not.

Caveats

The PDFs of Equation 9.15 and therefore of Equation 7.24 are only valid for PAM applications where the hydrophone is at the same depth as the animals. In fact, no vertical separation is used in the simulation of Figure 9.6. Introducing a vertical separation will change the PDF at short ranges, as not all ranges are possible and the minimal range is given by the vertical distance between the hydrophone and the layer in which the animals are foraging.

The off-axis angle of the individual echolocation clicks is also not considered in this simulation. Consequently, the results are only useful where sound is emitted equally in all directions, or where the off-axis angle statistics do not depend on the range of the animal. The latter is a reasonable assumption when the whale is searching not only horizontally but also vertically, or in other words, where the pitch angle of the whale is widely distributed and not constrained around the horizontal plane.

9.4.1   Stochastic sampling from an arbitrary probability density function

In the previous simulation, we generated a random variable by using a known or assumed PDF. The assumption was that the variation of the heading is circular and normally distributed according to Equation 8.38. The random variables were generated as described below.

Consider a random variable that may be characterized by the PDF $f(x)$. The cumulative density function (CDF) $F(y)$ is then given by

$$F(y) = \int_0^y f(x)\mathrm{d}x \tag{9.17}$$

where the CDF varies by definition from zero to one.

If we now draw a variable y from a uniform distribution U that is limited between zero and one, then

$$x = F^{-1}(y) \tag{9.18}$$

is a new random variable that is distributed with PDF $f(x)$.

Figure 9.7 shows the result of this operation for a random heading using the PDF given in Equation 8.38 with $\kappa = 5$ and $u = 1$.

The left panel shows the CDF $F(y)$ indicating that a random value $y = 0.8$ results in a random heading value of $22.2°$. The right panel shows as stars the histogram values of the simulated headings, which are in perfect agreement with the original PDF (circular normal distribution), shown as a solid line.

Matlab code for Figure 9.7

```
%Scr9_6
%
%heading distribution
dg=0.01;
gam=-pi:dg:pi;
k=5;
wg=1/(2*pi*besseli(0,k))*exp(k*cos(gam));
%estimate CDF
cwg=cumsum(wg)*dg;
%
```

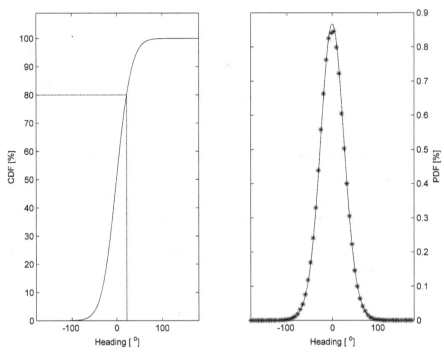

Figure 9.7   Simulation of random variables based on PDF models.

```
%simulate random heading
rand('state',1)
nsim=1000000;
arg=rand(nsim,1);
%
[i1,i2,i3]=unique(cwg);
gamx=gam(i3);
cwgx=cwg(i3);
head=interp1(cwgx,gamx,arg,'lin','extrap');
%
% obtain histogram (PDF) of simulated heading
hx=gam;
hy=histc(head,hx); hy=hy/sum(hy);

isi=0.8;
isl=find(cwg>=isi,1,'first');

figure(1)
set(gcf,'position',[300 400 700 420])
subplot(121)
plot(gam*180/pi,cwg*100,'k')
line([-180 gam(isl) gam(isl)]*180/pi, ...
     isi*[1 1 0]*100,'color','k','linestyle','--')
xlim([-180 180])
```

```
ylim([0 109])
xlabel('Heading [^o]')
ylabel('CDF [%]')
box on

subplot(122)
%bar(hx*180/pi,hy*100,'facecolor','w',' edgecolor','k','barwidth',
    1);
plot(hx(1:10:end)*180/pi,hy(1:10:end)*100,'k*');
line(gam*180/pi,wg*dg*100,'color','k')
set(gca,'yaxislocation','right')
xlim([-180 180])
xlabel('Heading [^o]')
ylabel('PDF [%]')
box on
```

Comment on Matlab code

Given the PDF wg(gam), the CDF cwg(gam) is in Matlab simply estimated by carrying out a cumulative sum

```
cwg=cumsum(wg)*dg;
```

The inversion of the CDF and the generation of random heading values is achieved by the line

```
head=interp1(cwgx,gamx,arg,'lin','extrap');
```

where arg are the uniformly distributed random numbers. The method works as long as cwgx (the first argument in interp1) is either a monotonically increasing or decreasing function. As the CDF may have multiple equal values due to rounding conditions, the inversion is executed only on unique CDF values.

9.5   Stochastic simulation of detection function

In the following, we obtain the detection function by simulating the random motion of a hypothetical acoustically active deep-diving whale, which we assume to be a beaked whale echolocating at a frequency of 40 kHz.

9.5.1   Distance and off-axis joint probability distribution

As mentioned above, the detection of echolocation clicks depends not only on the separation between whale and hydrophone but also in the off-axis angle, i.e. the angle between the acoustic axis of the click and the line connecting whale and hydrophone.

Being a function of two variables (range and off-axis angle), the resulting joint PDF now becomes a surface, as shown in Figure 9.8. To generate this figure, the script used for generating Figure 9.6 was extended by considering a vertical distance between whale and

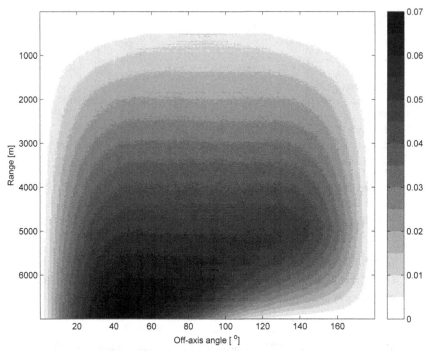

Figure 9.8   Mean click count as function of range and off-axis angle. The marginal sum over all off-axis angles results in the expected click count as a function of range (similar to Figure 10.6 but now with 10 m range bin).

sensor of 500 m, and by generating and accumulating a two-dimensional histogram. The number of clicks per dive was again assumed to be 3600 but the number of replicating detection experiments was increased to 100 000 so that the simulation results in a reasonably smooth two-dimensional histogram. In addition, the range bin was reduced from 500 m to 10 m to better cover short ranges, and the bin-width for the off-axis angles was selected to be 1°.

Figure 9.8 shows two particular features. First, no clicks are available for ranges below 500 m, owing to the vertical distance of the hydrophone and the foraging depth of the whale. Second, for ranges above 5000 m one may note a lack of clicks for larger off-axis angles. This is a side effect of the simulation, as for start ranges of the simulated animal tracks that are close to the limiting 7000 m, off-axis angles that are greater than 90° are more or less absent. This is because whales that are on the edge of the area of interest are only detectable if they move towards the observer (off-axis angles < 90°) and whales that move away (off-axis angles > 90°) move quickly out of the area of interest.

9.5.2   Detection function

Figure 9.8 shows for increasing ranges r the distribution of clicks as function of off-axis angle. This distribution can therefore be used to obtain a simulated detection function. For this, we start with Equation 8.24, which may be rewritten as

$$g(r) = \int_{DL_{max}}^{\infty} w_{SL}(u + F_0 + TL(r)) du$$

$$+ \int_0^{DL_{max}} w_{SL}(u + F_0 + TL(r)) \left(\frac{1}{w_{cl}(r)} \int_0^{DL^{-1}(u)} w_{OA}(\vartheta, r) d\vartheta \right) du \tag{9.19}$$

and replace $w_{OA}(\vartheta, r)$ by the simulated PDF of Figure 9.8 and $w_{cl}(r)$ is the click density as function of range as given by

$$w_{cl}(r) = \int_0^{\pi} w_{OA}(\vartheta, r) d\vartheta \tag{9.20}$$

As with Equation 8.28, the first integral of this equation depends only on the source level distribution and not on the off-axis angle. The second integral points out that the range-dependent off-axis PDF is to be integrated up to a critical angle $DL^{-1}(u)$, the result of which is range-dependent owing to the range-dependent off-axis PDF.

The integration (Equation 9.20) may be carried out for all ranges using the data of Figure 9.8 without further knowledge of the source level and transmission loss. This integration transforms the two-dimensional PDF of Figure 9.8 into a range-dependent partial CDF, where the integral over all off-axis angles results in a range-only marginal PDF counting the relative number of clicks at range r. To obtain the detection function, which is the conditional detection probability given the presence of clicks, we have to correct the two-dimensional PDF of Figure 9.8 by the range-dependent click count (Equation 9.20). It should be noted that the click count given by Equation 9.16 is not suited for use here, as the geometries are different in the two cases: Figure 9.6 and therefore Equation 9.16 assume that source and receiver are at the same depth, whereas for Figure 9.8 the receiver was chosen to be at a depth different from that of the whale.

To continue this estimation of the detection function, we assume further that the source level is normally distributed according to Equation 8.29 with a source level variation $\sigma_{SL} = 5$ dB, the off-axis attenuation given by Equation 8.14, and the transmission loss is due to spherical spreading.

Figure 9.9 shows a detection function that is similar to the theoretical one presented in Figures 8.1 and 8.2. The only visible difference is that the contribution of the second integral shows a maximum value of 0.14 in Figure 9.9 but 0.11 in Figure 8.1, indicating a slight difference in the simulated and theoretical off-axis distribution. However, the similarity between these two curves demonstrates the validity of the theoretical description of the off-axis distribution for detecting arbitrary echolocation clicks (Equation 8.16).

Common to both detection functions is that their first integrals (Equations 8.24 or 9.19), which cover the contributions that do not depend on the actual off-axis distribution, are already a very reasonable approximation to a passive acoustic detection function.

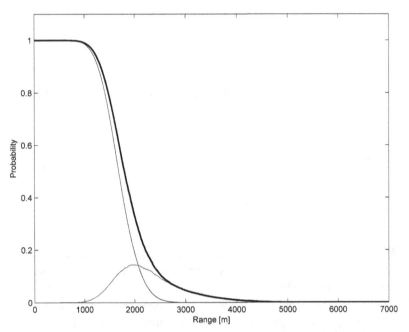

Figure 9.9   Simulated detection function as function of range. Assumed source level variation $\sigma_{\text{SL}} = 5$ dB.

The detection function is influenced by the implemented signal processing methods; in particular, it is important to specify the method used to obtain the detections. The example has assumed so far that all clicks emitted during a dive are of importance. In the following, we assume that only the near-on-axis clicks during each dive are used to decide on the presence of an animal. This approach makes sense, as for animals that are very far away, only the loudest click, i.e. when the whale is oriented towards the receiver, may be received by a PAM system. In other words, an animal cannot be detected if the loudest clicks of a whole dive are too weak to exceed the detection threshold. Fortunately, owing to the scanning nature of the foraging process, some clicks will be always significantly louder than the remainder, and clicks that are directed towards the receiver exceed the weakest clicks by about 30–40 dB.

The PDF of the near-on-axis clicks may be modelled according to a Weibull distribution

$$w_{\text{OA\_L}}(\vartheta) = \frac{\eta\vartheta^{\eta-1}}{\sigma_{\text{OA\_L}}{}^{\eta}}\exp\left\{-\left(\frac{\vartheta}{\sigma_{\text{AO\_L}}}\right)^{\eta}\right\} \tag{9.21}$$

where $\sigma_{\text{OA\_L}}$ and η are two parameters to be estimated from the data. For a shape parameter $\eta = 1$ this PDF would become a exponential distribution and for $\eta = 2$ the PDF would be a Rayleigh distribution with a slightly modified $\sigma_{\text{OA\_L}}$.

The Weibull-type PDF has also been chosen because it can be integrated analytically and the probability of having a click at angles below ϑ results in

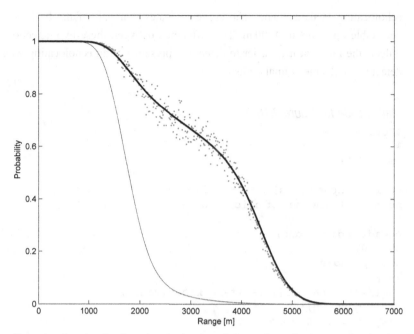

Figure 9.10   Detection function for detecting the loudest click per dive. The dots are the simulated detection function; the dashed line corresponds to Figure 9.9; the solid line is based on a modelled off-axis distribution with $\sigma_{OA\_L} = 14\,°$ and $\eta = 0.46$. Source level variation $\sigma_{SL} = 5$ dB for both curves.

$$P_{\mathrm{OA\_L}}(\vartheta) = \int_{0}^{\vartheta} w_{\mathrm{OA\_L}}(x)\mathrm{d}x = 1 - \exp\left\{ -\left(\frac{\vartheta}{\sigma_{\mathrm{OA\_L}}}\right)^{\eta} \right\} \tag{9.22}$$

Figure 9.10 shows the simulated detection function (dots) for the near-on-axis click of a dive, which may be considered as the detection function for detecting a dive. The same figure shows as a dashed line the detection function of all clicks, as shown in Figure 9.9. The solid line is based on a modelled off-axis distribution for clicks that are randomly distributed around the on-axis, fitting very well the simulated detection function (grey dots). The assumed source level variation for both curves is $\sigma_{SL} = 5$ dB.

As shown in Figure 9.10, a reasonable Weibull fit to the simulated data is obtained for parameters $\sigma_{OA\_L} = 14°$ and $\eta = 0.46$. Zimmer *et al.* (2008) also simulated the detection probability of the loudest click but used a model that differs from the present one, mainly by not considering the source-level statistics. As a result, their detection function resulted in a very sharp cut-off at large range, a feature that is not present in Figure 9.10 owing to the explicit consideration of the source-level statistics.

Figure 9.10 compares two detection functions, one of which (dashed line) is replicated from Figure 9.9, describing the probability of detecting all beaked whale clicks. Half of the emitted clicks are detected for ranges up to nearly 2000 m. The other detection function (bold line in Figure 9.10) expresses the probability of detecting beaked whales if one accepts that only a few near-on-axis clicks are required to detect the presence of an echolocating beaked whale. The curve indicates that half of the beaked whales are detectable up to about 4000 m. The difference between these two detection functions reflects the fact that it is easier to detect the presence of an echolocating whale than to detect all clicks the animal emits.

Matlab code for Figure 9.10

```
%Scr9_8
%
Rmax=7000;
%
vo=1.5; %speed [m/s]
nsim=3600; %number of clicks / dive
%
%heading distribution
dg=0.01;
gam=-pi:dg:pi;
k=5;
wg=1/(2*pi*besseli(0,k))*exp(k*cos(gam));
cwg=cumsum(wg)*dg;
icwg=cwg>0.0001 & cwg<0.9999;
%
% pitch distribution
bet=-pi/2:dg:pi/2;
k=5;
wb=1/(2*pi*besseli(0,k))*exp(k*cos(bet));
cwb=cumsum(wb)*dg;
icwb=cwb>0.0001 & cwb<0.9999;
%
% number of simulations
nrep=100000;
%
% needed for histograms
dx=10;
hx=0:dx:Rmax;
do=0.1;
ho=(0:do:180)/180*pi;
%
%off-axis attenuation
C1=47;
C2=0.218*17.8;
DL=@(x) C1*C2^2*x.^2./(1+C2*x+C2^2*x.^2);
%
DL_max=C1*C2^2/(1+C2+C2^2);
%
%transmission loss
TL=@(x) 20*log10(x)+9.5*x/1000;
%
if 0
    dcl=zeros(nrep,2);
    Ho=zeros(length(hx),length(ho));
```

```
tic
rand('state',1)
for ii=1:nrep
    if mod(ii,100)==0, ii, end
    % simulate initial track position
    % randomly located within circle
    ro=inf;
    while ro>Rmax
        rdo=rand(1,2);
        xo=2*Rmax*(rdo(1,1)-0.5);
        yo=2*Rmax*(rdo(1,2)-0.5);
        zo=500;
        ro=sqrt(xo^2+yo^2+zo^2);
    end
    %simulate initial direction
    gamo=2*pi*rand(1,1);
    %
    %simulate animal track
    ncl=rand(nsim,2);

    %simulate animal track
    arg=ncl(:,1);
    head=gamo+ ...
        interp1(cwg(icwg),gam(icwg),arg,'lin', 'extrap');
    %
    arg=ncl(:,2);
    pitch= ...
        interp1(cwb(icwb),bet(icwb),arg,'lin','extrap');
    %
    %assume constant speed between clicks ans 2 clicks/s
    dro=vo*0.5;
    dxa=dro*cos(head).*cos(pitch);
    dya=dro*sin(head).*cos(pitch);
    dza=dro*sin(pitch);

    %estimate track (cumsum) and range
    xx=xo+[0;cumsum(dxa)];
    yy=yo+[0;cumsum(dya)];
    zz=zo+[0;0*cumsum(dza)]; % stay at same depth
    rr=sqrt(xx.^2+yy.^2+zz.^2);
    TLr=TL(rr(2:end));
    %
    %off-axis
    oa1=[xx(2:end),yy(2:end),zz(2:end)];
    oa2=[dxa,dya,dza];
    oa=acos(sum(oa1.*oa2,2)./(dro*rr(2:end)));
    % accumulate range off-axis histograms
    hy=hist3([rr(2:end),oa],'edges',[hx,ho]};
    Ho=Ho+hy;
    %
    DLr=DL(abs(sin(min(oa,pi/2))));
```

```
        %
        [a,ia]=min(TLr+DLr); % select only one ping per dive
        dcl(ii,:)=[rr(ia),oa(ia)];
        %
    end
    toc
    save('Scr9_8dat','dcl','Ho','hx','ho','nsim', 'nrep')
    return
else
    load('Scr9_8dat');
end
%
%all clicks
%Ho=Ho/nrep; %get average
%partial CDF
CHo=cumsum(Ho,2);
%
% loudest click
Hol=hist3(dcl,'edges',{hx,ho});
%partial CDF
CHol=cumsum(Hol,2);
% click count as function of range
wcl=CHol(:,end);
%
%plot surfaces
hxx=hx(1:end-1)+dx/2;
hox=ho(1:end-1)+do/2/180*pi;

if 0
figure(1)
imagesc(hox*180/pi,hxx,Ho)
xlabel('Off-axis angle [^o]')
ylabel('Range [m]')
colorbar
%colormap(1-gray(7*2))%, caxis(0.07*[0 1])

figure(2)
imagesc(hox*180/pi,hxx,CHo)
xlabel('Off-axis angle [^o]')
ylabel('Range [m]')
colorbar
colormap(1-gray(8*2))%, caxis(8*[0 1])
end

% Estimation of the detection function
% nominal figure of merit
FOM=116;
%
%Gaussian SL weighting
sig=5;
Wx=@(x,xo) 1/(sqrt(2*pi)*sig)*exp(-1/2*((x-xo)/sig).^2);
```

```
%Transmission loss
rTL=max(1,hx(:));
TLr=20*log10(rTL)+9.5*rTL/1000;

% needed for integration
dx=0.1;
x=0:dx:(FOM+DL_max+4*sig); %FOM+DL_max+4*sig is assumed to be infinity
%
%warning verbose on % to check proper command
warning off MATLAB:divideByZero %switch off warning in next line
th=asin(1/(2*C2)*(x./(x-C1)).*real(-sqrt(1-4*((x-C1)./x))-1));
warning on all % reactivate warning
%
th(1)=0; % correct for NaN
th(x>=DL_max)=pi; % for x>=DL_max


%
% select integration support
u2=x(x<DL_max);
u3=x(x>=DL_max);
th2=th(x<DL_max);

gr0=0*TLr;
gr1=0*TLr;
gr2=0*TLr;
gr3=0*TLr;

wth1=0.5*(1-cos(th2));
%
sigth=14/180*pi;
eta=0.46;
woa2 = @(x,sigth,eta) (1-exp(-(x/sigth).^eta));
wth2=woa2(th2,sigth,eta);

%
for ii=1:length(TLr)
    W2=Wx(u2+TLr(ii),FOM);
    W3=Wx(u3+TLr(ii),FOM);
    %
    gr0(ii)=sum(W3)*dx;
    %
    gr1(ii)=sum(W2.*wth1)*dx;
    %
    gr2(ii)=sum(W2.*wth2)*dx;
    %
    wclx=interp1(ho,CHol(ii,:),th2);
    gr3(ii)=sum(W2.*wclx)*dx;
end
%correct for click PDF
iwcl=wcl>0;
gr3(iwcl)=gr3(iwcl)./wcl(iwcl);
```

```
gr3(1)=gr3(2);

eps = sum((gr2-gr3).^2)

figure(4)
plot(hx,gr0+gr1,'k--')
line(hx,gr0+gr3,'color','k', ...
    'linestyle','none','marker','.','color',0.7* [1 1 1])
line(hx,gr0+gr2,'color','k', ...
    'linestyle','-','linewidth',2)
ylim([0 1.1])
xlabel('Range [m]')
ylabel('Probability')
return
```

9.5.3   Advantage of model-based detection functions

Comparing the two detection functions in Figure 9.10, one may conclude that detection functions depend not only on the animal's behaviour but also on the PAM implementation. With hindsight, this conclusion is not a surprise; in general, it is also true for visual surveys. This may be the reason why detection data are traditionally fitted with some anonymous empirical functions without further interpretation of the hidden physical/ human detection model.

Being based on technical systems, PAM systems may be described in detail and it is therefore appropriate to use this knowledge to reduce the model uncertainty. Of course, these models have to be fitted to the data to be useful for abundance estimation, which may or may not be as simple as for empirical functions. As already mentioned, empirical detection functions are all that is needed for pure abundance estimation, but habitat analysis with the introduction of covariates will gain when system variations are explicitly taken care of by the detection function.

In summary, we may conclude that if abundance estimation is the only purpose of the survey then empirical functions (e.g. hazard function with multiplicative modifications) may be sufficient, as in the end only the effective area or the mean detection range is of relevance (e.g. Equation 7.22). If, however, the survey should be used for assessing the habitat preferences of the animals, then it is important that PAM system-related parameters are treated separately from habitat-specific covariates.

10 PAM systems

The last chapter of this book addresses some aspects of the PAM system that are fundamental in any PAM implementation. PAM systems are in general composed of hardware and very often also of software, so this chapter will first address the hardware and in particular describe and discuss hydrophones and their integration into a properly functioning PAM system.

10.1 Hardware

As humans are terrestrial mammals and not very well adapted to listening to underwater sound, they must rely on technical tools, or hardware, to implement passive underwater acoustic systems. All PAM systems are generally composed of hydrophones, support electronics and some sort of human interface.

10.1.1 Hydrophones

The purpose of a classical hydrophone is to convert (transduce) sound pressure into electrical tension or voltage. Consequently, piezoelectric materials are suitable materials for the construction of hydrophones. Piezoelectricity describes the property of some materials to produce an electric potential when mechanical stress is applied or external pressure changes the shape. Equivalently, piezoelectric materials change their form when an electric field is applied. Such an effect is used to generate sound in a transmitter.

Hydrophones in their simplest and most common form are designed to respond directly to the pressure of an incident sound wave. Hydrophones may, however, be constructed to respond to some aspect of the particle motion, such as displacement, velocity or acceleration. Hydrophones may further be designed to respond to the pressure gradient or acoustic intensity.

Currently, the most commonly used piezoelectric transducer materials are poly-crystalline ceramics. In their original state, ceramic materials are composed of randomly oriented small crystallites and as such, they are isotropic and possess no piezoelectric properties. By temporarily applying a high electric field to the ceramic material, the small crystallites will align and the ceramic material will become anisotropic and therefore piezoelectric. This procedure is referred to as the poling operation; the poling direction defines the z-axis of a three-dimensional orthogonal co-ordinate system.

The piezoelectric effect is in general quite small. For example, PZT (lead zirconate titanates) ceramics, which are the most widely used materials for electro-acoustical transducers, exhibit a maximum shape change of about 0.1% of the original dimension.

There are two quantities that one should know about hydrophones. The first parameter is the sensitivity of the hydrophone, which describes the generated electric tension as function of acoustic sound pressure. The second one is the natural resonance frequency, which limits the useful bandwidth of the hydrophone.

Receiving sensitivity

The receiving sensitivity M of a pressure-sensitive hydrophone is defined as the ratio of the open-circuit voltage V_{OC} to the free-field pressure P_f

$$M = \frac{V_{OC}}{P_f} \tag{10.1}$$

As the name indicates, the open-circuit voltage is the electrical tension generated by the hydrophone in absence of any electronic circuitry that would result in the flow of an electric current. The free-field pressure is the pressure without the hydrophones, that is, without interaction with the hydrophone. If the hydrophone is small compared with the wavelength of the pressure field, then the presence of the hydrophone disturbs the acoustic pressure field very little, but if the size of the hydrophone approaches or exceeds $\lambda/2$, then the acoustic pressure on the hydrophone surface may become measurably different from the free-field pressure. The actual open-circuit voltage of the hydrophone depends on the actual shape of the ceramic.

Taking the logarithm of the magnitude of the sensitivity, we obtain the sensitivity expressed in decibels

$$S = 20 \log_{10}(M) \tag{10.2}$$

The sensitivity is expressed in xxx dB re $1\,\text{V}/\mu\text{Pa}$ and may be interpreted in the following way: if one assumes a sensitivity of -200 dB re $1\,\text{V}/\mu\text{Pa}$, then in order to generate 1 V tension on the hydrophone output one needs a sound pressure of 200 dB re 1 μPa.

Resonance frequency

The second important quantity of a hydrophone is the resonance frequency, which for a thin walled spherical hydrophone is simply the circumferential or radial mode, where the wavelength fits the circumference of the hydrophone. The resonant frequency f_r is in general given in the data sheets of commercial hydrophones.

Diffraction effect

The sensitivity of the hydrophone as described above is only valid if the hydrophone is small enough with respect to the wavelength of the sound wave, so that its presence does not disturb the sound field. The diffraction effect describes this interaction for increasing hydrophone dimensions. It is convenient to modify the free-field sensitivity by a multiplicative diffraction constant D, which for a spherical hydrophone is reasonably approximated by

$$D = \frac{1}{\sqrt{1 + \left(\frac{f}{f_r} d_c\right)^2}} \quad (10.3)$$

where d_c is a constant that controls the diffraction effect at resonance.

10.1.2 Hydrophone frequency response

In order to obtain the frequency response of the hydrophone it is necessary to translate the mechanical ceramic transducer characteristics into an equivalent electric schematic, that is, we replace the hardware with a combination of voltage source, resistor, inductor and capacitor.

Figure 10.1 shows the equivalent circuit of an idealized hydrophone with only one frequency of mechanical resonance. R_m is the mechanical loss and C_m and L_m represent stiffness and mass of the mechanical vibrating system. C_0 is the electrical capacitance between the electrodes on the piezoelectric element (see e.g. Burdic, 1984).

The circuit shown in Figure 10.1 is divided into two parts. One contains the mechanical quantities R_m, L_m and C_m, which are connected in series; the other one consists of the electrical components C_0 and R_1, which are connected in parallel. The impedances of the individual resistance-less parts (ZL_m, ZC_m, ZC_0) are estimated according to

$$ZL = i\omega L \quad (10.4)$$

$$ZC = \frac{1}{i\omega C} \quad (10.5)$$

which we can combine in serial with the resistance R_m

$$Z_m = R_m + ZL_m + ZC_m \quad (10.6)$$

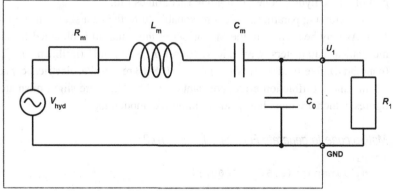

Hydrophone

Figure 10.1 Hydrophone equivalent circuit.

As we will use Matlab to take care of the complex numbers we continue the notation in terms of impedance and resistance.

We should note that the mechanical resonance frequency is related to the product of inductance L_m and capacitance C_m via

$$\omega_r^2 = \frac{1}{L_m C_m} \tag{10.7}$$

To estimate the admittance of the hydrophone at the output of the hydrophone we shorten the voltage source V_{hyd} and note that the capacitance C_0 is now parallel to the impedance Z_m so that the effective hydrophone admittance Y_H becomes

$$Y_H = \frac{1}{Z_H} = \left(\frac{1}{Z_m} + \frac{1}{ZC_0} \right) = \frac{Z_m + ZC_0}{Z_m ZC_0} \tag{10.8}$$

The conductance is the real part of Y_H while the susceptance is the imaginary part of the Y_H.

The output voltage, which is measured across C_0, becomes

$$V_1 = \left| \frac{ZC_0}{ZC_0 + Z_m} \right| V_{hyd} \tag{10.9}$$

where we model the hydrophone input voltage as

$$V_{hyd} = \gamma M D \tag{10.10}$$

with M the nominal sensitivity, D the diffraction effect, and $\gamma \approx 1 + \dfrac{C_0}{C_m}$, which is needed to correct the hydrophone sensitivity, which is defined at the hydrophone output and not at the hydrophone input.

In order to use this equivalent circuit one needs to quantify the values of the components. Unfortunately, these values are in most cases not readily available, and if at all, very often only in combined terms. However, most commercial hydrophones are described by an admittance plot near the resonance and a received sensitivity plot. These plots may be used to back-out the model parameters by fitting the model prediction to the published graphics.

Figure 10.2 shows the admittance plot and Figure 10.3 shows the received sensitivity plot of a D70 hydrophone of Neptune Electronics Ltd.

The following parameters are a reasonable fit to the data shown in Figures 10.2 and 10.3. As may be seen from the Matlab script the actual fit is obtained for the following mechanical parameters $R_m = 246\ \Omega$ (ohm), $L_m = 2.3$ mH (millihenry), $C_m = 2.27$ nF (nanofarad), free-field sensitivity of $M_0 = -198$ dB re $1\,\text{V}/\mu\text{Pa}$, electrical capacitance $C_0 = 7.5$ nF, and a diffraction effect constant $d_c = 3.5$. There are slight mismatches in both figures, which are most likely due to imperfect modelling.

Matlab code to generate Figures 10.2 and 10.3

```
%Scr10_1
f=10.^linspace(0,5.3,10000);
%
s=2*pi*f;
```

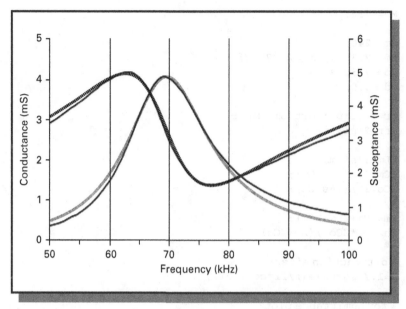

Figure 10.2 Modelled admittance plot of a hydrophone overlaid on a published commercial admittance plot of a D70 hydrophone. The dashed and dotted lines are predicted by the hydrophone model presented in Figure 10.1. Conductance, dashed line; susceptance, dotted line.

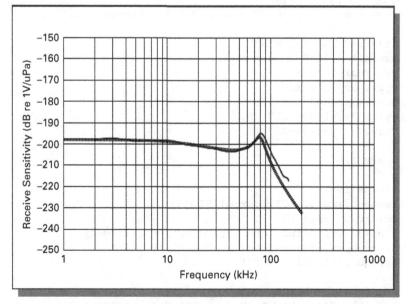

Figure 10.3 Sensitivity plot overlaid on a published commercial admittance plot of a D70 hydrophone. The dashed lines are predicted by the hydrophone model presented in Figure 10.1.

```
Mo=-198;
%
Rm = 246;
Lm = 2.3e-3; Cm = 2.27e-9;
Co = 7.5e-9;
cd = 3.5;
%
fr=1/(2*pi*sqrt(Lm*Cm));
sfr=2*pi*fr;
%
ZLm=i*s*Lm;
ZCm=1./(i*s*Cm);
ZCo=1./(i*s*Co);
%
Zm=(Rm+ZLm+ZCm);
Zh=Zm.*ZCo./(Zm+ZCo);
%
% diffraction effect
D=1./(sqrt(1+(f/fr*cd).^2));
%
% for admittance plot
Yh=1000./Zh;
T1=ZCo./(Zm+ZCo);
V1=T1./abs(T1(1)).*10.^(Mo/20).*D;

%graphics
%
% is from D/70 leaflet
img=imread('../PAM_BOOK_data/D70_Admittance_n.tif');

ifx=f>50000 & f<100000;
xx=f(ifx)/1000; xx=xx-xx(1);
yy1=real(Yh(ifx));
yy2=imag(Yh(ifx));
uu=57+xx*(407-57)/50;
vv1=257+yy1*(56-257)/5;
vv2=257+yy2*(56-257)/6;

figure(1)
clf
%image(img)
imagesc(img(:,:,2)),caxis([96 255])
colormap(gray)
line(uu,vv1,'linewidth',3,'Color','k','linestyle',':');
line(uu,vv2,'linewidth',3,'Color','k','linestyle','--');
set(gca,'visible','off')
%
% is from D/70 leaflet
img=imread('../PAM_BOOK_data/D70_Sensitivity_n.tif');

ifx=f>1000 & f<200000;
```

```
xx=log10(f(ifx)/1000); xx=xx-xx(1);
yy=20*log10(abs(V1(ifx)));

uv=[86.8 325;562.5 71];
xy=[0 -250; 3 -150];

uu=uv(1,1)+(xx-xy(1,1))*(uv(2,1)-uv(1,1))/(xy(2,1)-xy(1,1));
vv=uv(1,2)+(yy-xy(1,2))*(uv(2,2)-uv(1,2))/(xy(2,2)-xy(1,2));

figure(2)
clf
%image(img)
imagesc(img(:,:,2)),caxis([96 255])
colormap(gray)
line(uu,vv,'linewidth',3,'Color','k','linestyle','-');
set(gca,'visible','off')
```

10.1.3   Preamplifiers

The voltage generated by transducers is in general very low and requires amplification. Preamplifiers are special electronic devices that are matched to the hydrophone and designed to amplify very small voltages. Modern preamplifiers have as their first active element a field effect transistor (FET) to interface the hydrophone efficiently with the following amplifications, at the same time minimizing the impact of electronic noise.

Noise influences

Amplifying small signals will also amplify electronic noise from any source that is present at the input of the preamplifier. This electronic noise is in addition to the ambient noise, which is always present in the oceans. An acoustic signal will only be detectable if its pressure exceeds the total noise, which is composed of ambient and electronic noise. It is therefore necessary to chose electronic components and electronic circuits in such a way that they contribute as little as possible to the total noise floor; ideally, the electronic noise will be well below ocean ambient or environmental noise.

Thermal noise

Thermal noise is an important and hardly avoidable noise source, as it is generated when thermal energy causes free electrons to move randomly in a resistive material, an effect that is also called Johnson noise. The RMS thermal noise voltage across a resistor is given by

$$V_t(R) = \sqrt{4kTR\Delta f} \tag{10.11}$$

where k is the Bolzmann constant ($k = 1.38 \ 10^{-23}$ J/K), T is the absolute temperature ($T = 290$ K for 27 °C), R is the resistance, and Δf is bandwidth over which the noise is measured.

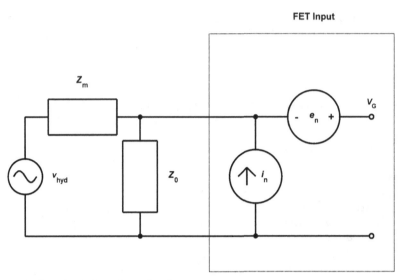

Figure 10.4   Equivalent circuit for noise estimation.

$V_n - I_n$ amplifier noise model

The noise at the output of an amplifier is a function of the noise at the input and the noise generated inside the amplifier. It is common to model amplifier noise by reflecting the internal noise sources to the input so they can easily be compared with the signal of interest at the input. The $V_n - I_n$ amplifier noise model uses two noise sources, a series voltage source e_n, and a shunt current source i_n. The shunt current noise source always needs a resistance to develop a noise voltage.

Figure 10.4 shows, for the input noise estimation, the simplified equivalent circuit of hydrophone and FET interface. The electrical description of the mechanical characteristics of the hydrophone (Figure 10.1) is combined in impedance Z_m, while the electrical capacitance C_0 and the load resistor R_1 form the impedance Z_0. As we see below, the shunt current noise i_n uses R_1 to develop noise voltage.

Electric equivalent environmental noise

The ambient noise level NL is, according to Equation 10.10, equivalent to a hydrophone input voltage V_{hyd}

$$V_{hyd} = \gamma M D 10^{NL/20} \tag{10.12}$$

which results at the hydrophone output in an environmental noise voltage V_E

$$V_E = \left| \frac{Z_0}{Z_0 + Z_m} \right| V_{hyd} \tag{10.13}$$

It is this noise level, which should be considered as a limit, which the system noise of a good hydrophone–preamplifier circuit should not exceed. Equation 10.13 says in principle that the voltage V_{hyd} is applied on Z_0 and Z_m functioning as a voltage divider, where the voltage V_E is read across Z_0.

System noise
The total system noise of hydrophone and preamplifier is composed of multiple components, which will be described next. Noise contributions are expressed in volts and where necessary converted to volts.

Thermal hydrophone noise
The resistance of the hydrophone R_m will generate thermal noise, which is given by

$$V_H = \left| \frac{Z_0}{Z_0 + Z_m} \right| V_t(R_m) \tag{10.14}$$

Thermal amplifier input resister noise
To estimate the thermal noise of the input resistor R_1, which is part of Z_0, we need to shorten the input voltage source, which effectively puts the capacitor C_0 in parallel to the mechanical impedance Z_m of the hydrophone, forming the effective hydrophone impedance Z_H (Equation 10.8). The output voltage is now measured across Z_H and is given by

$$V_R = \left| \frac{Z_H}{Z_H + R_1} \right| V_t(R_1) \tag{10.15}$$

Amplifier current noise
The amplifier current noise is generated by the current i_n flowing through resistor R_1. Again, V_{hyd} has to be shortened and the amplifier current noise becomes

$$V_i = \left| \frac{Z_H}{Z_H + R_1} \right| i_n R_1 \tag{10.16}$$

Total system noise
The total system noise V_{sys} is now given by taking the square root of the sum of all noise components squared

$$V_{\text{sys}} = \sqrt{e_n^2 + V_i^2 + V_R^2 + V_H^2} \tag{10.17}$$

where e_n is the amplifier voltage noise as obtained from the FET data sheets.

Example In addition to the previous example, we assume the following FET noise specifications

$$e_n = 0.3 \text{ nV}/\sqrt{\text{Hz}}$$
$$i_n = 0.2 \text{ pA}/\sqrt{\text{Hz}}$$

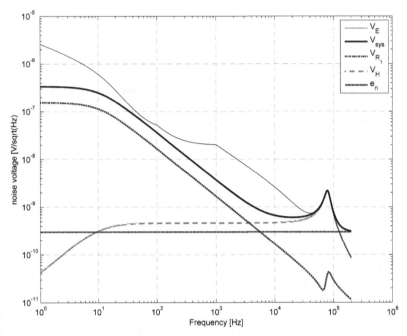

Figure 10.5   Noise contribution of hydrophone preamplifier configuration.

and estimate all the noise components, the total system noise, and the environmental noise for comparison.

Figure 10.5 shows the ambient noise level for a sea state zero and light ship traffic. It shows that the total system noise (solid thick line) is in general, for all frequencies, below the environmental noise, indicating that the assumed FET noise characteristics belong to a well-suited semiconductor. While not impossible, these noise values at higher frequencies are demanding, as most FETs feature noise voltage e_n values of 1 nV/$\sqrt{\text{Hz}}$ or higher. Nevertheless, the analysis presented allows the assessment of FETs that feature slightly different noise values.

We note further that the ratio between the strongest and weakest environmental noise is about 2500, which corresponds for wideband applications to a dynamic range of 68 dB. This dynamic range is somewhat reduced due to the C_0–R_1 highpass effect of the FET input stage, which limits the environmental noise below 20 Hz.

Matlab code to generate Figure 10.5

```
%Scr10_2
f=10.^linspace(0,5.3,10000);
%
s=2*pi*f;

Mo=-198;
%
```

```
Rm = 246;
Lm = 2.3e-3; Cm = 2.27e-9;
Co = 7.5e-9;
cd = 3.5;
%
fr=1/(2*pi*sqrt(Lm*Cm));
sfr=2*pi*fr;
%
ZLm=i*s*Lm;
ZCm=1./(i*s*Cm);
ZCo=1./(i*s*Co);
%
Zm=(Rm+ZLm+ZCm);
Zh=Zm.*ZCo./(Zm+ZCo);
%
% diffraction effect
D=1./(sqrt(1+(f/fr*cd).^2));

%
R1=1.5e+6; %hydrophone load resistance
%R1=1.5e+4; %alternative hydrophone load resistance

Zo=R1*ZCo./(R1+ZCo);
%
T1=Zo./(Zo+Zm);
T2=Zh./(Zh+R1);
%
%ambient noise
fkz=f/1000;
SD=0; %0:3
w=0; %m/s
%
T=13; %temperature
S=38; %salinity
z=100; %depth
c=1500; %sound speed
pH=7.8; %acidity
aa=FrancoisGarrison(fkz,T,S,z,c,pH);
dcorr=aa*z/1000;

NL1=17-30*log10(fkz);
NL2=44+23*log10(w+1)-17*log10(max(1,fkz));
NL3=30+10*SD-20*log10(max(0.1,fkz));
NL4=-15+20*log10(fkz);

NL3d=NL3-dcorr-10*log10(1+dcorr/8.686);

Na=10.^(NL1/10)+10.^(NL2/10)+10.^(NL3d/10)+10.^(NL4/10);
NL=10*log10(Na);

%FET noise model
```

```
en=0.3e-9;
in=0.2e-12;

%noise estimation
k=1.38e-23;
T=290;
vn= @(R) sqrt(4*k*T*R);
%
V_E=abs(T1).*(1+Co/Cm).*D.*10.^((Mo+NL)/20);
V_H=abs(T1).*vn(Rm);

V_R=abs(T2).*vn(R1);
V_I=abs(T2).*in*R1;
V_N=en+0*f;

V_sys=sqrt(V_N.^2+V_I.^2+V_R.^2+V_H.^2);

figure(1)
hp=plot(f,V_E,'k', f,V_sys,'k', ...
        f,V_R,'k-.', ...
        f,V_H,'k:', f,V_N,'k-');

set(hp(2:5),'linewidth',2)
set(gca,'xscale','log','yscale','log')
legend('V_E','V_[sys]','V_[R_1]','V_H','e_n',1)
xlabel('Frequency [Hz]')
ylabel('noise voltage [V/sqrt(Hz)')
grid on
set(gca,'xminorgrid','off','yminorgrid','off')
```

10.1.4   Analogue digital converter

Like most physical phenomena, sound pressure is also an analogue quantity, that is, sound is best described by a variable that in principle may take any value. Sound is an analogue signal. Hydrophones convert sound pressure into voltage, which is then amplified in multiple steps, but again, voltages are analogue quantities. Hydrophones plus some electronic amplifiers are in principle all that we need to listen to underwater sound and there are a significant number of field scientists doing exactly that. Analogue signals can be stored on analogue tape recorders, even though such recorders are difficult to find these days. Computers are now more generally used to interface the human to the underwater sound world. Even though there have been analogue computers in the past, modern computers are digital. It is therefore necessary to convert the analogue signal from the hydrophone into a digital form suited for further processing by digital computers.

Digital data

Digital data are usually represented as binary numbers, that is, as a collection of binary units (bits). A bit is a quantity that may take only two values, which are conveniently described as zero and one. Eight bits are combined into a byte, which for most computers is the reference quantity. Bits and bytes play an important role in modern computer applications. Memory storage is measured in bytes, or better millions or billions of bytes, (mega- or gigabytes, respectively); communication speed is expressed in megabits.

Digitizing analogue data is in principle nothing more than converting a real-valued into a discrete number, that is, we maintain only the relevant part of the real number and ignore all smaller contributions. We also do that continually in the real world where we limit, for example, the result of a division to one, two or three decimal places.

Decimal notation of a real number

The following number is given in three different notations. The first notation uses two decimals for the fraction smaller than one. The second notation also normalizes the whole number and uses four decimals, but adds a term that corrects for the normalization. The third term treats the real number as an integer and as a multiple of one hundred, that is, we scale the number to the minimal accepted decimal.

$$314.15 = 3.1415 \times 10^2 = \frac{31415}{100} \tag{10.18}$$

The example shows that in principle we can approximate a number by what is called floating-point notation (the second notation in Equation 10.18) or as an integer notation (the third notation in Equation 10.18). The constraints are only the number of digits available. Numbers that may vary over a large scale are best described by the floating-point notation, and numbers that are relatively constant or comparable are suited for integer notation.

Binary notation

In general, we may express any integer number I according to

$$I = \sum_{n=0}^{N-1} d_n 10^n = \sum_{m=0}^{M-1} b_m 2^m \tag{10.19}$$

where d_n are the different decimal digits, and b_m are the different binary bits.

The first encoding is the decimal notation taught in school, and the second encoding is the binary encoding used by computers. Obviously, the number of digits N needed to represent a given number is smaller than the number of bits M.

The maximal binary number is given when all bits are set to one and with

$$\sum_{m=0}^{M-1} 2^m = 2^M - 1 \tag{10.20}$$

Table 10.1 Decimal and binary format of integers

Decimal	Binary
0	00000000
1	00000001
127	01111111
−1	11111111
−128	10000000

We note that a single byte ($M = 8$ bits) can therefore describe 256 numbers between 0 and 255. Bit b_0 is also called the least significant bit and bit b_7, the last of the eight bits, is called the most significant bit.

The encoding shown in Equation 10.19 is fine if we only consider positive values, that is, unsigned integers, but in general, integer numbers may also be negative. A simple solution is to introduce a fixed offset, which is in the middle of the binary range, or given by the most significant bit

$$I = I_u - I_0 = \sum_{m=0}^{M-1} b_m 2^m - 2^{M-1} \tag{10.21}$$

To give a small example, Table 10.1 shows some numbers both in decimal and binary format.

We note that we still can encode 256 numbers but they vary now from -128 to 127, and that negative numbers are characterized by having the most significant bit always set to one.

The binary floating-point format is a little bit more complicated than the binary integer format and is defined by the IEEE 754–2008 standard, which in addition to the format also describes the set of standard floating-point operations.

Dynamic range
The result of analogue–digital conversion is a binary number having some M bits. In general, standard analogue–digital converters (ADC) generate integer numbers using 16 or 24 bits per sample (word size). The maximal number generated by the ADCs falls, for the 16 bit converter, within the −32 768 to +32 767 and for the 24 bit converter within the −8 388 608 to 8 388 607 range. As the smallest number is ±1, the theoretical dynamic range corresponds to about 2^{15} or 2^{23}, which translates to 90.3 dB or 138 dB, respectively.

A more realistic dynamic range (DR) is defined as the ratio of the RMS full-scale voltage to the RMS noise voltage of the converter in a certain bandwidth and is generally specified in dB according to

$$DR = 20 \times \log\left(\frac{V_{\text{fsc}}}{V_{\text{n}}}\right) \tag{10.22}$$

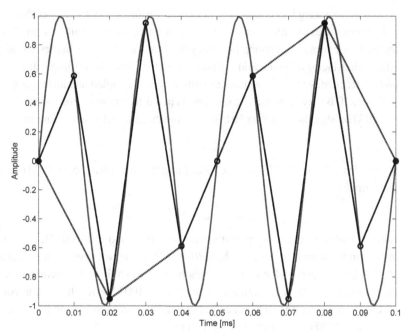

Figure 10.6   Effect of sampling below Nyquist critical frequency.

where V_{fsc} is the full-scale voltage and V_n is the noise voltage.

In general, the dynamic range of the ADC is below the maximal attainable value given by the word size of the ADC outputs. In particular, 24-bit ADC hardly ever exceeds a realistic dynamic range of 120 dB, indicating that only the 20 most significant bits may be relevant and the remaining bits are to be considered as electronic noise introduced by the ADC.

Data sampling

Analogue–digital conversion is in general carried out by sampling the analogue signal in a regular fashion. Whereas an analogue signal is defined for times that vary continuously, say $t_0 < t < t_1$, the output of the analogue–digital converter is defined for discrete times t_n, $n = 0, 1, \ldots, N-1$.

It seems logical that the sampled signal cannot show all the details of the original analogue signal. I have already mentioned in Section 2.4.1 the *Nyquist criterion*, which requires that the sampling frequency has to be higher than twice the maximum frequency in the signal and if necessary, the data have to be lowpass filtered to remove all higher-frequency content before sampling.

Figure 10.6 shows a simulated analogue sinusoidal signal of 40 kHz (grey solid line) that is sampled twice, first at a sampling frequency of 100 kHz (solid dotted line) and also at a sampling frequency of 50 kHz (dashed line). The original signal shows four periods, and so does the signal sampled at 100 kHz. In contrast, the signal sampled at 50 kHz shows only one period, that is, the resulting frequency of the under-sampled signal is what is called aliased: it does not correspond to the original frequency but to a frequency of 10 kHz.

To avoid aliasing it is common to design the analogue front-end in such a way that not only it introduces as little electronic noise as possible but also limits the bandwidth so that the ADC operates at a sampling frequency that is at least twice the highest signal frequency. Although a factor of two avoids aliasing, a better choice is to sample at three times the highest frequency to improve the similarity between sampled and analogue signal.

Figure 10.6 may, however, be interpreted in a different way. For a sampling frequency of 100 kHz, signals from 0 to 50 kHz are converted directly, according to

$$f_{out} = f_{in} \tag{10.23}$$

Signals from 50 to 100 kHz will be folded into the lower band of 0–50 kHz according to the formula

$$f_{out} = f_{samp} - f_{in} \tag{10.24}$$

Assume that the signals of interest are in the upper band (50–100 kHz) and we have no interest in the signals in the lower band (0–50 kHz); then we may bandpass the data with a 50–100 kHz, or somewhat narrower bandpass filter, so that all information in the lower band disappears, then a sampling frequency of 100 kHz results in a uncontaminated spectrum, with the only difference being that the spectrum is now frequency-inverted: what was 100 kHz now becomes 0 kHz, etc.

Equalization filter

In addition to noise considerations, the amplification of the analogue front-end should be also matched to the dynamic range of the ADC. This is of particular concern in wideband PAM applications, where the desire exists to detect low- and high-frequency signals simultaneously, as about 68 dB of the dynamic range is already consumed by the dynamic of the environmental noise, leaving only 20–50 dB for low-frequency signals. Stronger signals in the low frequency range will result in saturation of the ADC and will inhibit the detection of signals at higher frequencies, even if they were easily detectable otherwise.

One approach is to construct the amplifier stages in such a way that they gently suppress the low-frequency components of the environmental sound field to equalize the spectrum before entering the ADC stage. In this way, the full dynamic range of the ADC may be used for sound signals more or less independently of the frequency. As shown in the signal processing section (Section 4.2.1), a single-pole highpass filter (6 dB/octave) with a cut-off frequency of up to 45 kHz could be considered useful to equalize the spectrum on the hardware side. The implicit $R-C$ combination connecting the hydrophone with the preamplifier could, in principle, be used for implementing such a single-pole highpass filter. However, Figure 10.5 indicates that for the preamplifier a trade-off exists between suppressing environmental noise at lower frequencies and the sensitivity of the system to hydrophone system noise, so this equalization has to be designed carefully, using multiple amplification steps. For example, simply reducing R_1 from 1.5 MΩ to 15 kΩ to increase the highpass cut-off frequency by a factor of 100 would have the consequence that the thermal noise of R_1 would exceed the received environmental noise below 1 kHz.

10.1.5   Non-acoustic sensors

Whereas a simple underwater sound listening system may consist only of a hydrophone and some electronics, real PAM systems require, in addition to the acoustic sensors, some non-acoustic sensors. Depending on the PAM operation, if the hydrophone is moving or stationary, the sampling rate for digitizing these non-acoustic sensors may be low and will hardly ever exceed 10 Hz.

Depth sensor
The depth sensor is, after the hydrophone, the most important sensor. Measuring the depth of the hydrophone allows a better interpretation of the underwater sound. In particular, in the presence of multi-path arrivals of the sound (Section 6.3), knowledge of the hydrophone depth allows range and depth estimation of the acoustically active whale or dolphin.

Compass
Combining two or more hydrophones into an array requires orientation sensors to determine the actual orientation of the hydrophone array. Magnetic compasses are standard but deserve some care during the operation. Degaussing of the compasses should be done regularly before deployment, and ferrous materials should not be used in the vicinity of the compass to avoid disturbances of the earth's local magnetic field.

Inclination sensor
Compasses are not the only useful orientation sensors for PAM applications. In cases where the hydrophone array covers a vertical range, e.g. if we have a vertical or even a volumetric array, then we need a sensor that measures the inclination of the array. This may be done with accelerometers that measure the orientation of the sensor with respect to earth gravity. Inclination sensor and compass are very often integrated into a digital compass, allowing the estimation of pitch, roll and heading of the array.

Temperature sensor
Depth sensors, compasses and inclination sensors are important instruments for range and depth estimation of acoustically active whales and dolphins. While not strictly necessary for this task, an additional temperature sensor may be very useful to monitor the environmental conditions during PAM operation. We have seen earlier on that sound propagation is easily influenced by variations in water temperature. It therefore makes sense to complement the acoustic receiving system not only with orientation and depth sensors, but also with a temperature sensor.

10.2   Software

Using computers to acquire and process underwater sound also means using software. The development of PAM software is still a work in progress and is evolving from specific passive acoustic applications developed by researchers for their own fieldwork. As such, different approaches exist that use different hardware and computer languages. Recent developments in computer science and the dramatic improvements in processing power allow, in principle, the implementation of PAM software on an individual basis. Of course, adapting existing software, especially if developed as open source, to one's own needs is always more efficient than programming the whole software package from scratch.

All passive acoustic systems share, in theory, the same concept. Practical implementation may differ, mainly due to the possibilities for individual developers to program shortcuts in the otherwise general system concepts.

An important factor for the actual implementation is the computer language that is used to encode the PAM software. Whereas in the early years bioacoustic programs were coded in machine language, nowadays there is a variety of choices available. In the end, what counts is that the developer feels comfortable with the language and does not run into conceptual barriers. Small PAM systems may be programmed in any language, even using Matlab, which is efficient in signal processing but not so good for modular programming. Large modular programs are best programmed with languages that are intended for the development of large systems.

10.2.1   Data acquisition module

Most modern software is designed and developed to be modular and for PAM systems the data acquisition module plays a central role. It is the only module that has some strict time constraints, as it must guarantee that all data converted by the ADC are captured by the computer and made available to the other PAM modules.

The data acquisition module should be programmed as efficiently as possible to load all acoustic data into memory and to allow the computer to process the acquired data. Parallel processing is key to the implementation of a useful data acquisition module. Most data capture hardware, be it analogue sound cards or digital interfaces, feature their own DMA (direct memory access) that may transfer the input data directly into memory without use of the computer's CPU. Alternatively, the presence of multiple CPUs in the computer, or multiple networked computers, would allow one CPU, or one computer, to be dedicated to data acquisition.

In any case, a data acquisition module would concentrate on the transfer of input data into memory and would ignore what is done with the data. The interaction with the acquisition module should be as limited as possible to minimize unforeseen conflicts. Ideally, the data acquisition module would send messages to the PAM control module that a predefined amount of data is available for processing, and would only receive messages from the PAM control module that start and stop the data acquisition.

A key feature in the data acquisition module is the existence of multiple data buffers to allow concurrent acquisition and processing. Using two data buffers, where one is usually called ping and the other called pong, allows the processing of the data in the pong buffer while the acquisition module fills in the ping buffer. After having filled a buffer, the acquisition module signals to the PAM controller the availability of this buffer, and continues to fill the alternative buffer. The processing module and all other modules are only allowed to access the buffer that is not used by the acquisition module. To maintain real-time functionality, all data consumers, that is, all modules that read the acquisition buffers, must be finished with the processing before the acquisition module requires write access to the buffer. At any given time, an acquisition buffer is always in exclusive use, either by the acquisition module (the producer) or by the consumers, but not by both.

Using two acquisition buffers requires that the consumers always process the data faster than the producer can fill up the buffers. In general-purpose computers this may not be guaranteed, owing to activities that are not related to the PAM software and that may from time to time delay the processing significantly. To overcome this latency problem the double buffer concept is extended to a multiple circular buffer system. The acquisition module fills up one buffer after the other and signals to the PAM control module which one it has filled and which it is actually writing to.

As the data acquisition module should ideally run in parallel to the remainder of the PAM software, inter-module communication is of importance. Two types of information are relevant to the functionality of the PAM system. The first one is the identification of the buffer to which the acquisition module is currently writing, in order to avoid access by other PAM modules. The other type of information contains data that are specific to the acquisition module, but are relevant to the other PAM modules, such as timestamp of acquisition, or number of samples actually acquired. This additional information may be added as a header to the acquired input data, or be placed as a message in a message queue.

10.2.2   Data archiving module

While the data acquisition module interfaces with the data source and is therefore central to every PAM system, the data-archiving module ensures that the precious data are not lost in memory but are available after the PAM operation for further analysis. The quality and quantity of data that should be archived vary with the type and objective of PAM activity and the performance of the PAM system.

The simplest PAM system will archive all raw input data without any further processing. Such a system may run completely autonomously, and the absence of an operator eliminates the need for sophisticated signal and display processing. Signal detection and data analysis will then be done in a second phase after retrieval of this autonomous PAM system. Such an approach is easy to implement but finds its limit in the available storage capability. Fortunately, the capacity of permanent storage is increasing continuously, allowing more and more continuous archiving of raw input data.

The amount of storage required for continuous archiving depends mainly on the number of hydrophones, the sampling rate, and the word size of the ADC. Four hydrophones, sampled at 192 kHz with 24 bit ADC resolution, generate a continuous data flow of 2.3 Mb/s and require a permanent storage of 200 Gb/day. A continuous month-long PAM activity will result in an overall storage requirement of 6 Tb. This requirement is nowadays not out of reach, at least for desktop applications, but is still somewhat demanding for low-power long-endurance autonomous systems.

Traditionally, data were archived on tapes and other magnetic media. In recent years, hard disks have taken on the role of archiving media; they allow random and therefore easier access to the archived data than magnetic tapes, which can only be accessed serially. However, traditional hard disks require benign environments: they are in general shock-sensitive, and are to be handled with care. Solid-state disks are a good alternative as they have no moving parts but they still lack the huge capacity available on traditional hard disks.

The data-archiving module should handle not only the data from the data acquisition module, but also processed data from the signal processing modules. If required, the raw and processed data that are ready for archiving should be buffered in a similar way as done with the data acquisition. This would allow time to bridge delays due to temporary unavailability of disks, e.g. during initial disk spin-up.

When programming the data-archiving module it is appropriate to implement a replay capability, allowing seamless access to archived data during post-processing. Switching the data source from acquisition mode to replay mode is not only a convenient feature during software development, but also facilitates the analysis of results and performance assessment of the PAM system.

10.2.3 Signal processing modules

While some applications do not require elaborate signal processing, all modern PAM systems allow more or less complex signal processing as discussed in Part II of this book. Generally, we may characterize the signal processing according to the generality of the implemented algorithms. Very general-purpose algorithms, e.g. estimation of spectrograms, are found in all acoustic software packages. More specific algorithms, e.g. click detector, or whistle detector, are more specific to cetacean research; and some techniques, e.g. localization and tracking algorithms, are specific to particular PAM implementations and require particular hardware.

All signal-processing modules have in common that they transform a dataset from one form into another. With properly designed interfaces, the different modules can then be chained together to perform more or less complex tasks. This serial mode of operation is common in simple computer systems but requires a strict linear dependency. For example, implementing a complete DCLT algorithm (detection, classification, localization and tracking) in a serial fashion does not allow the use of multi-paths for detection or classification, at this information belongs to the localization module and is therefore only available after detection or classification.

Overall, the signal processing modules should be programmed in such a way that they do not block the PAM operation and that the signal processing is carried out in the allocated period. This is more or less equivalent to the requirement that all software modules must implement sufficient error handling for easy avoidance of undesired situations. For example, if the detection threshold of a click detector is too low, generating too many false alarms, so that the classifier or tracker will overrun the allocated processing time, then it may be better either to skip this dataset or to adapt the detection threshold to allow timely classification or tracking. The classification module must therefore handle the error situation (running out of time) and signal to the PAM controller to handle the missing classification results and if necessary to increase the detection threshold.

Using multiple CPUs for signal processing is interesting and challenging at the same time. This is especially due to the need to synchronize the different activities or processes, so that a process can wait for the availability of relevant data that are produced by another parallel process. Inter-process communication has always been important in parallel processing on mainframe computers and is nowadays also of interest in PAM software running on multi-core desktop or single-board computers.

It seems therefore intuitive that the implementation of complex signal processing algorithms in a general-purpose and flexible PAM system is very demanding and that most existing PAM software implementations tend to be specific to the applications or research preferences of the developer. As more and more software becomes available, programmatic consolidation of the signal processing techniques may be needed.

10.2.4 Display modules

Display software is both an art and a key to the success of human-oriented PAM software, as it is the graphic presentation of the PAM results that must meet the expectation of the human operator in order for a PAM system to be acceptable. Overall, displays should be intuitive and present only the information that is needed by the operator to carry out the PAM operation. Unfortunately, the information to be displayed depends on the functionality and complexity of the signal processing modules and the integration of the operator in PAM processing.

Most PAM displays feature a spectrogram display for one or more hydrophones, in which two modes are possible. Some displays keep the actual time always at the same screen location and scroll the history either horizontally or vertically, depending on the orientation of the time axis. Alternatively, the display keeps all data at a fixed screen location, with the consequence that the newest data are presented in a circular way, always overwriting the oldest data on the screen. Experience shows that if the operator needs to look carefully at spectral features, it is better to keep the spectrogram fixed and to have the latest update at a moving position. In addition, scrolling spectrograms require a fairly high update rate to avoid jerky movements of the whole spectrogram.

Although display modules are in general output-only, displaying results of the signal processing modules, some display activities require operator interaction. Two input modes are common with desktop computers: input using the mouse pointer and input via the keyboard. As with all MMIs (man–machine interfaces) the objective should be to provide the necessary flexibility with a minimum of confusion. In this sense, it is useful to follow the MMI concepts of successful commercial software, using similar menu setups, keystrokes and mouse operations.

A successful display will maximize the space for the most important graphics and will defer less frequently used information to secondary display panels that are in general hidden and only visualized on demand. Nevertheless, a PAM display differs from general acoustic analysis software and should not allow the operator to hide the primary PAM display completely with secondary graphics as long as the data acquisition continues and the risk exists that important events may be missed. For example, if one is trying to detect rare beaked whale events on the spectrogram, then covering the spectrogram with a map or a geographic information system (GIS) display only to see where the ship is, may jeopardize the PAM objective. Multiple screens on the same computer provide an easy solution to the increasing need for display space by complex PAM software.

10.2.5   Monitoring module

Although processing acoustic data and displaying the results of the signal processing modules are among the main tasks of a PAM system, the monitoring aspect adds an additional layer by relating the acoustic acquisition system to an overarching objective, the monitoring of whales and dolphins. It is therefore intuitive to add a special module to the basic acquisition, processing and display system that addresses the needs of the monitoring objective. Monitoring, in this context, is best described as the localization and tracking of cetaceans, that is, assigning time–space attributes to detected animals.

A convenient way to deal with this monitoring aspect is to use a common data layer, or format, which satisfies the monitoring requirement and is practicable for the PAM system. One such method is to geo-reference the PAM observations and to make them suitable for use in a standard GIS. Geo-referencing requires, in addition to the processing of acoustic data, also the acquisition of the location of the PAM system. The method of choice is to use GPS to determine the location of the PAM system and to use the different localization techniques (Chapter 6) to determine the absolute location of the species of interest. The approach seems similar to monitoring methods employed during visual monitoring. The only difference is that not all PAM implementations are capable of determining the absolute location of the acoustically active animal, complicating the use of PAM information for cetacean monitoring.

As a minimal requirement, PAM systems should provide for manual logging of acoustic events by the operator. A signal-processing module may support this task by carrying out routine calculations, either on demand or automatically. A useful aid in this context is a small geographic display that automatically shows possible locations, allowing the operator to select the most likely solution interactively.

We have seen that the detection function plays an important role in PAM applications. It is therefore very useful for improved PAM analysis to collect and present environmental data and to visualize the acoustic propagation conditions. A simple-to-use propagation model and the associated database of environmental parameters should therefore be part of the PAM system.

Finally, the actual state of the art in PAM systems still needs the presence of the operator for final decisions, and assigns a supporting role to the PAM software. The overall development goal, however, is that the different signal processing components evolve to provide not only robust DCLT but also reliable monitoring capabilities. Only when the need for operator intervention is minimized will automated PAM systems become state-of-the-art. For the time being, such operator-less systems still remain the goal of most PAM development.

10.2.6 PAM control module

At the end of the description of PAM software, it is important to mention the backbone of the entire PAM, the control module. This control module is generally an efficiently coded task controller ensuring that the overall system runs smoothly and that error conditions are assessed and handled properly. The control module handles further all interprocess communication and should be the only interface to start and stop the different PAM modules. If convenient, the implementation of the control module may use the functionality of an existing operating system.

10.3 PAM implementations

Passive acoustic monitoring is carried out with more or less complex systems. The systems may be adapted to human operators or even rely on the capacity of the human brain, or may function completely autonomously.

10.3.1 Lowered hydrophones

The simplest PAM system is a single hydrophone that is lowered from a boat to listen to whales and dolphins. Most researchers that are interested in cetacean sounds have used this approach at least once. It is easy to implement and requires only a minimal amount of hardware and hardly any software. A hydrophone with an electronic amplifier that will make the sound audible in headphones is all that it needs to get started, but is very often also all that one has available.

If the system is completely adapted to the human ear then it is clear that only audible sound may be detected. If one wants to extend the frequency coverage into the ultra- or infrasonic range, a computer is desirable to visualize the whole spectrum in the form of a spectrogram, or to convert ultra- or infrasound to an audible frequency range. A bat detector is one such electronic device that makes ultrasonic bat clicks audible; similar devices have been conceived for detecting the presence of harbour porpoises (e.g. Thomsen *et al.* 2005).

Single lowered hydrophones may be used in certain circumstances to determine the range of the acoustically active cetaceans, that is, ranging is possible when sufficient multi-path sound arrivals are detectable. If multiple hydrophones are lowered from a ship or boat, e.g. in the form of a vertical array or a compact volumetric array, than ranging becomes much easier, as demonstrated in Chapter 6. The presence of a depth sensor (and orientation sensors in arrays) are then important features of the hydrophone array.

10.3.2   Towed arrays

While lowering hydrophones over the side of a stationary ship or boat has minimal requirements on the shape of the system, towing hydrophones needs some hydrodynamic considerations. Bulky hydrophone configurations will generate significant drag, not only making it difficult to deploy the hydrophones at the desired depth, but also necessitating stronger tow cables to withstand the drag forces.

In general, towed arrays are implemented as line arrays, where the hydrophones are positioned serially inside an oil-filled plastic tube (usually made of polyurethane). The length of the towed array and the number of hydrophones is variable and application-dependent. Arrays that are optimized for low-frequency applications may be very long; arrays used typically by bioacoustic researchers tend to be relatively short. In general, array length is measured in terms of shortest useful wavelength, or highest frequency. This is mainly because the angular resolution of a line array is determined by the array length relative to the wavelength of the frequency of interest. The longer the array in terms of wavelength, the more precisely one may determine the direction from which the sound arrives. For sonic frequencies (say up to 15 kHz) an array of, say, 9 m length may achieve an angular resolution (3 dB beam width) of about 0.6° (53° λ/L). This maximal angular resolution is only achieved for this high frequency, and only if the sound arrives from broadside, that is, 90° with respect to the line that forms the array. It is interesting to note that the angular resolution does not depend on the number of hydrophones but only on the overall length perpendicular to the direction of sound propagation, with the consequence that the maximal angular resolution of the array may be achieved with only two hydrophones, one at the beginning and one at the end of the array. The purpose of using multiple hydrophones in a towed array is mainly to carry out beam-forming (Chapter 6.9) to filter the ambient noise spatially.

Conceptually, a towed array consists not only of the hydrophone section, but also of a tow cable that connects the hydrophone section to the ship. The tow cable is in general much longer than the hydrophone section in order to lower the hydrophones to the desired operational depth. The length of the tow cable is determined effectively by the desired array depth and the speed of the tow vessel, because the faster the array is towed, the longer the tow cable must be to maintain the same operational depth. As long tow cables tend to vibrate, it is important to decouple the tow cable from the acoustic section. This is usually done by inserting, between tow cable and hydrophone section, what is called a VIM (vibration isolation module). A typical VIM is made similarly to the hydrophone section, but without any acoustic sensors. Its only purpose is to pass all electric cables, or optical fibres, from the tow cable to the hydrophone section and to absorb all vibration that may come from the tow cable. To straighten the array, it is common to attach at the end of the acoustic section a rope drogue. Finally, a towed array is designed to be neutrally buoyant and to stay horizontal at most operational speeds.

So far most towed arrays are analogue arrays, that is, each hydrophone is pre-amplified and passed as an individual element through a multi-wire tow cable to the tow ship, where all hydrophone channels are interfaced to an analogue acquisition system. This is economical only for a small number of hydrophones, also because strengthened multi-channel tow cables and underwater connectors are rather expensive. An alternative solution is to add an ADC to each hydrophone or group of hydrophones to convert the analogue signal already inside the array to a digital data stream. These digital data streams may then easily be multiplexed and sent over a single line, copper or optical, to the tow ship. The price of this cost reduction, however, is the loss of flexibility in selecting the characteristics of the equalizing and anti-aliasing filters and the sampling frequency. While the flexibility of an analogue array is a desired feature in bioacoustic research, it may not be as important in PAM applications, allowing the implementation of cleverly designed digital arrays.

Towed arrays with at least two hydrophones allow direction finding. It is therefore desirable to have, close to the acoustic section, not only a depth sensor but also a heading sensor. Most depth and heading sensors are available with built- in ADC and when added to an analogue array care must be taken that standard high-voltage digital wires do not interfere electrically with the low-voltage analogue signal (normally of the order of 10^{-3} V). An all-digital approach could be used to avoid this electrical crosstalk in the tow cable.

10.3.3   Acoustic buoys and autonomous systems

Although towed arrays are the only viable implementation for PAM systems that use moving ships, acoustic buoys and autonomous systems are becoming more and more interesting for PAM applications. Compared with optical systems, PAM is a low-data-rate system and therefore suited for unattended operations. Furthermore, most optical systems require light, whereas acoustics work day and night. Night-vision devices, whether based on low-light amplification or on infrared, are in principle possible, but so far technologically more demanding than purely acoustic systems. Even if automatic PAM system are still in their infancy, the need for large-scale monitoring of whales and dolphins will inevitably result in autonomous acoustic systems, for which acoustic buoys are a promising tool.

Sound-receiving buoys (sonobuoys) have been used for quite some time by the military. The concept is straightforward: sonobuoys relay the sound received by a single hydrophone via radio to a nearby receiver. The analogue signal is then processed in the traditional way. At the end of the lifespan of the power supply, the sonobuoy sinks to the bottom of the ocean. In that sense, sonobuoys are disposable wireless hydrophones. Being produced for military applications, sonobuoys are not widely available or even specified in detail to be useful for civilian PAM applications.

The development of microelectronics in recent years allows the implementation of affordable multiple-use acoustic buoys with built-in signal processing capability. These systems are programmable and may be set up to execute standard DCLT tasks but may also be conceived to support the monitoring objective of the PAM application. An on-board DSP (digital signal processor) may accommodate all the modules that a desktop-based PAM system also implements. Only the display part is not necessary in an autonomous system, for obvious reasons.

In addition to the challenges when implementing a successful PAM system, the developer of autonomous systems is faced with some severe constraints. Foremost, processing power and data storage capacity are limited. This is mainly due to spatial limitations, due to reduced availability of energy, but also due to the adverse environmental conditions where the autonomous PAM system is employed.

Using PAM systems at depth, say 100 m or even 1000 m, either requires a pressure housing capable of withstanding 10 or 100 atmospheres, or needs to be oil-filled. Oil-filled systems are a clever way to mitigate the pressure issue, but require a careful selection of the components: for example, they cannot contain hard disks or devices that work only in air.

Acoustic buoys may be implemented in a variety of forms. They may be free-floating on the surface, or completely submerged without any connection to the surface, but they may be also bottom-moored. Submerged buoys may be designed to stay at a predefined depth or they may move up and down the water column.

Even if we are interested in underwater sound, there are a few reasons why a buoy should be on the surface. One reason is to be able to transmit data or results via radio or telephone to the user, another reason is to collect energy (e.g. solar or wave energy) to extend the lifespan of the on-board battery. However, any surface expression of a buoy poses a risk to the integrity of the system, not only of being overrun by surface ships, but also of being damaged by the forces of heavy seas.

Acoustic buoys are well suited to applications that prefer stationary PAM systems or allow the PAM system to drift with the current. If a larger part of the ocean should be monitored, and if survey time does not play a big role because the animals are assumed to stay in the area during the survey time, then the PAM system may be placed on slow-moving platforms that are optimized for very low power consumption. Underwater gliders, for example, are cost-effective underwater vehicles that move vertically up and down in such a way that, at the same time, they gain horizontal distance. Some surface systems, e.g. wave gliders, exploit wave energy to move around, or stay in a small area and provide at the same time solar energy to the PAM payload. Staying all the time on the surface, such platforms provide the opportunity to use GPS for precise localization of the platform and enable radio communication to transfer relevant acoustic and PAM status information.

All these new PAM platforms provide only limited space and energy compared with ship-based applications. They offer, however, the opportunity to carry out PAM operation over long periods and in environments that are not easily accessible. Even though they are not expendable, autonomous PAM systems are a low-cost alternative to ship-based PAM, which may become very expensive, especially for large-scale monitoring activities that last over weeks or months. A drawback of autonomous PAM systems with surface expression is the vulnerability to ship strikes, requiring that autonomous PAM systems are either affordable enough to be lost, or are equipped with radar reflectors or, better, an AIS (automated identification system) that allow easy localization by approaching vessels. Using AIS in combination with autonomous PAM systems allows the implementation of cetacean warning systems signalling the presence of cetaceans, when necessary.

10.4   The future of PAM

In recent years, passive acoustic monitoring of whales, dolphins and porpoises has received increased attention from scientists, funding agencies and policy makers. Much of this interest is based on the observation that cetaceans use sound to make their living; passive acoustics is an appropriate tool for detecting cetaceans and monitoring their behaviour. The interest is further driven by the expectation that PAM may in the near future be operated completely autonomously; this will reduce costs and overcome limitations that are usually attributed to visual detection that requires human observers.

Even though the term PAM is widely used in the scientific community, the monitoring aspect is not very well developed, at least when dealing with cetaceans, and it would be more appropriate to characterize the existing scientific activities as passive acoustic detection (PAD). PAD has always been a part of cetacean research where scientists have deployed hydrophones to characterize acoustic activities of whales and dolphins, or where scientists have used small towed arrays to detect submerged cetaceans to complement visual survey operations. Although the monitoring aspect of PAM is based also on detection, it is more demanding in terms of classification, localization and tracking. PAM requires persistent operation, which may be done by human operators but is better performed by autonomous systems, which are more cost-effective and whose performance can be objectively specified.

Recent advances in electronic engineering and signal processing have resulted in passive acoustic systems that are capable of archiving data of interest for further analysis of cetacean activity (archival buoys and tags), or that relay, in near real-time, simple cues to the behaviour of cetaceans (satellite tags). One should expect that in the future the different paths will merge together, resulting in an intelligent autonomous PAM system that not only detects but also classifies cetaceans, localizes and tracks their behaviour, and which may even be capable of adapting the mode of operations to the behaviour of monitored species.

An intelligent autonomous PAM system will be capable of persisting over multiple temporal scales of interest (at least a year), will report regularly on its status even without cetacean activity, will detect and classify multiple species and assess the animals' behaviour by localizing and tracking their motion, will report salient information using modern communication tools to allow human intervention when deemed necessary, and will be cost-effective to facilitate widespread, redundant and networked use, tolerating some failing systems. In other words, future PAM systems will be at the same level of technology as modern high-tech computerized cellphone devices, except that instead of inter-human chat they will collect and relay information on whales, dolphins and porpoises.

References and further reading

Note: Publications marked with an asterisk are not explicitly referred to in the book, but are offered here as further reading.

*Abbot, T. A., Premus, V. E., and Abbot, P. A. (2010). A real-time method for autonomous passive acoustic detection-classification of humpback whales. *J. Acoust. Soc. Am.* **127**: 2894–903.

Adam, O. (2006a). The use of the Hilbert-Huang transform to analyze transient signals emitted by sperm whales. *Appl. Acoust.* **67**: 1134–43.

Adam, O. (2006b). Advantages of the Hilbert-Huang transform for marine mammals signals analysis. *J. Acoust. Soc. Am.* **120**: 2965–73.

Adler-Fenchel, H. S. (1980). Acoustically derived estimate of the size distribution for a sample of sperm whales (*Physeter catodon*) in the Western North Atlantic. *Can. J. Fish. Aquat. Sci.* **37**: 2358–61.

Alling, A., Dorsey, E. M., and Gordon, J. C. D. (1991). Blue whales *Balaenoptera musculus* off the northeast coast of Sri Lanka: distribution, feeding and individual identification. *UNEP Mar. Mamm. Tech. Rep.* **3**: 247–58.

*Altes, R. A. (1980). Detection, estimation, and classification with spectrograms. *J. Acoust. Soc. Am.* **67**: 1232–46.

Amano, M., and Yoshioka, M. (2003). Sperm whale diving behavior monitored using a suction-cupattached TDR tag. *Mar. Ecol. Prog. Ser.* **258**: 291–5.

Amundin, M. (1991). Click repetition rate patterns in communicative sounds from the harbour porpoise, *Phocoena phocoena*. In *Sound Production in Odontocetes with Emphasis on the Harbour Porpoise, Phocoena phocoena*. Stockholm: Swede Publishing AB, pp. 91–111.

Anderson, R. C., Clark, R., Madsen, P. T. *et al.* (2006). Observation of Longman's beaked whale (*Indopacetus pacificus*) in the western Indean ocean. *Aquat. Mamm.* **32**: 223–234.

*Andrew, R. K., Howe, B. M., Mercer, J. A., and Dzieciuch, M. A. (2002). Ocean ambient sound: comparing the 1960s with the 1990s for a receiver off the California coast. *ARLO* **3**: 65–70.

Antunes, R., Rendell, L., and Gordon, J. (2010). Measuring inter-pulse intervals in sperm whale clicks: consistency of automatic estimation methods. *J. Acoust. Soc. Am.* **127**: 3239–47.

Akamatsu, T., Wang, D., Nakamura, K., and Wang, K. (1998). Echolocation range of captive and free-ranging baiji (*Lipotes vexillifer*), finless porpoise (*Neophocaena phocaenoides*), and bottlenose dolphin (*Tursiops truncatus*). *J. Acoust. Soc. Am.* **104**: 2511–16.

*Akamatsu, T., Wang, D., Wang, K., and Naito, Y. (2005). Biosonar behaviour of free-ranging porpoises. *Proc. Roy. Soc. Lond.* B **272**: 797–801.

Arnason, U., Gullberg, A., and Janke, A. (2004). Mitogenomic analyses provide new insights into cetacean origin and evolution. *Gene* **333**: 27–34.

Au, W. W. L. (1993). *The Sonar of Dolphins*. New York: Springer.

*Au, W. W. L. (1997). Echolocation in dolphins with a dolphin-bat comparison. *Bioacoustics* **8**: 137–62.

*Au, W. W. L., and Benoit-Bird, K. J. (2003). Automatic gain control in the echolocation system of dolphins. *Nature* **423**: 861–3.

Au, W. W. L., and Hastings, M. C. (2008). *Principles of Marine Bioacoustics*. New York: Springer.

Au, W. W. L., and Herzing, D. L. (2003). Echolocation signals of wild Atlantic spotted dolphin (*Stenella frontalis*). *J. Acoust. Soc. Am.* **113**: 598–604.

Au, W. W. L., Carder, D. A., Penner, R. H., and Scronce, B. L. (1985). Demonstration of adaptation in Beluga whale echolocation signals. *J. Acoust. Soc. Am.* **77**: 726–30.

Au, W. W. L., Floyd, R. W., Penner, R. J., and Murchison, E. (1974). Measurements of echolocation signals of the Atlantic bottlenose dolphin, *Tursiops truncatus* Montagu, in open waters. *J. Acoust. Soc. Am.* **56**: 1280–90.

Au, W. W. L., Ford, J. K. B., Horne, J. K., and Newman Allman, K. A. (2004). Echolocation signals of free-ranging killer whales (*Orcinus orca*) and modelling of foraging for Chinook salmon (*Oncorhynchus tshawytscha*). *J. Acoust. Soc. Am.* **115**: 901–9.

Au, W. W. L., Kastelein, R. A., Rippe, T., and Schooneman, N. M. (1999). Transmission beam pattern and echolocation signals of a harbor porpoise (*Phocoena phocoena*). *J. Acoust. Soc. Am.* **106**: 3699–705.

Au, W. W. L., Moore, P. W., and Pawloski, D. (1986). Echolocation transmitting beam of the Atlantic bottlenose dolphin. *J. Acoust. Soc. Am.* **80**: 688–91.

Au, W. W. L., Pack, A. A., Lammers, M. O. *et al.* (2006). Acoustic properties of humpback whale songs. *J. Acoust. Soc. Am.* **120**: 1103–10.

Au, W. W. L., Pawloski, J. L., Nachtigall, P. E., Blonz, M., and Gisner, R. C. (1995). Echolocation signals and transmission beam pattern of a false killer whale (*Pseudorca crassidens*). *J. Acoust. Soc. Am.* **98**: 51–9.

Au, W. W. L., Penner, R. H., and Kadane, J. (1982). Acoustic behavior of echolocating Atlantic Bottlenose Dolphins. *J. Acoust. Soc. Am.* **71**: 1269–75.

Au, W. W. L., Penner, R. H., and Turl, C. W. (1987). Propagation of beluga echolocation signals. *J. Acoust. Soc. Am.* **82**: 807–13.

Backus, R. H., and Schevill, W. E. (1966). *Physeter* clicks. In *Whales, Dolphins and Porpoises*, edited by K. Norris. Berkeley, CA: University of California Press, pp. 510–27.

*Baggenstoss, P. M. (2008) Joint localization and separation of sperm whale clicks. *Can. Acoust.* **36**: 125–31.

Baggeroer, A. B., Kuperman, W. A., and Mikhalevsky, P. N. (1993). An overview of matched field methods in ocean acoustics. *IEEE J. Ocean Eng.* **18**: 401–24.

Baggeroer, A. B., Kuperman, W. A., and Schmidt, H. (1988). Matched field processing: source localization in correlated noise as an optimum parameter estimation problem. *J. Acoust. Soc. Am.* **83**: 571–87.

Baird, R. W., Borsani, J. F., Bradley Hanson, M., and Tyack, P. L. (2002). Diving behavior of long-finned pilot whales in the Ligurian Sea. *Mar. Ecol. Prog. Ser.* **237**: 301–5.

Baird, R. W., Ligon, A. D., Hooker, S. K., and Gorgone, A. M. (2001). Subsurface and nighttime behaviour of pantropical spotted dolphins in Hawai'i. *Can. J. Zool.* **79**: 988–96.

Baker, C. S., Herman, L. M., Perry, A. *et al.* (1985). Population characteristics and migration of summer and late-season humpback whales *(Megaptera novaeangliae)* in southeastern Alaska. *Mar. Mamm. Sci.* **1**: 304–23.

Baraff, L. S., Clapham, P. J., and Mattila, D. K. (1991). Feeding behavior of a humpback whale in low-latitude waters. *Mar. Mamm. Sci.* **7**: 197–202.

*Barbarossa, S., Scaglione, A., and Giannakis, G. (1998). Product high-order ambiguity function for multicomponent polynomial-phase signal modeling. *IEEE Trans. Sig. Proc.* **46**: 691–708.

*Barlow, J. (1994). Abundance of large whales in California coastal waters: a comparison of ship surveys in 1979/80 and in 1991. *Rep. Int. Whal. Comm.* **44**: 399–406.

*Barlow, J. (1995). The abundance of cetaceans in California waters. Part I: Ship surveys in summer and fall of 1991. *Fish. Bull.* **93**: 1–14.

Barlow, J. (1999). Trackline detection probability for long-diving whales. In *Marine Mammal Survey and Assessment Methods*, edited by G. W. Garner *et al.* Rotterdam: A.A. Balkema, pp. 209–24.

*Barlow, J., and Gisiner, R. (2006). Mitigating, monitoring and assessing the effects of anthropogenic sound on beaked whales. *J. Cetacean Res. Manage.* **7**: 239–49.

Barlow, J., and Taylor, B. L. (2005). Estimates of sperm whale abundance in the northeastern temperate Pacific from combined acoustic and visual survey. *Mar. Mamm. Sci.* **21**: 429–45.

Barlow, J., Ferguson, M. C., Perrin, W. F. *et al.* (2006). Abundance and densities of beaked and bottlenose whales (family Ziphiidae). *J. Cetacean Res. Manage.* **7**: 263–70.

Baumgartner, M. F., and Fratantoni, D. M. (2008). Diel periodicity in both sei whale vocalization rates and the vertical migration of their copepod prey observed from ocean gliders. *Limnol. Oceanogr.* **53**: 2197–209.

Baumgartner, M. F., and Mate, B. R. (2003). Summertime foraging ecology of North Atlantic right whales. *Mar. Ecol. Prog. Ser.* **264**: 123–35.

Bazúan-Durán, C., and Au, W. W. L. (2004). Geographic variations in the whistles of spinner dolphins (*Stenella longirostris*) of the main Hawai'ian islands. *J. Acoust. Soc. Am.* **116**: 3757–69.

Bedholm, K., and Møhl, B. (2006). Directionality of sperm whale sonar clicks and its relation to piston radiation theory. *J. Acoust. Soc. Am.* **119**: EL 14–19.

Belikov, R. A., and Bel'kovich, V. M. (2003). Underwater vocalization of the white whales (*Delphinapterus leucas*) in a reproductive gathering during different behavioral situations. *Okeanologiya* **43**: 118–26.

*Benson, S. R., Croll, D. A., Marinovic, B. B., Chavez, F. P., and Harvey, J. T. (2002). Changes in the cetacean assemblage of a coastal upwelling ecosystem during El Niño 1997–98 and La Niña 1999. *Progr. Oceanog.* **54**: 279–91.

Bernal, R., Olavarría, C., and Moraga, R. (2003). Occurrence and long-term residency of two long-beaked common dolphins, *Delphinus capensis* (Gray 1828), in adjacent small bays on the Chilean central eoast. *Aquat. Mamm.* **29**: 396–9.

Blackwell, S. B., Richardson, W. J., Greene, C. R. Jr., and Streever, B. (2007). Bowhead whale (*Balaena mysticetus*) migration and calling behavior in the Alaskan Beaufort Sea, autumn 2001–04: an acoustic localization study. *Arctic* **60**: 255–70.

*Booth, N. O., Baxley, P. A., Rice, J. A. *et al.* (1996). Source localization with broad-band matched-field processing in shallow water. *IEEE J. Ocean Eng.* **21**: 402–12.

*Boisseau, O. (2005). Quantifying the acoustic repertoire of a population: the vocalizations of free-ranging bottlenose dolphins in Fiordland, New Zealand. *J. Acoust. Soc. Am.* **117**: 2318–29.

*Bowen, W. D. (1997). Role of marine mammals in acquatic ecosystems. *Mar. Ecol. Prog. Ser.* **158**: 267–74.

*Bowen, W. D., Tully, D., Boness, D. J., Bulheier, B. M., and Marshall, G. J. (2002). Prey-dependent foraging tactics and prey profitability in a marine mammal. *Mar. Ecol. Prog. Ser.* **244**: 235–45.

*Bradbury, J. W., and Vehrencamp, S. L. (1998). *Principles of Animal Communication.* Sunderland, MA: Sinauer.

*Brickley, P. J., and Thomas, A. C. (2004). Satellite-measured seasonal and interannual chlorophyll variability in the Northeast Pacific and Coastal Gulf of Alaska. *Deep-Sea Res. II* **51**: 229–45.

Brigham, E. O. (1974). *The Fast Fourier Transform*. Englewood Cliffs, NJ: Prentice-Hall.

*Brown, J. C., and Smaragdis, P. (2009). Hidden Markov and Gaussian mixture models for automatic call classification. *J. Acoust. Soc. Am.* **125**: EL 221–4.

*Brown, J. C., Hodgind-Davis, A., and Miller, P. J. O. (2006). Classification of vocalizations of killer whales using dynamic time warping. *J. Acoust. Soc. Am.* **119**: EL 34–40.

Brownell, R. L., Clapham, P. J., Miyashita, T., Kasuya, T. (2001). Conservation status of North Pacific right whales. *J. Cetacean Res. Manage.* **2**: 269–86.

Brueggeman, J. J., Newby, T. C., and Grotefendt, R. A. (1985). Seasonal abundance, distribution and population characteristics of blue whales reported in the 1917 to 1939 catch records of two Alaska whaling stations. *Rep. Int. Whal. Comm.* **35**: 405–11.

Buck, J. R., and Tyack, P. L. (1993). A quantitative measure of similarity for *Tursiops truncatus* signature whistles. *J. Acoust. Soc. Am.* **94**: 2497–506.

*Buck, J. R., Morgenbesser, H. B., and Tyack, P. L. (2000). Synthesis and modification of the whistles of the bottlenose dolphin, *Tursiops truncatus*. *J. Acoust. Soc. Am.* **108**: 407–16.

*Buckland, S. T., and Garthwaite, P. H. (1991). Quantifying precision of mark–recapture estimates using the bootstrap and related methods. *Biometrics* **47**: 255–68.

Buckland, S. T., Anderson, D. R., Burnham, K. P. *et al.* (2001). *Introduction to Distance Sampling: Estimating Abundance of Biological Populations*. Oxford: Oxford University Press.

Buckland, S. T., Anderson, D. R., Burnham, K. P. *et al.* (2004). *Advanced Distance Sampling. Estimating Abundance of Biological Populations*. Oxford: Oxford University Press.

Burdic, W. S. (1984). *Underwater Acoustic System Analysis*. Englewood Cliffs, NJ: Prentice-Hall.

Burgess, W. C., Tyack, P. L., Le Boeuf, B. J., and Costa, D. P. (1998). A programmable acoustic recording tag and first results from free-ranging northern elephant seals. *Deep-Sea Res. II* **45**: 1327–51.

*Calambokidis, J., and Barlow, J. (2004). Abundance of blue and humpback whales in the Eastern North Pacific estimated by capture–recapture and line-transect methods. *Mar. Mamm. Sci.* **20**: 63–85.

*Calambokidis, J., Steiger, G. H., Cubbage, J. C. *et al.* (1990). Sightings and movements of blue whales off central California 1986–88 from photo-identification of individuals. *Rep. Int. Whal. Comm.* (special issue) **12**: 343–48.

*Calambokidis, J., Steiger, G. H., Evenson, J. R. *et al.* (1996). Interchange and isolation of humpback whales off California and other North Pacific feeding grounds. *Mar. Mamm. Sci.* **12**: 215–26.

*Caldwell, D. K., and Caldwell, M. C. (1971). Sounds produced by two rare cetaceans stranded in Florida. *Cetology* **4**: 1–6.

Caldwell, M. C., Caldwell, D. K., and Tyack, P. L. (1990). Review of the signature-whistle hypothesis for the Atlantic bottlenose dolphin. In *The Bottlenose Dolphin*, edited by S. Leatherwood and R. R. Reeves. New York: Academic Press, pp. 199–234.

Cañadas, A., and Sagarminaga, R. (2000). The northeastern Alboran Sea, an important breeding and feeding ground for the long-finned pilot whale (*Globicaephala melas*) in the Mediterranean Sea. *Mar. Mamm. Sci.* **13**: 513–29.

*Cañadas, A., Sagarminaga, R., and García-Tiscar, S. (2002). Cetacean distribution related with depth and slope in the Mediterranean waters off south Spain. *Deep-Sea Res. I* **49**: 2053–73.

Carlström, J. (2005). Diel variation in echolocation behavior of wild harbor porpoises. *Mar. Mamm. Sci.* **21**: 1–12.

Cato, D. (1991). Songs of humpback whales: the Australian perspective. *Mem. Queensl. Mus.* **30**: 277–90.

*Cato, D. H., and McCauley R. D. (2002). Australian research in ambient sea noise. *Acoust. Australia* **30**: 13–20.

Cato, D. H., Paterson, R. A., and Paterson, P. (2001). Vocalization rates of migrating humpback whales over 14 years. *Mem. Queensl. Mus.* **47**: 481–9.

*Cerchio, S., and Dahlheim, M. (2001). Variation in feeding vocalizations of humpback whales *Megaptera novaeangliae* from southeast Alaska. *Bioacoustics* **11**: 277–95.

Cerchio, S., Jacobsen, J. K., and Norris, T. F. (2001). Temporal and geographical variation in songs of humpback whales, *Megaptera novaeangliae*: synchronous change in Hawaiian and Mexican breeding assemblages. *Anim. Behav.* **62**: 313–29.

Chabot, D. (1988). A quantitative technique to compare and classify humpback whale (*Megaptera novaeangliae*) sounds. *Ethology* **77**: 89–102.

*Chan, Y. T., and Ho, K. C. (1994). A simple and efficient estimator for hyperbolic location. *IEEE Trans. Sig. Proc.* **42**: 1905–15.

Charif, R. A., Clapham, P. J., and Clark, C. W. (2001). Acoustic detections of singing humpback whales in deep waters off the British Isles. *Mar. Mamm. Sci.* **17**: 751–68.

Charif, R. A., Mellinger, D. K., Dunsmore, K. J., Fristrup, K. M., and Clark, C. W. (2002). Estimated source levels of fin whale (*Balaenoptera physalus*) vocalizations: adjustments for surface interference. *Mar. Mamm. Sci.* **18**: 81–98.

*Chow, R. K., and Browning, D. G. (1983). Low-frequency attenuation in the Northeast Pacific Subarctic transition zone. *J. Acoust. Soc. Am.* **74**: 1635–8.

*Chow, R. K., and Turner, R. G. (1982). Attenuation of low-frequency sound in the Northeast Pacific Ocean. *J. Acoust. Soc. Am.* **72**: 888–91.

Clapham, P. J., and Mattila, D. K. (1990). Humpback whale songs as indicators of migration routes. *Mar. Mamm. Sci.* **6**: 155–60.

Clark, C. W. (1982). The acoustic repertoire of the southern right whale, a quantitative analysis. *Anim. Behav.* **30**: 1060–71.

Clark, C. W. (1983). Acoustic communication and behavior of the Southern Right Whale (*Eubalaena australis*). In *Communication and Behavior of Whales*, edited by R. Payne. Boulder, CO: Westview, pp. 163–98.

Clark, C. W. (1989). Call tracks of bowhead whales based on call characteristics as an independent means of determining tracking parameters. *Rep. Intl. Whal. Comm.* **39**: 111–12.

*Clark, C. W. (1990). Acoustic behavior of mysticete whales. In *Sensory Abilities of Cetaceans*, edited by J. A. Thomas and R. A. Kastelein. New York: Plenum, pp. 571–83.

Clark, C. W., and Clapham, P. J. (2004). Acoustic monitoring on a humpback whale (*Megaptera novaeangliae*) feeding ground shows continual singing into late spring. *Proc. R. Soc. Lond.* B **271**: 1051–7.

Clark, C. W., and Ellison, W. T. (2000). Calibration and comparison of the acoustic location methods used during the spring migration of the bowhead whale, *Balaena mysticetus*, off Pt. Barrow, Alaska, 1984–1993. *J. Acoust. Soc. Am.* **107**: 3509–17.

*Clark, C. W., and Ellison, W. T. (2004). Potential use of low-frequency sounds by baleen whales for probing the environment: evidence from models and empirical measurements. In *Advances in the Study of Echolocation in Bats and Dolphins*, edited by J. A. Thomas and R. A. Kastelein. New York: Plenum, pp. 564–89.

Clark, C. W., and Fristrup, K. M. (1997). Whales '95: a combined visual and acoustic survey of blue and fin whales off Southern California. *Rep. Intl. Whal. Comm.* **47**: 583–600.

Clark, C. W., and Johnson, J. H. (1984). The sounds of the bowhead whale, *Balaena mysticetus*, during the spring migrations of 1979 and 1980. *Can. J. Zool.* **62**: 1436–41.

Clark, C. W., Borsani, F., and Notarbartolo-di-Sciara, G. (2002). Vocal activity of fin whales, *Balaenoptera physalus*, in the Ligurian Sea. *Mar. Mamm. Sci.* **18**: 286–95.

*Clark, C. W., Bower, J. B., and Ellison, W. T. (1991). Acoustic tracks of migrating bowhead whales, *Balaena mysticetus*, off Point Barrow, Alaska, based on vocal characteristics. *Rep. Intl. Whal. Comm.* **40**: 596–7.

Clark, C. W., Ellison, W. T., and Beeman, K. (1986). Acoustic tracking of migrating bowhead whales. In *Proc. IEEE Oceans '86*. New York: IEEE, pp. 341–6.

Clarke, M. R., Matins, H. R., and Pascoe, P. (1993). The diet of sperm whales *(Physeter macrocephalus* Linnaeus 1758) off the Azores. *Phil. Trans. R. Soc. Lond.* B **339**: 67–82.

Cosens, S. E., and Blouw, A. (2003). Size- and age-class segregation of bowhead whales summering in northern Foxe Basin: A photogrammetric analysis. *Mar. Mamm. Sci.* **19**: 284–96.

Cranford, T. W. (1999). The sperm whale's nose: sexual selection on a grand scale? *Mar. Mamm. Sci.* **15**: 1133–57.

*Cranford, T. W., and Amundin, M. (2004). Biosonar pulse production in Odontocetes: the state of our knowledge. In *Echolocation in Bats and Dolphins*, edited by J. A. Thomas, C. F. Moss, and M. Vater. Chicago, IL: University of Chicago Press, pp. 27–35.

Cranford, T. W., Amundin, M., and Norris, K. S. (1996). Functional morphology and homology in the odontocete nasal complex: implications for sound generation. *J. Morphol.* **228**: 223–85.

Crane, N. L., and Lashkari, K. (1996). Sound production of gray whales, *Eschrichtius robustus*, along their migration route: a new approach to signal analysis. *J. Acoust. Soc. Am.* **100**: 1878–86.

Crease, R. P. (2008) *The Great Equations*. New York: W.W. Norton & Company.

Croll, D. A. *et al.* (2002). Only male fin whales sing loud songs. *Nature* **417**: 809.

Cummings, W. C., and Holliday, D. V. (1987). Sounds and source levels from bowhead whales off Pt. Barrow, Alaska. *J. Acoust. Soc. Am.* **82**: 814–21.

Cummings, W. C., and Thompson, P. O. (1971). Underwater sounds from the blue whale, *Balaenoptera musculus*. *J. Acoust. Soc. Am.* **50**: 1193–8.

Cummings, W. C., Thompson, P. O., and Cook, R. (1968). Underwater sounds of migrating gray whales, *Eschrichtius glaucus* (Cope). *J. Acoust. Soc. Am.* **44**: 1278–81.

*Curtis, K. R., Howe, B. M., and Mercer, J. A. (1999). Low-frequency ambient sound in the North Pacific: long time series observations. *J. Acoust. Soc. Am.* **106**: 3189–200.

Dalebout, M. L., Hooker, S. K., and Christensen, I. (2001). Genetic diversity and population structure among northern bottlenose whales, *Hyperoodon ampullatus*, in the western North Atlantic. *Can. J. Zool.* **79**: 478–84.

*D'Amico, A., Bergamasco, A., Zanasca, P. *et al.* (2003). Qualitative correlation of marine mammals with physical and biological parameters in the Ligurian Sea. *IEEE J. Ocean. Eng.* **28**: 29–43.

*Datta, S., and Sturtivant, C. (2002). Dolphin whistle classification for determining group identities. *Signal Proc.* **82**: 251–8.

Dawbin, W. H. (1966). The seasonal migratory cycle of humpback whales. In *Whales, Dolphins and Porpoises*, edited by K. S. Norris. Berkeley CA: University of California Press, pp. 145–70.

Dawbin, W. H., and Cato, D. H. (1992). Sounds of a pygmy right whale (*Caperea marginata*). *Mar. Mamm. Sci.* **8**: 213–19.

Dawson, S. M. (1991). Clicks and communication: the behavioural and social contexts of Hector's dolphin vocalizations. *Ethology* **88**: 265–76.

Dawson, S. M., and Thorpe, C. W. (1990). A quantitative analysis of the sounds of Hector's dolphin. *Ethology* **86**: 131–45.

Dawson, S. M., Barlow, J., and Ljungblad, D. (1998). Sounds recorded from Baird's beaked whale, *Berardius bairdii*. *Mar. Mamm. Sci.* **14**: 335–44.

*Deecke, V. B., and Janik, V. M. (2006). Automated categorization of bioacoustic signals: avoiding perceptual pitfalls. *J. Acoust. Soc. Am.* **119**: 645–53.

Deecke, V. B., Ford, J. K. B., and Slater, P. J. B. (2005). The vocal behaviour of mammal-eating killer whales: communicating with costly calls. *Anim. Behav.* **69**: 395–405.

Deecke, V. B., Ford, J. K. B., and Spong, P. (1999). Quantifying complex patterns of bioacoustic variation: use of a neural network to compare killer whale (*Orcinus orca*) dialects. *J. Acoust. Soc. Am.* **105**: 2499–507.

Dolphin, W. F. (1987). Ventilation and dive pattern of humpback whales, *Megaptera novaeangliae*, on their Alaskan feeding grounds. *Can. J. Zool.* **65**: 83–90.

Douglas, L. A., Dawson, S. M., and Jaquet, N. (2005). Click rates and silences of sperm whales at Kaikoura, New Zealand. *J. Acoust. Soc. Am.* **118**: 523–9.

Drouot, V., Gannier, A., and Goold, J. C. (2004a). Diving and feeding behavior of sperm whales (*Physeter macrocephalus*) in the northwestern Mediterranean Sea. *Aquat. Mamm.* **30**: 419–26.

Drouot, V., Gannier, A., and Goold, J. C. (2004b). Summer social distribution of sperm whales (*Physeter macrocephalus*) in the Mediterranean Sea. *J. Mar. Biol. Assoc. U.K.* **84**: 675–80.

Dunphy-Daly, M. M., Heithaus, M. R., and Claridge, D. E. (2008). Temporal variation in dwarf sperm whale (*Kogia sima*) habitat use and group size off Great Abaco Island, Bahamas. *Mar. Mamm. Sci.* **24**: 171–82.

Edds, P. L. (1982). Vocalisations of the blue whale, *Balaenoptera musculus* in the St. Lawrence River. *J. Mammal.* **63**: 345–7.

*Erbe, C. (2000). Detection of whale calls in noise: performance comparison between a beluga whale, human listeners, and a neural network. *J. Acoust. Soc. Am.* **108**: 297–303.

Erbe, C., and Farmer, D. M. (1998). Masked hearing thresholds of a beluga whale (*Delphinapterus leucas*) in icebreaker noise. *Deep-Sea Res. II* **45**: 1373–88.

*Erbe, C., and King, A. R. (2008). Automatic detection of marine mammals using information entropy. *J. Acoust. Soc. Am.* **124**: 2833–40.

Esch, H. C., Sayigh, L. S., and Wells, R. S. (2009). Quantifying parameters of bottlenose dolphin signature whistles. *Mar. Mamm. Sci.* **25**: 976–86.

*Evans, W. E. (1967). Vocalizations among marine mammals. In *Marine Bioacoustics*, edited by W. N. Tavolga, Vol. 2. New York: Pergamon Press, pp. 159–86.

Evans, W. E., and Awbrey, F. T. (1984). High frequency pulses of Commerson's dolphin and Dall's porpoise. *Am. Zool.* **24**, 2A.

Fertl, D., Jefferson, T. A., Moreno, I. B., Zerbini, A. N., and Mullin, K. D. (2003). Distribution of the Clymene dolphin *Stenella clymene*. *Mamm. Rev.* **33**: 253–71.

Fiedler, P. C., Reilly, S. B., Hewitt, R. P. *et al.* (1998). Blue whale habitat and prey in the California Channel Islands. *Deep-Sea Res. II* **45**: 1781–801.

Fisher, N. I. (1993). *Statistical Analysis of Circular Data*. Cambridge, UK: Cambridge University Press.

Fish, J. F., Sumich, J. L., and Lingle, G. L. (1974). Sounds produced by the gray whale, *Eschrichtius robustus*. *Mar. Fish. Rev.* **36**: 38–45.

Fish, M. P. and Mowbray, W. H. (1962). Production of underwater sound by the white whale or beluga, *Delphinapterus leucas*. *J. Mar. Res.* **20**: 149–62.

Flandrin, P., Rilling, G., and Gonçalvés, P. (2004). Empirical mode decomposition as a filter bank. *IEEE Signal Process. Lett.* **11**: 112–14.

*Folse, L. J., Packard, J. M., and Grant, W. E. (1989). AI modelling of animal movements in a heterogeneous habitat. *Ecol. Model.* **46**: 57–72.

Ford, J. K. B., and Fisher, H. D. (1978). Underwater acoustic signals of the narwhal *Monodon monoceros. Can. J. Zool.* **56**: 552–60.

Ford, J. K. B., and Fisher, H. D. (1982). Killer whale (*Orcinus orca*) dialects as an indicator of stocks in British Columbia. *Rep. Int. Whal. Comm.* **32**: 671–9.

*Forney, K. A. (2000). Environmental models of cetacean abundance: reducing uncertainty in population trends. *Conserv. Biol.* **14**: 1271–86.

*Forney, K. A., and Barlow, J. (1998). Seasonal patterns in the abundance and distribution of California cetaceans, 1991–1992. *Mar. Mamm. Sci.* **14**: 460–89.

Forney, K. A., Hanan, D. A., and Barlow, J. (1991). Detecting trends in harbor porpoise abundance from aerial surveys using analysis of covariance. *Fish. Bull.* **89**: 367–77.

*Fox, C. G., Matsumoto, H., and Lau, T. K. A. (2001). Monitoring Pacific Ocean seismicity from an autonomous hydrophone array. *J. Geophys. Res.* **106**: 4183–206.

François, R. E., and Garrison, G. R. (1982a). Sound absorption based on ocean measurements: Part I: Pure water and magnesium sulfate contribution. *J. Acoust. Soc. Am.* **72**: 896–907.

François, R. E., and Garrison, G. R. (1982b). Sound absorption based on ocean measurements: Part II: Boric acid contribution and equation for total absorption. *J. Acoust. Soc. Am.* **72**: 1879–90.

*Franke, A., Caelli T., and Hudson, R. J. (2004). Analysis of movements and behavior of caribou (*Rangifer tarandus*) using hidden Markov models. *Ecol. Model.* **173**: 259–70.

Frankel, A. S., and Yin, S. (2010). A description of sounds recorded from melon-headed whales (*Peponocephala electra*) off Hawai'i. *J. Acoust. Soc. Am.* **127**: 3248–55.

Frankel, A. S., Clark, C. W., Herman, L. M., and Gabriele, C. M. (1995). Spatial distribution, habitat utilization, and social interactions of humpback whales, *Megaptera novaeangliae*, off Hawai'i, determined using acoustic and visual techniques. *Can. J. Zool.* **73**: 1134–46.

Frantzis, A., and Alexiadou, P. (2008). Male sperm whale (*Physeter macrocephalus*) coda production and coda-type usage depend on the presence of conspecifics and the behavioural context. *Can. J. Zool.* **86**: 62–75.

Frantzis, A., Goold, J. C., Sharsoulis, E. K., Taroudakis, M. I., and Kandia, V. (2002). Clicks from Cuvier's beaked whales, *Ziphius cavirostris* (L). *J. Acoust. Soc. Am.* **112**: 34–7.

Freitag, L. E., and Tyack, P. L. (1993). Passive acoustic localization of the Atlantic bottlenose dolphin using whistles and echolocation clicks. *J. Acoust. Soc. Am.* **93**: 2197–205.

*Freitas, C., Kovacs, K. M., Lydersen, C., and Ims, R. A. (2008). A novel method for quantifying habitat selection and predictive habitat use. *J. Appl. Ecol.* **45**: 1213–20.

Gabriele, C., and Frankel, A. (2003). The occurrence and significance of humpback whale songs in Glacier Bay, southeast Alaska. *Arctic Res.* **16**: 42–7.

Gaetz, W., Jantzen, K., Weinberg, H., Spong, P., and Symonds, H. (1993). A neural network mechanism for recognition of individual *Orcinus orca* based on their acoustic bahavior: Phase 1. In *Proc. IEEE Oceans '93.* New York: IEEE, pp. 455–7.

Gardner, S. C., and Chávez-Rosales, S. (2000). Changes in the relative abundance and distribution of gray whales (*Eschrichtius robustus*) in Magdalena Bay, Mexico during an El Niño event. *Mar. Mamm. Sci.* **16**: 728–38.

Gedamke, J., Costa, D. P., and Dunstan, A. (2001). Localization and visual verification of a complex minke whale vocalization. *J. Acoust. Soc. Am.* **109**: 3028–47.

George, J.C., Zeh, J., Suydam, R., and Clark, C.W. (2004). Abundance and population trend (1978–2001) of western Arctic bowhead whales surveyed near Barrow, Alaska. *Mar. Mamm. Sci.* **20**: 755–73.

Gerard, O., Carthel, C., Coraluppi, S., and Willett, P. (2008). Feature-aided tracking for marine mammal detection and classification. *Can. Acoust.* **36**: 13–19.

*Ghosh, J., Deuser, L.M., and Beck, S.D. (1992). A neural network-based hybrid system for detection, characterization, and classification of short duration oceanic signals. *IEEE J. Ocean Eng.* **17**: 351–63.

Gillespie, D. (1997). An acoustic survey for sperm whales in the Southern Ocean Sanctuary conducted from *RSV Aurora Australis*. *Rep. Int. Whal. Comm.* **47**: 897–906.

Gillespie, D. (2004). Detection and classification of right whale calls using an edge detector operating on a smoothed spectrogram. *Can. Acoust.* **32**: 39–47.

*Gillespie, D., and Caillat, M. (2008). Statistical classification of odontocete clicks. *Can. Acoust.* **36**: 20–6.

Gillespie, D., and Chappell, O. (2002). An automatic system for detecting and classifying the vocalizations of harbour porpoises. *Bioacoustics* **13**: 37–61.

Gillespie, D., Dunn, C., Gordon, J. *et al.* (2009). Field recordings of Gervais' beaked whales *Mesoplodon europaeus* from the Bahamas. *J. Acoust. Soc. Am.* **125**: 3428–33.

*Gómez de Segura, A., Hammond, P.S., Cañadas, A., and Raga, J.A. (2007). Comparing cetacean abundance estimates derived from spatial models and design-based line transect methods. *Mar. Ecol. Prog. Ser.* **329**: 289–99.

Goodson, A.D., and Sturtivant, C.R. (1996). Sonar characteristics of the harbour porpoise (*Phocoena phocoena*): source levels and spectrum. *ICES J. Mar. Sci.* **53**: 465–72.

*Goold, J.C. (1996). Signal processing techniques for acoustic measurement of sperm whale body lengths. *J. Acoust. Soc. Am.* **100**: 3431–41.

Goold, J.C. (2009). Acoustic assessment of populations of Common Dolphin *Delphinus delphis* in conjunction with seismic surveying. *J. Mar. Biol. Ass. U.K.* **76**: 811–20.

Goold, J.C., and Jones, S.E. (1995). Time and frequency domain characteristics of sperm whale clicks. *J. Acoust. Soc. Am.* **98**: 1279–91.

*Goold, J.C., Bennell, J.D., and Jones, S.E. (1996). Sound velocity measurements in spermaceti oil under the combined influences of temperature and pressure. *Deep-Sea Res. I* **43**: 961–9.

Gordon, J.C.D. (1987). *The behaviour and ecology of sperm whales off Sri Lanka*. Ph.D. thesis, Darwin College, University of Cambridge.

Gordon, J.C.D. (1991). Evaluation of a method for determining the length of sperm whales (*Physeter catodon*) from their vocalizations. *J. Zool. Lond.* **224**: 301–14.

Gordon, J., and Steiner, L. (1992). Ventilation and dive patterns in sperm whales, *Physeter macrocephalus*, in the Azores. *Rep. Int. Whal. Comm.* **42**: 561–5.

*Gordon, J., and Tyack P.L. (2001a). Sound and cetaceans. In *Marine Mammals. Biology and Conservation*, edited by P.G.H. Evans and J.A. Raga. New York: Kluwer Academic, pp. 139–96.

*Gordon, J., and Tyack P.L. (2001b). Acoustic techniques studying cetaceans. In *Marine Mammals. Biology and Conservation*, edited by P.G.H. Evans and J.A. Raga. New York: Kluwer Academic, pp. 293–325.

*Gordon, J.C.D., Matthews, J.N., Panigada, S. *et al.* (2000). Distribution and relative abundance of striped dolphins, and distribution of sperm whales in the Ligurian Sea cetacean sanctuary: results from a collaboration using acoustic techniques. *J. Cetacean Res. Manage.* **2**: 27–36.

*Gowans, S., Dalebout, M. L., Hooker, S. K., and Whitehead, H. (2000a). Reliability of photographic and molecular techniques for sexing northern bottlenose whales (*Hyperoodon ampullatus*). *Can. J. Zool.* **78**: 1224–9.

Gowans, S., Whitehead, H., Arch, J. K., and Hooker, S. K. (2000b). Population size and residency patterns of northern bottlenose whales (*Hyperoodon ampullatus*) using the Gully, Nova Scotia. *J. Cetacean Res. Manage.* **2**: 201–10.

Gowans, S., Whitehead, H., and Hooker, S. K. (2001). Social organization in northern bottlenose whales (*Hyperoodon ampullatus*): not driven by deep water foraging? *Anim. Behav.* **62**: 369–77.

Griffin, D. R. (1958). *Listening in the Dark: The Acoustic Orientation of Bats and Men*. New York: Cornell University Press.

Grewal, M. S., and Andrews, A. P. (2008). *Kalman Filtering. Theory and Practice using MATLAB®*. 3rd edn. Hoboken, NJ: John Wiley & Sons.

Hamming, R. W. (1977). *Digital Filters*. Englewood Cliffs, NJ: Prentice-Hall.

Hamming, R. W. (1986). *Numerical Methods for Scientists and Engineers*. 2nd Edn. New York: Dover.

*Hammond, P. S. (1986). Estimating the size of naturally marked whale populations using capture–recapture techniques. *Rep. Int. Whal. Comm.* (Special Issue) **8**: 253–82.

*Hammond, P. S. (1990). Heterogeneity in the Gulf of Maine? Estimating humpback whale population size when capture probabilities are not equal. *Rep. Int. Whal. Comm.* (Special Issue) **12**: 135–9.

Hansen, M., Wahlberg, M., and Madsen, P. T. (2008). Low-frequency components in harbor porpoise (*Phocoena phocoena*) clicks: communication signal, by-products, or artifacts? *J. Acoust. Soc. Am.* **124**: 4059–68.

*Harwood, J., and Wilson, B. (2001). The implications of developments on the Atlantic Frontier for marine mammals. *Cont. Shelf Res.* **21**: 1073–93.

Hastie, G. D., Swift, R., Gordon, J. C. D., Slesser, G., and Turrell, W. R. (2003). Sperm whale distribution and seasonal density in the Faroe Shetland Channel. *J. Cetacean Res. Manage.* **5**: 247–52.

*Hastie, G. D., Swift, R. J., Slesser, G., Thompson, P. M., and Turrell, W. R. (2005). Environmental models for predicting oceanic dolphin habitat in the Northeast Atlantic. *ICES J. Mar. Sci.* **62**: 760–70.

Hastie, T. J., and Tibshirani, R. J. (1990). *Generalised Additive Models*. London: Chapman and Hall.

Hawkins, E. R., and Gartside, D. F. (2009). Patterns of whistles emitted by wild Indo-Pacific bottlenose dolphins (*Tursiops aduncus*) during a provisioning program. *Aquat. Mamm.* **35**: 171–86.

Hawkins, E. R., and Gartside, D. F. (2010). Whistle emissions of Indo-Pacific bottlenose dolphins (*Tursiops aduncus*) differ with group composition and surface behaviors. *J. Acoust. Soc. Am.* **127**: 2652–63.

*Hayes, S. A., Mellinger, D. K., Croll, D., Costa, D. P., and Borsani, J. F. (2000). An inexpensive passive acoustic system for recording and localizing wild animal sounds. *J. Acoust. Soc. Am.* **107**: 3552–5.

Heide-Jørgensen, M. P., Bloch, D., Stefansson, E., *et al.* (2002). Diving behavior of long-finned pilot whales *Globicephala melas* around the Faroe Islands. *Wildl. Biol.* **8**: 307–13.

Heide-Jørgensen, M. P., Laidre, K. L., Borchers, D., and Samarra, F. I. P. (2007). Increasing abundance of bowhead whales in West Greenland. *Biol. Lett.* **3**: 577–80.

Heide-Jørgensen, M. P., Laidre, K. L., Jensen, M. V., Dueck, L., and Postma, L. D. (2006). Dissolving stock discreteness with satellite tracking: bowhead whales in Baffin Bay. *Mar. Mamm. Sci.* **22**: 34–45.

Heide-Jørgensen, M. P., Laidre, K. L., Wiig, O. *et al.* (2003). From Greenland to Canada in two weeks: movements of bowhead whales, *Balaena mysticetus*, in Baffin Bay. *Arctic* **56**: 21–31.

*Heller, J. R., and Pinezich, J. D. (2008). Automatic recognition of harmonic bird sounds using a frequency track extraction algorithm. *J. Acoust. Soc. Am.* **124**: 1830–7.

Helweg, D. A., Herman, L. M., Yamamoto, S., and Forestell, P. H. (1990). Comparison of songs of humpback whales (*Megaptera novaeangliae*) recorded in Japan, Hawaii, and Mexico during the winter of 1989. *Sci. Rep. Cetacean Res.* **1**: 1–20.

Henshaw, M. D., LeDuc, R. G., Chivers, S. J., and Dizon, A. E. (1997). Identifying beaked whales (Family Ziphiidae) using mtDNA sequences. *Mar. Mamm. Sci.* **13**: 495–8.

*Herman, L. M., and Tavolga, W. N. (1980). The communication systems of cetaceans. In *Cetacean Behavior: Mechanism and Function*, edited by L. M. Herman. New York: Wiley, pp. 149–209.

Herzing, D. L. (1996). Vocalizations and associated underwater behavior of free-ranging Atlantic spotted dolphins, *Stenella frontalis* and bottlenose dolphins, *Tursiops truncatus. Aquat. Mamm.* **22**: 61–79.

Hobson, R. P., and Martin, A. R. (1996). Behaviour and dive times of Arnoux's beaked whales, *Berardius arnuxii*, at narrow leads in fast ice. *Can. J. Zool.* **74**: 388–93.

Hooker, S. K., and Baird, R. W. (1999a). Deep-diving behaviour of the northern bottlenose whale, *Hyperoodon ampullatus* (Cetacea: Ziphiidae). *Proc. R. Soc. Lond.* B **266**: 671–6.

Hooker, S. K., and Baird, R. W. (1999b). Observations of Sowerby's beaked whales, *Mesoplodon bidens*, in The Gullys, Nova Scotia. *Can. Field-Nat.* **113**: 273–7.

*Hooker, S. K., and Baird, R. W. (2001). Diving and ranging behaviour of odontocetes: a methodological review and critique. *Mamm. Rev.* **31**: 81–105.

Hooker, S. K., and Whitehead, H. (2002). Click characteristics of northern bottlenose whales (*Hyperoodon ampullatus*). *Mar. Mamm. Sci.* **18**: 69–80.

*Horn, J. S., Garton, E. O., and Rachlow, J. L. (2008). A synoptic model of animal space use: Simultaneous estimation of home range, habitat selection, and inter/intra-specific relationships. *Ecol. Model.* **214**: 338–48.

*Houser, D. S. (2006). A method for modeling marine mammal movement and behavior for environmental impact assessment. *IEEE J. Ocean Eng.* **31**: 76–81.

*Houser, D. S., Helweg, D. A., and Moore, P. W. (1999). Classification of dolphin echolocation clicks by energy and frequency distributions. *J. Acoust. Soc. Am.* **106**: 1579–85.

Huang, N. E., Shen, Z., Long, S. R. *et al.* (1998). The empirical mode decomposition and the Hilbert transform spectrum for nonlinear and nonstationary time series analysis. *Proc. R. Soc. Lond.* A **454**: 903–95.

*Iona, C., and Quinquis, A. (2005). Time-frequency analysis using warped-based high-order phase modeling. *EURASIP J. Appl. Signal Proc.* **17**: 2856–73.

*Janik, V. M. (1999). Pitfalls in the categorization of behaviour: a comparison of dolphin whistle classification methods. *Anim. Behav.* **57**: 133–43.

Janik, V. M. (2000). Source levels and estimated active space of bottlenose dolphin (*Tursiops truncatus*) whistles in the Moray Firth, Scotland. *J. Comp. Physiol.* A **186**: 673–80.

Janik, V. M., Dehnhardt, G., and Todt, D. (1994). Signature whistle variations in a bottlenosed dolphin, *Tursiops truncatus. Behav. Ecol. Sociobiol.* **35**: 243–8.

Jaquet, N., and Whitehead, H. (1999). Movements, distribution and feeding success of sperm whales in the Pacific Ocean, over scales of days and tens of kilometers. *Aquat. Mamm.* **25**: 1–13.

Jaquet, N., Dawson, S., and Douglas, L. (2001). Vocal behavior of male sperm whales: why do they click? *J. Acoust. Soc. Am.* **109**: 2254–9.

Jaquet, N., Gendron, D., and Coakes, A. (2003). Sperm whales in the Gulf of California: residency, movements, behavior, and the possible influence of variation in food supply. *Mar. Mamm. Sci.* **19**: 545–62.

*Jefferson, T. A., Leatherwood, S., and Webber, M. A. (1993). *Marine Mammals of the World*. Rome: FAO.

Jefferson, T. A., Newcomer, M. W., Leatherwood, S., and van Waerebeek, K. (1994). Right whale dolphins *Lissodelphis borealis* (Peale, 1848) and *Lissodelphis peronii* (Lacépède, 1804). In *Handbook of Marine Mammals*, Vol. 5, *The First Book of Dolphins*, edited by S. H. Ridgway and R. Harrisan. London: Academic Press, pp. 335–62.

*Jensen, F. B., and Kuperman, W. A. (1983). Optimum frequency of propagation in shallow water environments. *J. Acoust. Soc. Am.* **73**: 813–19.

Jensen, F. B., Kuperman, W. A., Porter, M. B., and Schmidt, H. (2000) *Computational Ocean Acoustics*. New York: AIP.

Johnson, M., and Tyack, P. L. (2003). A digital acoustic recording tag for measuring the response of wild marine mammals to sound. *IEEE J. Ocean. Eng.* **28**: 3–12.

Johnson, M., Madsen, P. T., Zimmer, W. M. X., Aguilar de Soto, N., and Tyack, P. L. (2004). Beaked whales echolocate on prey. *Proc. R. Soc. Lond.* B **271**: S383–6.

Johnson, M., Madsen, P. T., Zimmer, W. M. X., Aguilar de Soto, N., and Tyack, P. L. (2006). Foraging Blainville's beaked whales (*Mesoplodon densirostris*) produce distinct click types matched to different phases of echolocation. *J. Exp. Biol.* **209**: 5038–50.

Kaiser, J. F. (1990). On a simple algorithm to calculate the "Energy" of a signal. In *Proc. IEEE ICASSP-90*. Albuquerque, NM: IEEE, pp. 381–4.

Kandia, V., and Stylianou, Y. (2006). Detection of sperm whale clicks based on the Teager–Kaiser energy operator. *Appl. Acoust.* **67**: 1144–63.

Karlsen, J. D., Bisther, A., Ludersen, D., Haug, T., and Kovacs, K. M. (2002). Summer vocalisations of adult male white whales (*Delphinapterus leucas*) in Svalbard, Norway. *Polar Biol.* **25**: 808–17.

Kasamatsu, F., and Joyce, G. G. (1995). Current status of odontocetes in the Antarctic. *Antarctic Sci.* **7**: 365–79.

*Kawamura, A. (1975). A consideration on an available source of energy and its cost for locomotion in fin whales with special reference to the seasonal migrations *Sci. Rep. Whales Res. Inst.* **27**: 61–79.

*Kawamura, A. (1980). A review of food of Balaenopterid whales. *Sci. Rep. Whales Res. Inst.* **32**: 155–97.

*Kawamura, A. (1982). Food habits and prey distributions of three rorqual species in the North Pacific Ocean. *Sci. Rep. Whales Res. Inst.* **34**: 59–91.

Kemper C. M. (2002). Distribution of the pygmy right whale, *Caperea marginata*, in the Australasian region. *Mar. Mamm. Sci.* **18**: 99–111.

*Kibblewhite, A. C., Bedfordand, N. R., and Mitchell, S. K. (1977). Regional dependence of low-frequency attenuation in the North Pacific Ocean. *J. Acoust. Soc. Am.* **61**: 1169–77.

*Kibblewhite, A. C., Denham, R. N., and Barnes, D. J. (1967). Unusual low frequency signals observed in New Zealand waters. *J. Acoust. Soc. Am.* **41**: 644–55.

Kinsler, L. E., Frey, A. R., Coppens, A. B., and Sanders, J. V. (2000). *Fundamentals of Acoustics*. New York: Wiley.

Knudsen, V. O., Alford, R. S., and Emling, J. W. (1948). Underwater ambient noise. *J. Mar. Res.* **7**: 410–29.

Kraus, S. D., Prescott, J. H., Knowlton, A. R., and Stone, G. S. (1986). Migration and calving of right whales (*Eubalaena glacialis*) in the western North Atlantic. *Rep. Int. Whal. Comm.* (Special Issue) **10**: 139–44.

Kyhn, L. A., Jensen, F. H., Beedholm, K. *et al.* (2010). Echolocation in sympatric Peale's dolphins (*Lagenorhynchus australis*) and Commerson's dolphins (*Cephalorhynchus commersonii*) producing narrow-band high-frequency clicks. *J. Exp. Biol.* **213**: 1940–9.

*Laidre, K. L., Heide-Jørgensen, M. P., and Nielsen, T. G. (2007). The role of the bowhead whale as a predator in West Greenland. *Mar. Ecol. Prog. Ser.* **346**: 285–97.

Lammers, M. O., and Au, W. W. L. (2003). Directionality in the whistles of Hawaiian spinner dolphins (*Stenella longirostris*): a signal feature to cue direction of movement? *Mar. Mamm. Sci.* **19**: 249–64.

Lammers, M. O., Au, W. W. L., Aubauer, R., and Nachtigall, P. E. (2004). A comparative analysis of echolocation and burst-pulse click trains in *Stenella longirostris*. In *Echolocation in Bats and Dolphins*, edited by J. A. Thomas, C. F. Moss, and M. M. Vater. Chicago, IL: University of Chicago Press, pp. 414–19.

*Lammers, M. O., Brainard, R. E., Au, W. W. L., Mooney, T. A., and Wong, K. B. (2008). An ecological acoustic recorder (EAR) for long-term monitoring of biological and anthropogenic sounds on coral reefs and other marine habitats. *J. Acoust. Soc. Am.* **123**: 1720–8.

Lammers, M. O., Schotten, M., and Au, W. W. L. (2006). The spatial context of free-ranging Hawaiian spinner dolphins (*Stenella longirostris*) producing acoustic signals. *J. Acoust. Soc. Am.* **119**: 1244–50.

Leaper, R., Chappell, O., and Gordon, J. (1992). The development of practical techniques for surveying sperm whale populations acoustically. *Rep. Intl. Whal. Comm.* **42**: 549–60.

*Leaper, R., Gillespie, D., and Papastavrou, V. (2000). Results of passive acoustic surveys for odontocetes in the Southern Ocean. *J. Cetacean Res. Manage.* **2**: 187–96.

Leroy, C., Robinson, S. P., and Goldsmith, M. J. (2008). A new equation for the accurate calculation of sound speed in all oceans *J. Acoust. Soc. Am.* **124**: 2774–83.

Leroy, C., Robinson, S. P., and Goldsmith, M. J. (2009). Erratum: A new equation for the accurate calculation of sound speed in all oceans. *J. Acoust. Soc. Am.* **126**: 2117.

Ljungblad, D. K., Leatherwood, S., and Dahlheim, M. E. (1980). Sounds recorded in the presence of an adult and calf bowhead whales. *Mar. Fish. Rev.* **42**: 86–87.

Ljungblad, D. K., Moore, S. E., and VanSchoik, D. R. (1986). Seasonal patterns of distribution, abundance, migration and behavior of the western Arctic stock of bowhead whales, *Balaena mysticetus*, in Alaskan Seas. *Rep. Int. Whal. Comm.* **8**: 177–205.

Ljungblad, D. K., Thompson, P. O., and Moore, S. E. (1982). Underwater sounds recorded from migrating bowhead whales, *Balaena mysticetus*, in 1979. *J. Acoust. Soc. Am.* **71**: 477–82.

*Lopatka, M., Adam, O., Laplanche, C., Zarzycki, J., and Motsch, J.-F. (2005). An attractive alternative for sperm whale click detection using the wavelet transform in comparison to the Fourier spectrogram. *Aquat. Mamm.* **31**: 463–467.

Lucifredi, I., and Stein, P. J. (2007). Gray whale target strength measurements and the analysis of the backscattered response. *J. Acoust. Soc. Am.* **121**: 1383–91.

*Lucke, K., and Goodson, A. D. (1997). Characterising wild dolphins echolocation behaviour: off line analysis. *Proc. Inst. Acoust.* **19**: 179–83.

Lurton, X. (2002). *An Introduction to Underwater Acoustics*. London: Springer.

Lynn, S. K., and Reiss, D. L. (1992). Pulse sequence and whistle production by two captive beaked whales, *Mesoplodon* species. *Mar. Mamm. Sci.* **8**: 299–305.

*Madsen, P. T. (2002a). Sperm whale sound production – in the acoustic realm of the biggest nose on record. In *Sperm whale sound production*, Ph.D. dissertation, University of Aarhus, Denmark.

Madsen, P. T. (2002b). Morphology of the sperm whale head: a review and some new findings. In *Sperm whale sound production*, Ph.D. dissertation, University of Aarhus, Denmark.

Madsen, P. T., and Wahlberg, M. (2007). Recording and quantitative analysis of clicks from echolocating toothed whales in the wild. *Deep-Sea Res. II* **54**: 1421–44.

Madsen, P. T., Kerr, I., and Payne, R. (2004a). Source parameter estimates of echolocation clicks from wild pygmy killer whales (*Feresa attenuata*). *J. Acoust. Soc. Am.* **116**: 1909–12.

Madsen, P. T., Kerr, I., and Payne, R. (2004b). Echolocation clicks of two free-ranging delphinids with different food preferences: false killer whales (*Pseudorca crassidens*) and Risso's dolphin (*Grampus griseus*). *J. Exp. Biol.* **207**: 1811–23.

*Madsen, P. T., Carder, D. A., Au, W. W. *et al.* (2003). Sound production in neonate sperm whales. *J. Acoust. Soc. Am.* **113**: 2988–91.

Madsen, P. T., Carder, D. A., Bedholm, K., and Ridgway, S. H. (2005a). Porpoise clicks from a sperm whale nose – convergent evolution of 130 kHz pulses in toothed whale sonars? *Bioacoustics* **15**: 195–206.

Madsen, P. T., Johnson, M., Aguilar de Soto, N., Zimmer, W. M. X. and Tyack, P. L. (2005b). Biosonar performance of foraging beaked whales (*Mesoplodon densirostris*). *J. Exp. Biol.* **208**: 181–94.

*Madsen, P. T., Payne, R., Kristiansen, N. U. *et al.* (2002a). Sperm whale sound production studied with ultrasound time/depth-recording tags. *J. Exp. Biol.* **205**: 1899–906.

Madsen, P. T., Wahlberg, M., and Møhl, B. (2002). Male sperm whale (*Physeter macrocephalus*) acoustics in a high latitude habitat: implications for echolocation and communication. *Behav. Ecol. Sociobiol.* **53**: 31–41.

*Mann, J. (1999). Behavioral sampling methods for cetaceans: a review and critique. *Mar. Mamm. Sci.* **15**: 102–22.

Marques, T. A., Thomas, L., Ward., J., DiMarzio, N., and Tyack, P. L. (2009). Estimating cetacean population density using fixed passive acoustic sensors: an example with Blainville's beaked whales. *J. Acoust. Soc. Am.* **125**: 1982–94.

Marten, K. (2000). Ultrasonic analysis of pygmy sperm whale (*Kogia breviceps*) and Hubb's beaked whale (*Mesoplodon carlhubbsi*) clicks. *Aquat. Mamm.* **1**: 45–48.

Martin, A. R., Katona, S. K., Matilla, D., Hembree, D., and Waters, T. D. (1984). Migration of humpback whales between the Caribbean and Iceland. *J. Mammal.* **65**: 330–33.

*Martin, J., Calenge, C., Quenette, P. Y., and Allainé, D. (2008). Importance of movement constraints in habitat selection studies. *Ecol. Model.* **213**: 257–62.

Mate, B. R., Gisiner, R., and Mobley, J. (1998). Local and migratory movements of Hawaiian humpback whales tracked by satellite telemetry. *Can. J. Zool.* **76**: 863–8.

Mate, B. R., Lagerquist, B. A., and Calambokidis, J. (1999). Movements of North Pacific blue whales during the feeding season off Southern California and their southern fall migration. *Mar. Mamm. Sci.* **15**: 1246–57.

*Matthews, J. (2004). Detection of frequency-modulated calls using a chirp model. *Can. Acoust.* **32**: 66–75.

Matthews, J. N., Brown, S., Gillespie, D. *et al.* (2001). Vocalization rates of the North Atlantic right whale (*Eubalena glacialis*). *J. Cet. Res. Manage.* **3**: 271–82.

May-Collado, L. J., and Wartzok, D. (2007). The Freshwater dolphin *Inia geoffrensis geoffrensis* produces high frequency whistles. *J. Acoust. Soc. Am.* **121**: 1203–12.

McDonald, M. A., and Fox, C. G. (1999). Passive acoustic methods applied to fin whale population density estimation. *J. Acoust. Soc. Am.* **105**: 2643–51.

McDonald, M. A., and Moore, S. E. (2002). Calls recorded from North Pacific right whales (*Eubalaena japonica*) in the eastern Bering Sea. *J. Cetacean Res. Manage.* **4**: 261–6.

McDonald, M. A., Hildebrand, J. A., and Webb, S. C. (1995). Blue and fin whales observed on seafloor array in the northeast Pacific. *J. Acoust. Soc. Am.* **98**: 712–21.

*McDonald, M. A., Hildebrand, J. A., and Wiggins, S. (2006). Increases in deep ocean ambient noise in the Northeast Pacific west of San Nicolas Island, California. *J. Acoust. Soc. Am.* **120**: 711–18.

McGregor, P. K., Dabelsteen, T., Clark, C. W. *et al.* (1997). Accuracy of a passive acoustic location system: empirical studies in terrestrial habitats. *Ethol. Ecol. Evol.* **9**: 269–86.

McSweeney, D. J., Chu, K. C., Dolphin, W. F., and Guinee, L. N. (1989). North Pacific humpback whale songs: a comparison of southeast Alaskan feeding ground songs with Hawaiian wintering ground songs. *Mar. Mamm. Sci.* **5**: 139–48.

*Mead, J. G. (1989). Beaked whales of the genus *Mesoplodon*. In *Handbook of Marine Mammals*, Vol. 4, *River Dolphins and the Large Toothed Whales*, edited by S. Ridgway and R. Harrison. London: Academic Press, pp. 349–64.

*Medwin, H. (1975). Speed of sound in water: a simple equation for realistic parameters *J. Acoust. Soc. Am.* **58**: 1318–19.

Medwin, H. and Clay, C. S. (1998). *Fundamentals of Acoustical Oceanography*. Boston, MA: Academic Press.

Medwin, H., *et al.* (2005). *Sound in the Sea. From Ocean Acoustics to Acoustical Oceanography*. Cambridge: Cambridge University Press.

*Mellen, R. H. (1952). The thermal-noise limit in the detection of underwater acoustic signals. *J. Acoust. Soc. Am.* **24**: 478–80.

*Mellinger, D. K. (1993). Handling time variability in bioacoustic transient detection. In *Proc. IEEE Oceans '93*. New York: IEEE, pp. 116–21.

*Mellinger, D. K. (2004). A comparison of methods for detecting right whale calls. *Can. Acoust.* **32**: 55–65.

*Mellinger, D. K., and Clark, C. W. (1993). A method for filtering bioacoustic transients by spectrogram image convolution. In *Proc. IEEE Oceans '93*. New York: IEEE, pp. 122–7.

*Mellinger, D. K., and Clark, C. W. (1997). Methods for automatic detection of mysticete sounds. *Mar. Freshw. Behav. Physiol.* **29**: 163–81.

*Mellinger, D. K., and Clark, C. W. (2000). Recognizing transient low-frequency whale sounds by spectrogram correlation. *J. Acoust. Soc. Am.* **107**: 3518–29.

Mellinger, D. K., and Clark, C. W. (2003). Blue whale (*Balaenoptera musculus*) sounds from the North Atlantic. *J. Acoust. Soc. Am.* **114**: 1108–19.

Mellinger, D. K., Stafford, K. M., and Fox, C. G. (2004a). Seasonal occurrence of sperm whale (*Physeter macrocephalus*) sounds in the gulf of Alaska, 1999–2001. *Mar. Mamm. Sci.* **20**: 48–62.

Mellinger, D. K., Stafford, K. M., Moore, S. E., Munger, L., and Fox, C. G. (2004b). Detection of North Pacific right whale *(Eubalaena japonica)* calls in the Gulf of Alaska. *Mar. Mamm. Sci.* **20**: 872–9.

*Mellinger, D. K., Stafford, K. M., Moore, S. E., Dziak, R. P., and Matsumoto, H. (2007). An overview of fixed passive acoustic observation methods for cetaceans. *Oceanography* **20**: 36–45.

*Mignerey, P. C., and Finette, S. (1992). Multichannel deconvolution of an acoustic transient in an oceanic waveguide. *J. Acoust. Soc. Am.* **92**: 351–64.

*Miller, B., and Dawson, S. (2009). A large-aperture low-cost hydrophone array for tracking whales from small boats. *J. Acoust. Soc. Am.* **126**: 2248–56.

Miller, L. A., Pristed, J., Møhl, B., and Surlykke, A. (1995). Click sounds from narwhals (*Monodon monoceros*) in Inglefield Bay, Northwest Greenland. *Mar. Mamm. Sci.* **11**: 491–502.

Miller, P. J. O. (2002). Mixed-directionality of killer whale stereotyped calls: a direction of movement cue? *Behav. Ecol. Sociobiol.* **52**: 262–70.

*Miller, P. J. O., and Tyack, P. L. (1998). A small towed beamforming array to identify vocalizing *resident* killer whales (*Orcinus orca*) concurrent with focal behavioral observations. *Deep-Sea Res. II* **45**: 1389–405.

Miller, P. J. O., Johnson, M., and Tyack, P. L. (2004a). Sperm whale behaviour indicates the use of echolocation click buzzes 'creaks' in prey capture. *Proc. R. Soc. Lond.* B **271**: 2239–47.

Miller, P. J. O., Johnson, M. P., Tyack, P. L., and Terray, E. A. (2004b). Swimming gaits, passive drag and buoyancy of diving sperm whales *Physeter macrocephalus*. *J. Exp. Biol.* **207**: 1953–67.

Minkler, G., and Minkler, J. (1993). *Theory and Application of Kalman Filtering*. Palm Bay, FL: Magellan Book Company.

Møhl, B. (2001). Sound transmission in the nose of the sperm whale *Physeter catodon*. A post mortem study. *J. Comp. Physiol. (A)* **187**: 335–40.

Møhl, B., and Andersen, S. (1973). Echolocation: high-frequency component in the click of the Harbour Porpoise (*Phocoena ph. L.*). *J. Acoust. Soc. Am.* **54**: 1369–72.

Møhl, B., Surlykke, A., and Miller, L. A. (1990). High intensity narwhal click. In *Sensory Abilities of Cetaceans*, edited by J. Thomas and R. Kastelein. New York: Plenum Press, pp. 295–304.

Møhl, B., Wahlberg, M., Madsen, P. T., Heerfordt, A., and Lund, A. (2003). The monopulsed nature of sperm whale clicks. *J. Acoust. Soc. Am.* **114**: 1143–54.

Møhl, B., Wahlberg, M., Madsen, P. T., Miller, L. A., and Surlykke, A. (2000). Sperm whale clicks: directionality and source level revisited. *J. Acoust. Soc. Am.* **107**: 638–48.

Møhl, B., Madsen, P. T., Wahlberg, M. *et al.* (2002). Sound transmission in the spermaceti complex of a recently expired sperm whale calf. *ARLO* **4**: 19–24.

Monestiez, P., Dubroca, L., Bonnin, E., Durbec, J.-P., and Guinet, C. (2006). Geostatistical modelling of spatial distribution of *Balaenoptera physalus* in the Northwestern Mediterranean Sea from sparse count data and heterogeneous observation efforts. *Ecol. Model.* **193**: 615–28.

Moore, K. E., Watkins, W. A., and Tyack, P. L. (1993). Pattern similarity in shared codas from sperm whales (*Physeter catodon*). *Mar. Mamm. Sci.* **9**: 1–9.

*Moore, P. W. B., Roitblat, H. L., Penner, R. H., and Nachtigall, P. E. (1991). Recognizing successive dolphin echoes with an integrator gateway network. *Neural Networks* **4**: 701–9.

Moore, S. E., and Ljungblad, D. K. (1984). Gray whales in the Beaufort, Chukchi and Bering Seas: distribution and sound production. In *The Gray Whale, Eschrichtius robustus*, edited by M. L. Jones, S. L. Swartz, and S. Leatherwood. New York: Academic Press, pp. 543–59.

*Moore, S. E., Stafford, K. M., Dahlheim, M. E. *et al.* (1998). Seasonal variation in reception of fin whale calls at five geographic areas in the North Pacific. *Mar. Mamm. Sci.* **14**: 617–627.

*Moore, S. E., Stafford, K. M., Mellinger, D. K., and Hildebrand, J. A. (2006). Listening for large whales in the offshore waters of Alaska. *BioScience* **56**: 49–55.

*Moore, S. E., Watkins, W. A., Daher, M. A., and Davis, J. R. (2002). Blue whale habitat associations in the northwest Pacific: analysis of remotely sensed data using a geographic information system. *Oceanography* **15**: 20–5.

Morisaka, T., and Connor, R. C. (2007). Predation by killer whales (*Orcinus orca*) and the evolution of whistle loss and narrow-band high frequency clicks in odontocetes. *J. Evol. Biol.* **20**: 1439–58.

Moulins, A., Rosso, M., Nani, B., and Würtz, M. (2006). Aspects of distribution of Cuvier's beaked whale (*Ziphius cavirostris*) in relation to topographic features in the *Pelagos* Sanctuary (northwestern Mediterranean sea). *J. Mar. Biol. Ass. U.K.* **86**: 1–10.

Mullins, J., Whitehead, H., and Weilgart, L. S. (1988). Behaviour and vocalizations of two single sperm whales, *Physeter macrocephalus*, off Nova Scotia. *Can. J. Fish. Aquat. Sci.* **45**: 1736–43.

*Munger, L. M., Mellinger, D. K., Wiggins, S. M., Moore, S. E., and Hildebrand, J. A. (2005). Performance of spectrogram cross-correlation in detecting right whale calls in long-term recordings from the Bering Sea. *Can. Acoust.* **33**: 25–34.

Murchison, A. E. (1980). Maximum detection range and range resolution in echolocating bottlenose porpoise (*Tursiops truncatus*). In *Animal Sonar Systems*, edited by R. G. Busnel and J. F. Fish. New York: Plenum Press, pp. 43–70.

Murison, L. D., and Gaskin, D. E. (1989). The distribution of right whales and zooplankton in the Bay of Fundy, Canada. *Can. J. Zool.* **67**: 1411–20.

Niemann, H. (1974). *Methoden der Mustererkennung*. Frankfurt am Main: Akademische Verlagsanstalt.

*Nieukirk, S. L., Stafford, K. M., Mellinger, D. K., and Fox, C. G. (2004). Low-frequency whale sounds recorded from the mid-Atlantic Ocean. *J. Acoust. Soc. Am.* **115**: 1832–43.

*Nishiwaki, M. (1966). Distribution and migration of the larger cetaceans in the North Pacific as shown by Japanese whaling results. In *Whales, Dolphins and Porpoises*, edited by K. S. Norris. Berkeley, CA: University of California Press, pp. 171–91.

Nores, C., and Pérez, C. (1988). Overlapping range between *Globicephala macrorhynchus* and *Globicephala melaena* in the northeastern Atlantic. *Mammalia* **52**: 51–6.

Norris, K. S., and Harvey, G. W. (1972). A theory for the function of the spermaceti organ of the sperm whale (*Physter catodon L.*). In *Animal Orientation and Navigation*, edited by S. R. Galler, K. Schmidt-Koenig, G. J. Jacobs, and R. E. Belleville, SP-262. Washington, DC: NASA, pp. 397–417.

Norris, K. S., and Harvey, G. W. (1974). Sound transmission in the porpoise head. *J. Acoust. Soc. Am.* **56**: 659–64.

Norris, K. S., Prescott, J. H., Asa-Dorian, P. V., and Perkins, P. (1961). An experimental demonstration of echolocation behavior in the porpoise, *Tursiops truncatus* (Montagu). *Biol. Bull.* **120**: 163–76.

Norris, T. F., McDonald, M. A., and Barlow, J. (1999). Acoustic detections of singing humpback whales (*Megaptera novaeangliae*) in the eastern North Pacific during their northbound migration. *J. Acoust. Soc. Am.* **106**: 506–14.

*Northrup, J., Cummings, W. C. and Morrison, M. F. (1971). Underwater 20-Hz signals recorded near Midway Island. *J. Acoust. Soc. Am.* **49**: 1909–10.

*Northrop, J., Cummings, W. C., and Thompson, P. O. (1968). 20-Hz signals observed in the central Pacific. *J. Acoust. Soc. Am.* **43**: 383–4.

Nosal, E. M., and Frazer, L. N. (2006). Track of a sperm whale from delays between direct and surface-reflected clicks. *Appl. Acoust.* **67**: 1187–201.

Nosal, E. M., and Frazer, L. N. (2007). Sperm whale three-dimensional track, swim orientation, beam pattern, and click levels observed on bottom-mounted hydrophones. *J. Acoust. Soc. Am.* **122**: 1969–78.

*Oppenheim, A. V., and Schafer, R. W. (1975). *Digital Signal Processing*. Englewood Cliffs, NJ: Prentice-Hall.

*Oswald, J. N., Barlow, J., and Norris, T. F. (2003). Acoustic identification of nine delphinid species in the eastern tropical Pacific Ocean. *Mar. Mamm. Sci.* **19**: 20–37.

*Oswald, J. N., Rankin, S., and Barlow, J. (2004). The effect of recording and analysis bandwidth on acoustic identification of delphinid species. *J. Acoust. Soc. Am.* **116**: 3178–85.

*Oswald, J. N., Rankin, S., Barlow, J., and Lammers, M. O. (2007). A tool for real-time acoustic species identification of delphinid whistles. *J. Acoust. Soc. Am.* **122**: 587–95.

Page, S. E. (1954). Continuous inspection schemes. *Biometrika* **41**: 100–15.

*Palka, D. L., and Hammond, P. S. (2001). Accounting for responsive movement in line transect estimates of abundance. *Can. J. Fish. Aquat. Sci.* **58**: 777–87.

Panigada, S., Zanardelli, M., MacKenzie, M. *et al.* (2008). Modelling habitat preferences for fin whales and striped dolphins in the Pelagos Sanctuary (Western Mediterranean Sea) with physiographic and remote sensing variables. *Rem. Sens. Environ.* **112**: 3400–12.

Papastavrou, V., Smith, S. C., and Whitehead, H. (1989). Diving behaviour of the sperm whale, *Physeter macrocephalus*, off the Galapagos Islands. *Can. J. Zool.* **67**: 839–46.

Papoulis, A. (1962). *The Fourier Integral and its Applications*. New York, NY: McGraw-Hill.

Parks, S. E., and Tyack, P. L. (2005). Sound production by North Atlantic right whales (*Eubalaena glacialis*) in surface active groups. *J. Acoust. Soc. Am.* **117**: 3297–306.

Parra, G. J., Schick R., and Corkeron, P. J. (2006) Spatial distribution and environmental correlates of Austalian snubfin and Indo-Pacific humpback dolphins. *Ecography* **29**: 1–11.

*Parsons, S., and Jones, G. (2000). Acoustic identification of twelve species of echolocating bat by discriminant function analysis and artificial neural networks. *J. Exp. Biol.* **203**: 2641–56.

Pavan, G., Hayward, T. J., Borsani, J. F. *et al.* (2000). Time patterns of sperm whale codas recorded in the Mediterranean Sea 1985–1996. *J. Acoust. Soc. Am.* **107**: 3487–95.

Pavan, G., Priano, M., Manghi, M., and Fossati, C. (1997). Software tools for real-time IPI measurements on sperm whale sounds. *Proc. Institute Acoust.* **19**: 157–64.

Payne, R. S., and Guinee, L. N. (1983). Humpback whale (*Megaptera novaeangliae*) songs as an indicator of stocks. In *Communication and Behavior of Whales*, edited by R. S. Payne. Boulder, CO: Westview, pp. 333–58.

Payne, R. S., and McVay, S. (1971). Songs of humpback whales. *Science* **173**: 585–97.

Payne, R. S., and Payne, K. B. (1971). Underwater sounds of southern right whales. *Zoologica* **58**: 159–65.

Payne, K., Tyack, P., and Payne, R. (1983). Progressive changes in the songs of humpback whales (*Megaptera novaeangliae*): a detailed analysis of two seasons in Hawaii. In *Communication and Behavior of Whales*, edited by R. S. Payne. Boulder, CO: Westview, pp. 9–57.

*Pearce, J., and Freeier, S. (2000). Evaluating the predictive performance of habitat models developed using logistic regression. *Ecol. Model.* **133**: 225–45.

Perrin, W. F., Best, P. B., Dawbin, W. H. *et al.* (1973). Rediscovery of Fraser's Dolphin *Lagenodelphis hosei. Nature* **241**: 345–50.

Peterson, J. T., and Bayley, P. B. (2004). A Bayesian approach to estimating presence when a species is undetected. In *Sampling Rare or Elusive Species. Concepts, Designs, and Techniques for Estimating Population Parameters*, edited by W. L. Thompson. Washington D.C.: Island Press, pp. 173–88.

*Pledger, S., Pollock, K. H., and Norris, J. L. (2003). Open capture – recapture models with heterogeneity: I. Cormack–Jolly–Seber model. *Biometrics* **59**: 786–94.

Podos, J., da Silva, V. M. F., and Rossi-Santos, M. R. (2002). Vocalizations of Amazon river dolphins, Inia geoffrensis: insights into the evolutionary origins of delphinid whistles. *Ethology* **108**: 601–12.

*Pollock, K. H. (1982). A capture–recapture design robust to unequal probability of capture. *J. Wildl. Manage.* **46**: 752–7.

Pollock, K. H. (2000). Capture-recapture models. *J. Am. Stat. Ass.* **95**: 293–6.

Porter, M. B., and Bucker, H. P. (1987). Gaussian beam tracing for computing ocean acoustic fields. *J. Acoust. Soc. Am.* **82**: 1349–59.

*Potter, J. R., Mellinger, D. K., and Clark, C. W. (1994). Marine mammal call discrimination using artificial neural networks. *J. Acoust. Soc. Am.* **96**: 1255–62.

Raftery, A. E., and Zeh, J. E. (1998). Estimating bowhead whale population size and rate of increase from the 1993 census. *J. Am. Stat. Ass.* **93**: 451–63.

Randall, R. H. (1951). *An Introduction to Acoustics.* Cambridge, MA: Addison-Wesley Press. Republication 2005, Mineola, NY: Dover.

Rankin, S., and Barlow, J. (2007). Sounds recorded in the presence of Blainville's beaked whales, *Mesoplodon densirostris*, near Hawai'i (L). *J. Acoust. Soc. Am.* **122**: 42–5.

Rankin S., Oswald, J., Barlow, J., and Lammers, M. (2007). Patterned burst-pulse vocalizations of the northern right whale dolphin, *Lissodelphis borealis. J. Acoust. Soc. Am.* **121**: 1213–18.

Rasmussen, M. H., Lammers, M., Beedholm, K., and Miller, L. A. (2006). Source levels and harmonic content of whistles in white-beaked dolphins (*Lagenorhynchus albirostris*). *J. Acoust. Soc. Am.* **120**: 510–17.

Rasmussen, M. H., Miller, L. A., and Au, W. W. L. (2002). Source levels of clicks from free-ranging white-beaked dolphins (*Lagenorhynchus albirostris* Gray 1846) recorded in Icelandic waters. *J. Acoust. Soc. Am.* **111**: 1122–5.

Rasmussen, M. H., Wahlberg, M., and Miller, L. A. (2004). Estimated transmission beam pattern of clicks recorded from free-ranging white-beaked dolphins (*Lagenorhynchus albirostris*). *J. Acoust. Soc. Am.* **116**: 1826–31.

Rauch, H. E., Tung,. F., and Striebel, C. T. (1965). Maximum likelihood estimates of linear dynamic systems. *AIAA Journal* **3**: 1445–50.

Rebull, O. G., Cusí, J. D., Fernández, M. R., and Muset, J. G. (2006). Tracking fin whale calls offshore the Galicia Margin, North East Atlantic Ocean. *J. Acoust. Soc. Am.* **120**: 2077–85.

Reeves, R. R., Smith, T. D., Josephson, E. A., Clapham, P. J., and Woolmer, G. (2004). Historical observations of humpback and blue whales in the North Atlantic Ocean: clues to migratory routes and possibly additional feeding grounds. *Mar. Mamm. Sci.* **20**: 774–86.

*Redfern, J. V., Ferguson, M. C., Becker, E. A. *et al.* (2006). Techniques for cetacean–habitat modeling. *Mar. Ecol. Prog. Ser.* **310**: 271–95.

*Ren, Y., Johnson, M. T., Clemins, P. J. *et al.* (2009). A framework for bioacoustic vocalization analysis using hidden Markov models. *Algorithms* **2**: 1410–28.

Renaud, D., and Popper, A. N. (1975). Sound localization by the bottlenose porpoise *Tursiops truncatus. J. Exp. Biol.* **63**: 569–85.

Rendell, L., and Whitehead, H. (2004). Do sperm whales share coda vocalizations? Insights into coda usage from acoustic size measurements. *Anim. Behav.* **67**: 865–74.

*Reysenbach de Haan, F. W. (1966). Listening underwater: thoughts on sound and cetacean hearing. In *Whales, Dolphins, and Porpoises*, edited by K. S. Norris. Berkeley, CA: University of California Press, pp. 583–96.

Rhinelander, M. Q., and Dawson, S. M. (2004). Measuring sperm whales from their clicks: stability of interpulse intervals and validation that they indicate whale length. *J. Acoust. Soc. Am.* **115**: 1826–31.

Ribeiro, S., Viddi, F. A., Cordeiro, J. L., and Freitas, T. R. O. (2007). Fine-scale habitat selection of Chilean dolphins (*Cephalorhynchus eutropia*): interaction with aquaculture activities in southern Chiloé Island, Chile. *J. Mar. Biol. Ass. U.K.* **87**: 119–28.

*Rice, D. W. (1974). Whales and whale research in the eastern North Pacific. In *The Whale Problem: a Status Report*, edited by W. E. Schevill. Cambridge, MA: Harvard University Press, pp. 170–95.

Rice, D. W. (1998). *Marine Mammals of the World: Systematics and Distribution*. Lawrence, KS: Allen Press.

*Rice, D. W., and Wolman, A. A. (1982). Whale census in the Gulf of Alaska, June to August 1980. *Rep. Int. Whal. Comm.* **32**: 491–7.

Richardson, W. J., Finley, K. J., Miller, G. W., Davis, R. A., and Koski, W. R. (1995b). Feeding, social and migration behavior of bowhead whales, *Balaena mysticetus*, in Baffin Bay vs the Beaufort Sea – regions with different amounts of human activity. *Mar. Mamm. Sci.* **11**: 1–45.

*Richardson, W. J., Greene, C. R. Jr., Malme, C. I., and Thomson, D. H. (1995a). *Marine Mammals and Noise*. San Diego, CA: Academic Press.

Rivers, J. (1997). Blue whale, *Balaenoptera musculus*, vocalizations from the waters off central California. *Mar. Mamm. Sci.* **13**: 186–95.

*Roch, M. A., Soldevilla, M. S., Burtenshaw, J. C., Henderson, E. E., and Hildebrand, J. A. (2007). Gaussian mixture model classification of odontocetes in the Southern California Bight and the Gulf of California. *J. Acoust. Soc. Am.* **121**: 1737–48.

Rogers, T. L. (1999). Acoustic observations of Arnoux's beaked whale (*Berardius arnuxii*) off Kemp Land, Antarctica. *Mar. Mamm. Sci.* **15**: 198–204.

*Ross, D. (2005). Ship sources of ambient noise. *IEEE J. Ocean. Eng.* **30**: 257–61.

Rossi-Santos, M. R., Wedekin, L. L., and Monteiro-Filho, E. L. A. (2007). Residence and site fidelity of *Sotalia guianensis* in the Caravelas river estuary, eastern Brazil. *J. Mar. Biol. Ass. U. K.* **87**: 207–12.

*Sánchez-García, A., Bueno-Crespo, A., and Sancho-Gómez, J. L. (2010). An efficient statistics-based method for the automated detection of sperm whale clicks. *Appl. Acoust.* **71**: 451–9.

*Santos, M. B., Martín, V., Arbelo, M., Fernández, A., and Pierce, G. J. (2007). Insights into the diet of beaked whales from the atypical mass stranding in Canary Islands in September 2002. *J. Mar. Biol. Ass. U.K.* **87**: 243–51.

Santos, M. C. de O., Rosso, S., Siciliano, S. *et al.* (2000). Behavioral observations of the marine tucuxi dolphin (*Sotalia fluviatilis*) in Sao Paulo estuarine waters, Southeastern Brazil. *Aquat. Mamm.* **26**: 260–7.

Sayigh, L. S., Esch, H. C., Wells, R. S., and Janik, V. M. (2007). Facts about signature whistles of bottlenose dolphins, *Tursiops truncatus. Anim. Behav.* **74**: 1631–42.

*Sayigh, L. S., Tyack, P. L., and Wells, R. S. (1993). Recording underwater sound of free-ranging dolphins while underway in a small boat. *Mar. Mamm. Sci.* **9**: 209–13.

Sayigh, L. S., Tyack, P. L., Wells, R. S., Scott, M. D., and Irvine, A. B. (1975). Sex difference in signature whistle production of free-ranging bottlenose dolphins, *Tursiops truncatus. Behav. Ecol. Sociobiol.* **36**: 171–7.

*Schau, H. C., and Robinson, A. Z. (1987). Passive source localization intersecting spherical surfaces from Time-Of-Arrival differences. *IEEE Trans. ASSP* **35**: 1223–5.

*Schevill, W. E. (1964). Underwater sounds of cetaceans. In *Marine Bioacoustics*, edited by W. N. Tavolga. New York: Pergamon, pp. 307–16.

Schevill, W. E., and Lawrence, B. (1949). Underwater listening to the white porpoise (*Delphinapterus leucas*). *Science* **109**: 143–4.

*Schevill, W. E., and Watkins, W. A. (1966). Sound structure and directionality in *Orcinus* (killer whale). *Zoologica (N.Y.)* **51**: 70–6.

Schevill, W. E., Watkins, W. A., and Backus, R. H. (1964). The 20-cycle signals and *Balaenoptera* (fin whales). In *Marine Bioacoustics*, edited by W. N. Tavolga. New York: Pergamon, pp. 147–52.

*Schmidt, R. O. (1972). A new approach to geometry of range difference location. *IEEE Trans. AES* **8**: 821–35.

Schorr, G. S., Baird, R. W., Hansen, M. B. *et al.* (2009). Movements of satellite-tagged Blainville's beaked whales off the island of Hawai'i. *Endang. Species Res.* **10**: 203–13.

Schotten, M., Au, W. W. L., Lammers, M. O., and Aubauer, R. (2003). Echolocation recordings and localization of wild spinner dolphins (*Stenella longirostris*) and pantropical spotted dolphins (*Stenella attenuata*) using a four-hydrophone array. In *Echolocation in Bats and Dolphins*, edited by J. Thomas, C. F. Moss, and M. Vater. Chicago, IL: University of Chicago Press, pp. 383–400.

Schulz, T. M., Whitehead, H., and Rendell, L. (2009). Off-axis effects on the multi-pulse structure of sperm whale coda clicks. *J. Acoust. Soc. Am.* **125**: 1768–73.

*Schwager, M., Anderson, D. M., Butler, Z., and Rus, D. (2007). Robust classification of animal tracking data. *Comp. Electron. Agricult.* **56**: 46–59.

*Schwarz, C. J., and Seber, G. A. F. (1999). Estimating animal abundance: review III. *Stat. Sci.* **14**: 427–56.

*Seber, G. A. F. (1982). *The Estimation of Animal Abundance and Related Parameters*, 2nd edn. London: Griffin.

*Seber, G. A. F. (1992). A review of estimating animal abundance II. *Int. Stat. Rev.* **60**: 129–66.

Secchi, E. R., Ott, P. H., and Danilewicz, D. (2002). Report on the fourth workshop for the coordinated research and conservation of the Franciscana dolphin (*Pontoporia blainvillei*) in the western south Atlantic. *LAJAM* **1**: 11–20.

Selzer, L. A., and Payne, P. M. (1988). The distribution of white-sided (*Lagenorhynchus acutus*) and common dolphins (*Delphinus delphis*) vs. environmental features of the continental shelf of the Northeastern United States. *Mar. Mamm. Sci.* **4**: 141–53.

Shapiro, A. D. (2006). Preliminary evidence for signature vocalizations among free-ranging narwhals (*Monodon monoceros*). *J. Acoust. Soc. Am.* **120**: 1695–705.

*Shapiro, A. D., and Wang, C. (2009) A versatile pitch tracking algorithm: from human speech to killer whale vocalizations. *J. Acoust. Soc. Am.* **126**: 451–9.

Shelden, K. E. W., Moore, S. E., Waite, J. M., Wade, P. R., and Rugh, D. J. (2005). Historic and current habitat use by North Pacific right whales Eubalaena japonica in the Bering Sea and Gulf of Alaska. *Mamm. Rev.* **35**: 129–55.

Shinha, R. K., and Sharma, G. (2003). Current status of the Ganges river dolphin, *Platanista gangetica* in the rivers Kosi and Son, Bihar, India. *J. Bombay Nat. Hist. Soc.* **100**: 27–37.

Silber, G. K. (1986). The relationship of social vocalizations to surface behavior and aggression in the Hawaiian humpback whale *Megaptera novaeangliae*. *Can. J. Zool.* **64**: 2075–80.

Silber, G. K. (1991). Acoustic signals of the Vaquita (*Phocoena sinus*). *Aquat. Mamm.* **17**: 130–3.

Simar, P., Hibbard, A. L., McCallister, K. A. *et al.* (2010). Depth dependent variation of the echolocation pulse rate of bottlenose dolphins (*Tursiops truncatus*). *J. Acoust. Soc. Am.* **127**: 568–78.

Simon, M., Wahlberg, M., and Miller, L. A. (2007). Echolocation clicks from killer whales (*Orcinus orca*) feeding on herring (*Clupea harengus*) (L). *J. Acoust. Soc. Am.* **121**: 749–52.

Širović, A., Hildebrand, J. A., and Wiggins, S. M. (2007). Blue and fin whale call source levels and propagation range in the Southern Ocean. *J. Acoust. Soc. Am.* **122**: 1208–15.

Širović, A., Hildebrand, J. A., Wiggins, S. M. *et al.* (2004). Seasonality of blue and fin whale calls and the influence of sea ice in the Western Antarctic Peninsula. *Deep-Sea Res. II* **51**: 2327–44.

Sjare, B. L., and Smith., T. G. (1986). The vocal repertoire of white whales, *Delphinapterus leucas*, summering in Cunningham Inlet, Northwest Territories. *Can. J. Zool.* **64**: 407–15.

*Sjare, B. L., Stirling, I., and Spencer, C. (2003). Structural variation in the songs of Atlantic walruses breeding in the Canadian High Arctic. *Aquat. Mamm.* **29**: 297–318.

*Skarsoulis, E. K., and Kalogerakis, M. A. (2005). Ray-theoretic localization of an impulsive source in a stratified ocean using two hydrophones. *J. Acoust. Soc. Am.* **118**: 2934–43.

*Skarsoulis, E. K., and Piperakis, G. S. (2009). Use of acoustic navigation signals for simultaneous localization and sound-speed estimation. *J. Acoust. Soc. Am.* **125**: 1384–93.

Smeek, C., Addink, M. J., van den Berg, A. B., Bosman, C. A. W., and Cadée, G. C. (1996). Sightings of *Delphinus* cf. *tropicalis* Van Bree, 1971 in the Red Sea. *Bonn. Zool. Beitr.* **46**: 389–98.

*Smith, J. O., and Bell, J. S. (1987). Closed form least-squares source location estimation from range-difference measurements. *IEEE Proc. ASSP* **35**: 1661–9.

Soldevilla, M. S., Henderson, E. E., Campbell, G. S. *et al.* (2008). Classification of Risso's and Pacific white-sided dolphins using spectral properties of echolocation clicks *J. Acoust. Soc. Am.* **124**: 609–24.

Soldevilla, M. S., Wiggins, S. M., and Hildebrand, J. A.(2010). Spatio-temporal comparison of Pacific white-sided dolphin echolocation click types. *Aquat. Biol.* **9**: 49–62.

Southwood, T. R. E., and Henderson, P. A. (2006). *Ecological Methods.* 3rd edn. Malden, MA: Blackwell.

*Spencer, S. J. (2007) The two-dimensional source location problem for time differences of arrival at minimal element monitoring arrays. *J. Acoust. Soc. Am.* **121**: 3579–94.

*Spiesberger, J. L. (1998). Linking auto- and cross-correlation functions with correlation equations: application to estimating the relative travel times and amplitudes of multipath. *J. Acoust. Soc. Am.* **104**: 300–12.

Spiesberger, J. L. (1999). Locating animals from their sounds and tomography of the atmosphere: experimental demonstration. *J. Acoust. Soc. Am.* **106**: 837–46.

*Spiesberger, J. L. (2000). Finding the right cross-correlation peak for locating sounds in multipath environments with a fourth-moment function. *J. Acoust. Soc. Am.* **108**: 1349–52.

*Spiesberger, J. L. (2001). Hyperbolic location errors due to insufficient numbers of receivers. *J. Acoust. Soc. Am.* **109**: 3076–9.

Spiesberger, J., and Fristrup, K. (1990). Passive localization of calling animals and sensing of their acoustic environment using acoustic tomography. *Am. Nat.* **125**: 107–53.

Stafford, K. M. (2003). Two types of blue whale calls recorded in the Gulf of Alaska. *Mar. Mamm. Sci.* **19**: 682–3.

Stafford, K. M., Bohnenstiehl, D. R., Tolstoy, M. *et al.* (2004). Antarctic-type blue whale calls recorded at low latitudes in the Indian and the eastern Pacific Oceans. *Deep-Sea Res. I* **51**: 1337–46.

Stafford, K. M., Fox, C. G., and Clark, D. S. (1998). Long-range acoustic detection and localization of blue whale calls in the Northeast Pacific Ocean. *J. Acoust. Soc. Am.* **104**: 3616–25.

*Stafford, K. M., Mellinger, D. K. Moore, S. E., and Fox, C. G. (2007a). Seasonal variability and detection range modeling of baleen whale calls in the Gulf of Alaska, 1999–2002. *J. Acoust. Soc. Am.* **122**: 3378–90.

Stafford, K. M., Moore, S. E., Laidre, K. L., and Heide-Jørgensen, M. P. (2008). Bowhead whale springtime song off West Greenland. *J. Acoust. Soc. Am.* **124**: 3315–23.

Stafford, K. M., Moore, S. E., Spillane, M., and Wiggins, S. (2007b). Gray whale calls recorded near Barrow, Alaska, throughout the winter of 2003–04. *Arctic* **60**: 167–72.

Stafford, K. M., Nieukirk, S. L., and Fox, C. G. (1999a). An acoustic link between blue whales in the eastern tropical Pacific and the northeast Pacific. *J. Acoust. Soc. Am.* **15**: 1258–68.

*Stafford, K. M., Nieukirk, S. L., and Fox, C. G. (1999b). Low-frequency whale sounds recorded on hydrophones moored in the eastern tropical Pacific. *J. Acoust. Soc. Am.* **106**: 3687–98.

Stafford, K. M., Nieukirk, S. L., and Fox, C. G. (2001). Geographic and seasonal variation of blue whale calls in the North Pacific. *J. Cetacean Res. Manage.* **3**: 65–76.

Stearns, S. D., and David, R. A. (1988). *Signal Processing Algorithms*. Englewood Cliffs, NJ: Prentice-Hall.

Steinhausen, D., and Langer, K. (1977). *Clusteranalyse. Einführung in Methoden und Verfahren der automatischen Klassifikation*. Berlin: Walter de Gruyter.

*Stewart, B. S., Karl, S. A., Yochem, P. K., Leatherwood, S., and Laake, J. L. (1987). Aerial surveys for cetaceans in the former Akutan, Alaska, whaling grounds. *Arctic* **40**: 33–42.

Stimpert, A. K., Wiley, D. N., Au, W. W. L., Johnson, M. P., and Arsenault, R. (2007). 'Megapclicks': acoustic click trains and buzzes produced during night-time foraging of humpback whales (*Megaptera novaeangliae*). *Biol. Lett.* **3**: 467–70.

*Store, R., and Jokimäki, J. (2003). A GIS-based multi-scale approach to habitat suitability modeling. *Ecol. Model.* **169**: 1–15.

*Strindberg, S., and Bickland, S. (2004). ZigZag survey design in line transect sampling. *J. Agric. Biol. Environ. Stat.* **9**: 443–61.

*Sturtivant, C., and Datta, S. (1997). Automatic dolphin whistle detection, extraction, encoding, and classification. *Proc. Inst. Acoust.* **19**: 259–66.

Teloni, V., (2005). Patterns of sound production in diving sperm whales in the northwestern Mediterranean. *Mar. Mamm. Sci.* **21**: 446–57.

Teloni, V., Zimmer, W. M. X., and Tyack P. L. (2005). Sperm whale trumpet sounds. *Bioacoustics* **15**: 163–74.

Teloni, V., Zimmer, W. M. X., Wahlberg, M., and Madsen, P. T. (2007). Consistent acoustic size estimation of sperm whales using clicks recorded from unknown aspects. *J. Cetacean Res. Manage.* **9**: 127–36.

Teranishi, A. M., Hildebrand, J. A., McDonald, M. A., Moore, S. E., and Stafford, K. M. (1997). Acoustic and visual studies of blue whales near the California Channel Islands. *J. Acoust. Soc. Am.* **102**: 3121.

Thode, A. (2004). Tracking sperm whale (*Physeter macrocephalus*) dive profiles using a towed passive acoustic array. *J. Acoust. Soc. Am.* **116**: 245–53.

Thode, A. M., D'Spain, G. L., and Kuperman, W. A. (2000). Matched-field processing, geoacoustic inversion, and source signature recovery of blue whale vocalizations. *J. Acoust. Soc. Am.* **107**: 1286–300.

*Thomas, J. A., Fisher, S. R., and Awbrey, F. A. (1986). Acoustic techniques in studying whale behavior. *Rep. Int. Whal. Comm.* **8**: 121–38.

*Thomas, R. E., Fristrup, K. M., and Tyack, P. L. (2002). Linking the sounds of dolphins to their locations and behavior using video and multichannel acoustic recordings. *J. Acoust. Soc. Am.* **112**: 1692–901.

Thompson, P. O. (1992). 20-Hz pulses and other vocalizations of fin whales, *Balaenoptera physalus*, in the Gulf of California, Mexico. *J. Acoust. Soc. Am.* **92**: 3051–7.

*Thompson, P. O., and Friedl, W. A. (1982). A long-term study of low frequency sound from several species of whales off Oahu, Hawaii. *Cetology* **45**: 1–19.

Thompson, P. O., Cummings, W. C., and Ha, S. J. (1986). Sounds, source levels and associated behavior of humpback whales, southeast Alaska. *J. Acoust. Soc. Am.* **80**: 735–40.

Thompson, P. O., Findley, L. T., Vidal, O., and Cummings, W. C. (1996). Underwater sounds of blue whales, *Balaenoptera musculus*, in the Gulf of California, Mexico. *Mar. Mamm. Sci.* **12**: 288–93.

*Thompson, T. J., Winn, H. E., and Perkins, P. J. (1979). Mysticete sounds. In *Behavior of Marine Mammals: Current Perspectives in Research*, edited by H. E. Winn and B. L. Olla. New York: Plenum, pp. 403–31.

Thompson, W. L. (ed.) (2004). *Sampling Rare or Elusive Species. Concepts, Designs, and Techniques for Estimating Population Parameters*. Washington, D.C.: Island Press.

Thomsen, F., van Elk, N., Brock, V., and Piper, W. (2005). On the performance of automated porpoise-click-detectors in experiments with captive harbor porpoises (*Phocoena phocoena*) (L). *J. Acoust. Soc. Am.* **118**: 37–40.

*Thorp, W. H. (1965). Deep-ocean sound attenuation in the sub and low-kilocycle-per-second range. *J. Acoust. Soc. Am.* **38**: 648–54.

Tiemann, C. O. (2008). Three-dimensional single-hydrophone tracking of a sperm whale demonstrated using workshop data from the Bahamas. *Can. Acoustics* **36**: 67–73.

Tiemann, C. O., and Porter, M. (2003). Automated model-based localization of sperm whale clicks. In *IEEE Proc. Oceans '03*. New York: IEEE, pp. 821–7.

Tiemann, C. O., Thode, A. M., Straley, J., O'Connell, V., and Folkert, K. (2006). Three-dimensional localization of sperm whales using a single hydrophone. *J. Acoust. Soc. Am.* **120**: 2355–65.

*Tokuda, I., Riede, T., Neubauer, J., Owren, M. J., and Herzel, H. (2002). Nonlinear analysis of irregular animal vocalizations. *J. Acoust. Soc. Am.* **111**: 2908–19.

*Trifa, V. M., Kirschel, A. N. G., Taylor, C. E., and Vallejo, E. E., (2008). Automated species recognition of antbirds in a Mexican rainforest using hidden Markov models. *J. Acoust. Soc. Am.* **123**: 2424–31.

*Tyack, P. L. (1986). Population biology, social behavior, and communication in whales and dolphins. *Trends Ecol. Evol.* **1**: 144–50.

*Tyack, P. (1998). Acoustic communication under the sea. In *Animal Acoustic Communication. Sound Analysis and Research Methods*, edited by S. L. Hopp, M. J. Owren, and C. S. Evans. New York: Springer, pp. 163–220.

Tyack, P. (1999). Communication and cognition. In *Biology of Marine Mammals*, edited by J. E. Reynolds II and S. A. Rommel. Washington, D.C.: Smithsonian Institution Press, pp. 287–323.

*Tyack, P. L., and Clark, C. W. (2000). Communication and acoustic behavior of dolphins and whales. In *Hearing by Whales and Dolphins*, edited by W. W. L. Au, A. N. Popper, and R. R. Fay. New York: Springer-Verlag, pp. 156–224.

Tyack, P. L., Johnson, M., Soto, N. A., Sturlese, A., and Madsen, P. T. (2006). Extreme diving of beaked whales. *J. Exp. Biol.* **209**: 4238–53.

Urazghildiiev, I. R., and Clark, C. W. (2006). Acoustic detection of North Atlantic right whale contact calls using the generalized likelihood ratio test. *J. Acoust. Soc. Am.* **120**: 1956–63.

*Urazghildiiev, I. R., and Clark, C. W. (2007a). Detection performances of experienced human operators compared to a likelihood ratio based detector. *J. Acoust. Soc. Am.* **122**: 200–4.

Urazghildiiev, I. R., and Clark, C. W. (2007b). Acoustic detection of North Atlantic right whale contact calls using spectrogram-based statistics. *J. Acoust. Soc. Am.* **122**: 769–76.

Urban, H. G. (2002). *Handbook of Underwater Acoustic Engineering*. Bremen: STN Atlas.

*Urick, R. J. (1962). Generalized form of the sonar equations. *J. Acoust. Soc. Am.* **34**: 547–50.

Urick, R. J. (1983). *Principles of Underwater Sound*, 3rd edn. New York: McGraw-Hill.

van der Schaar, M., Delory, E., and André, M. (2009). Classification of sperm whale clicks (*Physeter macrocephalus*) with Gaussian-kernel-based networks. *Algorithms* **2**: 1232–47.

van der Schaar, M., Delory, E., Català, A., and André, M. (2007). Neural network-based sperm whale click classification. *J. Mar. Biol. Ass. U.K* **87**: 35–8.

Van Parijs, S. M., Lydersen, C., and Kovacs, K. M. (2003). Sounds produced by individual white whales, *Delphinapterus leucas*, from Svalbard during capture. *J. Acoust. Soc. Am.* **113**: 57–60.

Van Parijs, S. M., Parra, G., and Corkeron, P. J. (2000). Sounds produced by Australian Irrawaddy dolphins *Orcaella brevirostris*. *J. Acoust. Soc. Am.* **108**: 1938–40.

Van Parijs, S. M., Smith, J., and Corkeron, P. J. (2002). Using calls to estimate the abundance of inshore dolphins: a case study with Pacific humpback dolphins *Sousa chinensis*. *J. Appl. Ecol.* **39**: 853–64.

van Waerebeek, K., Barnett, L., Camara, A. *et al.* (2004). Distribution, status, and biology of the Atlantic humpback dolphin, *Sousa teuszii* (Kükenthal, 1892). *Aquat. Mamm.* **30**: 56–83.

Verfuss, U. K., Miller, L. A., and Schnitzler, H. U. (2005). Spatial orientation in echolocating harbour porpoises (*Phocoena phocoena*) *J. Exp. Biol.* **208**: 3385–94.

*Wagstaff, R. A. (2005). An ambient noise model for the Northeast Pacific ocean basin. *IEEE J. Ocean. Eng.* **30**: 286–94.

Wahlberg, M. (2002). The acoustic behaviour of diving sperm whales observed with a hydrophone array. *J. Exp. Mar. Biol. Ecol.* **281**: 53–62.

*Wahlberg, M., Møhl, B., and Madsen, P. T. (2001). Estimating source position accuracy of a larger-aperture hydrophone array for bioacoustics. *J. Acoust. Soc. Am.* **109**: 397–406.

*Waite, A. D. (2002). *Sonar for Practising Engineers*, 3rd edn. Chichester: Wiley

*Waite, J. M., Wynne, K., and Mellinger, D. K. (2003). Documented sighting of a North Pacific right whale in the Gulf of Alaska and post-sighting acoustic monitoring. *Northwestern Naturalist* **84**: 38–43.

Wald, A. (1947). *Sequential Analysis*. New York: Wiley.

Wang, K., Wang, D., Akamatsu, T., Fujita, K., and Shiraki, R. (2006). Estimated detection distance of a baiji's (Chinese river dolphin, *Lipotes vexillifer*) whistles using a passive acoustic survey method. *J. Acoust. Soc. Am.* **120**: 1361–5.

Wang, K., Wang, D., Akamatsu, T., Li, S., and Xiao, J. (2005). A passive acoustic monitoring method applied to observation and group size estimation of finless porpoises. *J. Acoust. Soc. Am.* **118**: 1180–5.

Ward, J., Morrissey, R., Moretti, D. *et al.* (2008). Passive acoustic detection and localization of *Mesoplodon densirostris* (Blainville's beaked whale) vocalizations using distributed bottom-mounted hydrophones in conjunction with a digital tag (DTAG) recording. *Can. Acoust.* **36**: 60–6.

*Waring, G. T., Hamazaki, T., Sheehan, D., Wood, G., and Baker, S. (2001). Characterization of beaked whale (Ziphiidae) and sperm whale (*Physeter macrocephalus*) summer habitat in shelf-edge and deeper waters off the Northeast U.S. *Mar. Mamm. Sci.* **17**: 703–17.

Watkins, W. A. (1967). The harmonic interval: fact or artifact in spectral analysis of pulse trains? *Mar. Bioacoust.* **2**: 15–43.

Watkins, W. A. (1980). Acoustics and the behavior of sperm whales. In *Animal Sonar Systems*, edited by R. Busnel and J. F. Fish. New York: Plenum, pp. 291–7.

Watkins, W. A. (1981). Activities and underwater sounds of fin whales. *Sci. Rep. Whal. Res. Inst.* **33**: 83–117.

Watkins, W. A., and Daher, M. A. (2004). Variable spectra and nondirectional characteristics of clicks from near-surface sperm whales (*Physeter catodon*). In *Echolocation in Bats and Dolphins*, edited by J. A. Thomas, C. F. Moss, and M. Vater. Chicago, IL: The University of Chicago Press, pp. 410–13.

Watkins, W. A., and Moore, K. E. (1982). An underwater acoustic survey for sperm whales (*Physeter catodon*) and other cetaceans in the southeast Caribbean. *Cetology* **46**: 1–7.

*Watkins, W. A., and Schevill, W. E. (1972). Sound source location by arrival times on a non-rigid three-dimensional hydrophone array. *Deep-Sea Res.* **19**: 691–706.

Watkins, W. A., and Schevill, W. E. (1977). Sperm whale codas. *J. Acoust. Soc. Am.* **62**: 1485–90.

Watkins W. A., Daher, M. A., DiMarzio, N. A. *et al.* (1999). Sperm whale surface activity from tracking by radio and satellite tags. *Mar. Mamm. Sci.* **15**: 1158–80.

Watkins, W. A., Daher, M. A., DiMarzio, N. A. *et al.* (2002). Sperm whale dives tracked by radio tag telemetry. *Mar. Mamm. Sci.* **18**: 55–68.

*Watkins, W. A., Daher, M. A., George, J. E., and Rodriguez, D. (2004). Twelve years of tracking 52-Hz whale calls from a unique source in the North Pacific. *Deep-Sea Res. I* **51**: 1889–901.

Watkins, W. A., Daher, M. A., Fristrup, K. M., Howald, T. J., and Notarbartolo di Sciara, G. (1993). Sperm whales tagged with transponders and tracked underwater by sonar. *Mar. Mamm. Sci.* **9**: 55–67.

*Watkins, W. A., Daher, M. A., Reppucci, G. M. *et al.* (2000). Seasonality and distribution of whale calls in the North Pacific. *Oceanography* **13**: 62–7.

*Watkins, W. A., Schevill, W. E., and Best, P. B. (1977) Underwater sound of *Cephalorhynchus heavisidii* (Mammalia: Cetacea). *J. Mamm.* **58**: 316–20.

Watkins, W. A., Schevill, W. E., and Ray, C. (1971). Underwater sounds of *Monodon* (Narwhal). *J. Acoust. Soc. Am.* **49**: 595–9.

Watkins, W. A., Tyack, P., Moore, K. E., and Bird, J. E. (1987). The 20-Hz signals of finback whales (*Balaenoptera physalus*). *J. Acoust. Soc. Am.* **82**: 1901–12.

*Watkins, W. A., Tyack, P., Moore, K. E., and Notarbartolo di Sciara, G. (1987). Steno bredanensis in the Mediterranean Sea. *Mar. Mamm. Sci.* **3**: 78–82.

Watwood, S. L., Miller, P. J. O., Johnson, M., Madsen, P. T., and Tyack, P. L. (2006). Deep-diving foraging behavior of sperm whales (*Physeter macrocephalus*). *J. Anim. Ecol.* **75**: 814–25.

Weilgart, L., and Whitehead, H. (1997). Group-specific dialects and geographical variation in coda repertoire in South Pacific sperm whales. *Behav. Ecol. Sociobiol.* **40**: 277–85.

Weinrich, M. A., Martin, M., Griffiths, R., Bove, J., and Schilling, M. (1997). A shift in distribution of humpback whales, *Megaptera novaeangliae*, in response to prey in the southern Gulf of Maine. *Fish. Bull.* **95**: 826–36.

*Weir, C. R., Pollok, C., Cronin, C., and Taylor, S. (2001). Cetaceans of the Atlantic Frontier, north and west of Scotland. *Cont. Shelf. Res.* **21**: 1047–71.

Wenz, G. M. (1962). Acoustic ambient noise in the ocean: spectra and sources. *J. Acoust. Soc. Am.* **34**: 1936–56.

*White, P. R., Leighton, T. G., Finfer, D. C., Powles, C., and Baumann, O. N. (2006). Localisation of sperm whales using bottom-mounted sensors. *Appl. Acoust.* **67**: 1074–90.

*Whitehead, H. (1990). Mark–recapture estimates with emigration and re-immigration. *Biometrics* **46**: 473–9.

*Whitehead, H. (2009). Estimating abundance from one-dimensional passive acoustic surveys. *J. Wildl. Manage.* **73**: 1000–9.

Whitehead, H., and Weilgart, L. (1990). Click rates from sperm whales. *J. Acoust. Soc. Am.* **87**: 1798–806.

Whitehead, H., and Weilgart L. (1991). Patterns of visually observable behaviour and vocalizations in group of female sperm whales. *Behaviour* **118**: 275–96.

Whitehead, H., and Wimmer, T. (2005). Heterogeneity and the mark–recapture assessment of the Scotian Shelf population of northern bottlenose whales (*Hyperoodon ampullatus*). *Can. J. Fish. Aquat. Sci.* **62**: 2573–85.

*Whitehead, H., Faucher, A., Gowans, S., and McCarrey, S. (1997). Status of the Northern Bottlenose Whale, *Hyperoodon ampullatus*, in the Gully, Nova Scotia. *Can. Field-Nat.* **111**: 287–92.

Whitehead, H., Gowans, S., Faucher, A., and McCarrey, S. W. (1997). Population analysis of northern bottlenose whale in the Gully, Nova Scotia. *Mar. Mamm. Sci.* **13**: 173–85.

Whitehead, H., Weilgart, L., and Waters, S. (1989). Seasonality of sperm whales off the Galapagos Islands, Ecuador. *Rep. Int. Whal. Commn* **39**: 207–10.

*Wille, P. C., and Geyer, D. (1984). Measurements on the origin of the wind-dependent ambient noise variability in shallow water. *J. Acoust. Soc. Am.* **75**: 173–85.

*Wilson, B., Hammond, P. S., and Thompson, P. M. (1999). Estimating size and assessing trends in coastal bottlenose dolphin populations. *Ecol. Applic.* **9**: 288–300.

*Wilson, R. P., Grant, S., and Duffy, D. C. (1986). Recording devices on free-ranging marine mammals: does measurement affect foraging performance? *Ecology* **67**: 1091–3.

*Wilson, R. P., Liebsch, N., Davies, I. M. *et al.* (2006). All at sea with animal tracks; methodological and analytical solutions for the resolution of movement. *Deep-Sea Res. II* **54**: 193–210.

Wimmer, T., and Whitehead, H. (2004). Movements and distribution of northern bottlenose whales, *Hyperoodon ampullatus*, on the Scotian Slope and in adjacent waters. *Can. J. Zool.* **82**: 1782–94.

Winkler, G. (1977). *Stochastische Systeme. Analyse und Synthese.* Wiesbaden: Akademische Verlagsgesellschaft.

*Winn, H. E., and Perkins, P. J. (1976). Distribution and sounds of the minke whale, with a review of mysticete sounds. *Cetology* **19**: 1–12.

Winn, H. E., Thompson, T. J., Cummings, W. C., Hain, J., Hudnall, J., Hays, H., and Steiner, W. W. (1981). Song of the humpback whale – population comparisons. *Behav. Ecol.* **8**: 41–6.

*Witteveen, B. H., Straley, J. M., von Ziegesar, O., Steel, D., and Baker, C. S. (2004). Abundance and mtDNA differentiation of humpback whales (*Megaptera novaeangliae*) in the Shumagin Islands, Alaska. *Can. J. Zool.* **82**: 1352–9.

*Wu, H., Li, B. L., Springer, T. A., and Neill, W. H. (2000). Modelling animal movement as a persistent random walk in two dimensions: expected magnitude of net displacement. *Ecol. Model.* **132**: 115–24.

*Würsig, B., and Clark, C. W. (1993). Behavior. In *The Bowhead Whale*, edited by J. J. Burns, J. J. Montague, and C. J. Cowles. Lawrence, KS: Allen, pp. 157–199.

Würsig, B., and Würsig, M. (1980). Behavior and ecology of the Dusky dolphin, *Langenorhynchus obscurus*, in the south Atlantic. *Fish. Bull.* **77**: 871–90.

Würsig, B., Dorsey, E. M., Fraker, M. A., Payne, R. S., and Richardson, W. J. (1985). Behavior of bowhead whales, *Balaena mysticetus*, summering in the Beaufort Sea: a description. *Fish. Bull.* **83**: 357–77.

Yochem, P. K., and Leatherwood, S. (1985). Blue whale – *Balaenoptera musculus* (Linnaeus, 1758). In *Handbook of Marine Mammals*, Vol. 3, *The Sirenians and Baleen Whales*, edited by S. H. Ridgway and R. Harrison. London and Orlando, FL: Academic Press, pp. 193–240.

*Zeh, J. E., Clark, C. W., George, J. C. *et al.* (1993). Current population size and dynamics. In *The Bowhead Whale*, edited by J. J. Burns, J. J. Montague, and C. J. Cowles. Lawrence, KS: Allen, pp. 409–89.

*Zerbini, A. N., Waite, J. M., Laake, J. L., and Wade, P. R. (2006). Abundance, trends and distribution of baleen whales in western Alaska and the central Aleutian Islands. *Deep-Sea Res. I* **53**: 1772–90.

Zimmer, W. M. X., Harwood, J., Tyack, P. L., Johnson, M. P., and Madsen, P. T. (2008). Passive acoustic detection of deep diving beaked whales. *J. Acoust. Soc. Am.* **124**: 2823–32.

Zimmer, W. M. X., Johnson, M. P., D'Amico, A., and Tyack, P. L. (2003). Combining data from a multisensor tag and passive sonar to determine the diving behavior of a sperm whale (*Physeter macrocephalus*). *IEEE J. Ocean. Eng.* **28**: 13–28.

Zimmer, W. M. X., Johnson, M. P., Madsen, P. T., and Tyack, P. L. (2005a). Echolocation clicks of free-ranging Cuvier's beaked whales (*Ziphius cavirostris*). *J. Acoust. Soc. Am.* **117**: 3919–27.

Zimmer, W. M. X., Madsen, P. T., Teloni, V., Johnson, M. P., and Tyack, P. L. (2005b). Off-axis effects on the multipulse structure of sperm whale usual clicks with implications for sound production. *J. Acoust. Soc. Am.* **118**: 3337–45.

Zimmer, W. M. X., Tyack, P. L., Johnson, M. P., and Madsen, P. T. (2005c). Three-dimensional beam pattern of regular sperm whale clicks confirms bent-horn hypothesis. *J. Acoust. Soc. Am.* **117**: 1473–85.

Index

Printed in the United States
By Bookmasters